NEW MONOMERS
AND POLYMERS

POLYMER SCIENCE AND TECHNOLOGY

Recent volumes in the series:

A Continuation Order Plan is available for this series. A continuation order will bring delivery of each new volume immediately upon publication. Volumes are billed only upon actual shipment. For further information please contact the publisher.

NEW MONOMERS AND POLYMERS

Edited by
Bill M. Culbertson
Ashland Chemical Company
Columbus, Ohio

and
Charles U. Pittman, Jr.
Mississippi State University
Mississippi State, Mississippi

PLENUM PRESS • NEW YORK AND LONDON

Library of Congress Cataloging in Publication Data

Main entry under title:

New monomers and polymers.

  (Polymer science and technology; v. 25)
  Proceedings of a symposium, New monomers and polymers, held Sept. 13–15,
1982, at the American Chemical Society meeting, Kansas City, Mo.
  Includes bibliographical references and index.
  1. Polymers and polymerization—Congresses. 2. Monomers—Congresses. I.
Culbertson, B. M., 1929–   . II. Pittman, Charles U. III. Series.
QD380.N48 1983   668.9   83-16045
ISBN-13: 978-1-4684-4621-0   e-ISBN-13: 978-1-4684-4619-7
DOI: 10.1007/978-1-4684-4619-7

Proceedings of a symposium on New Monomers and Polymers,
held September 13–15, 1982, at the American Chemical Society
meeting, in Kansas City, Missouri

© 1984 Plenum Press, New York
Softcover reprint of the hardcover 1st edition 1984
A Division of Plenum Publishing Corporation
233 Spring Street, New York, N.Y. 10013

PREFACE

    Interest in preparing new polymers peaked about 1966. Since that time, industrial and government support for the synthesis and study of new polymers has steadily declined. Gone are the good days when government funds supported a great push to attain ultimate thermal stability for organic polymeric materials. Gone are the good days when many chemical companies, encouraged by the obvious potential for rewards, had great interest and provided support for preparing new polymers. We now often hear managers say "we have enough polymers" or "all we need to do is find additional and better ways to use existing polymers." The latter often includes the statement, "we can get the new materials that are wanted from polymer alloys or blends."

    Interest in preparing new monomers has also waned, even though it is well recognized that monomers with special functionality are greatly needed to fine-tune existing polymers for specific tasks.

    Shrinkage of interest in new monomer and polymer research has not come about solely as a result of the obvious maturity of the polymers industry. Since uses for polymers continue to grow and there is still room for good concepts to study, lack of market growth and fields of study have probably not significantly contributed to that shrinkage. For the most part, loss of interest has come about due to the high cost of coming up with a totally new polymer and then doing all that is needed to market the new materials. Preoccupation with raw materials cost, market size, energy needs, and health and environmental concerns has also contributed to support shrinkage, especially for discovery and introduction of new monomers. However, management should keep in mind the rapid development and commercialization of a variety of new water-soluble coating polymers which followed the government's imposition of solvent emission standards in the late 1970's. A motivated management will find that a tremendous potential for synthetic development still exists.

    A report (1981) entitled "Polymer Science and Engineering: Challenges, Needs, and Opportunities," issued by the National

remarked "I am inclined to think that the development of
polymerization is perhaps the biggest thing chemistry has done,
where it has had the biggest effect on everyday life. The world
would be a totally different place without artificial fibers,
plastics, elastomers, etc. Even in the field of electronics, what
would you do without insulation, and there you come back to polymers again." Thinking about Lord Todd's remarks, one could add
that the story is far from finished. Polymeric materials, as they
are constantly being modified and improved, fine-tuned for current
and additional needs, and more readily accepted by the public, will
have an ever expanding influence on everyday life. However, lack
of long-term support of meaningful size for basic research on all
facets of polymer chemistry and engineering could well stunt the
growth of high technology in our country and help insure that we
won't have materials with performance profiles to meet requirements of emerging technologies and national needs. For example,
in the high-tech microelectronics area, many opportunities exist
for polymers to serve as improved dielectrics, improved plasma
etch resistance barriers, improved lithographic resists, etc.,
where, simultaneously, many other properties such as adhesion to a
variety of inorganic substances, water permeability, and temperature stability must not be limiting. Polymers for biomedical
applications, likewise, provide a rich opportunity for synthetic
innovation.

In consideration of the above and to foster more interest in
new monomers and new polymers, a symposium was scheduled for the
Kansas City meeting, titled "New Monomers and Polymers." At the
same time, it was recognized that such a symposium would help meet
the need and interest of many members of the ACS's Division of
Polymer Chemistry. It is well known that many members of the
Division have great interest in polymer synthesis and desire timely
symposia to present and discuss their studies.

The Kansas City symposium on New Monomers and Polymers consisted of twenty-six papers, reasonably representative of the
current interest in, direction of, and opportunities for research
on new monomers and polymers. The papers clearly showed that
great opportunities exist for new work in both step-growth and
chain-growth polymerization and in the introduction of new monomers.
For a variety of reasons, not all the papers from that symposium
are included in this book, even though each author was invited to
contribute. However, additional papers were solicited and/or
accepted to broaden the scope of the book and give readers a
better picture of the many opportunities and challenges that exist
for preparing new monomers and polymers.

Although all the chapters in this book cannot be neatly cate-
gorized, except as representative of either step-growth or chain-
growth polymerization studies, the subject matter falls into one

Research Council's ad hoc Panel on Polymer Science and Engineering, clearly recognized the need and urgency of increased commitment to basic studies on polymers. The Panel's list of needs and opportunities included challenges for basic/applied research on monomers, polymerization methods, specialty polymers, high-performance materials, and in situ polymerization methods for direct conversion of monomers to useful shapes. Clearly, as substantiated by the NRC report, industry and government must not neglect support of basic research on polymers. Further, a significant part of this support should be directed to new monomers and polymers.

At the fall 1982 National Meeting of the American Chemical Society in Kansas City, a special Polymer Education Award Honoring Herman F. Mark was presented. It was followed by a panel discussion concerning the state of polymer education at U.S. universities. While the award to Professor Mark was, indeed, a joyous event, the discussion that followed was sobering. Speaker after speaker pointed out a serious decline in the position of polymer chemistry within the chemistry departments of U.S. universities. It is true that several interdisciplinary polymer programs or institutes exist, but there is an increasing dominance by engineering and materials science aspects. Within chemistry departments the position of polymer chemistry, which has always been tenuous at best, was thought to be declining. We were told that it was unlikely that polymer chemists such as P.J. Flory (Stanford), W. Bailey (Maryland), and C. Marvel (Arizona) will be replaced by polymer chemists. Perhaps the most depressing aspect of this situation is that it will affect synthetic polymer chemistry most seriously because it is most difficult to develop and conduct innovative frontier syntheses outside a chemistry department where other strong synthetic programs already exist. Students who want to become synthetic chemists will seldom enroll in a program outside a chemistry department.

The discussion pointed out that very few of the "top twenty" chemistry departments have a polymer chemist (much less a synthetic polymer chemist) on their faculties, and there appear to be no plans to change this situation. The number of synthetic polymer chemists in the "top eighty" chemistry departments is also not growing and may be declining. There seems to be a reluctance within traditional chemistry departments to recognize and promote the elegance in synthetic methodology (organic, inorganic, biochemical, organometallic) which exists within the field of macromolecules. We all realize that several reasons exist for this prejudice and that many of these are less noble than many chemists care to admit. It appears that we could learn profitable lessons from the Japanese, who have placed a higher value on synthetic polymer chemistry within their "ivied halls."

Lord Todd, President of the Royal Society of London, recently or more of the following categories:

- High-Performance Polymers
- Aromatic Heterocyclic Monomers
- Thermally Stable Polymers
- High-Performance Composites
- Reactive Oligomers
- Aromatic Polyformals
- Star and Comb Shape Copolymers
- Polymers in Phase Transfer Catalysis
- Organometallic Polymers
- Polymers in Medical Applications
- Nitrogen-Containing Polymers
- Vinyl Organometallic Monomers
- Functionalized Polymers
- Anionic Copolymerizations
- Macromers
- Functional Methacrylate Polymers
- Block Copolymers
- Ring Opening Polymerizations
- Cationic Polymerizations
- Polymers Containing Transition Metals
- Conductive Polymers
- Diyne Polymers
- Ziegler-Natta Polymerizations
- Quinodimethane Polymers
- Charge-Transfer Copolymerizations
- Reactions on Polymers
- Polymeric Dyes
- p-Methylstyrene Monomers and Polymers

The work discussed in the various chapters clearly shows that great opportunities exist for research and development on new monomers and polymers. In addition, the work done at Mobil on poly(p-methylstyrene) decisively illustrates that great opportunities still exist for finding routes to new monomers that need not be tied to small specialty markets. After all, in a few years p-methylstyrene could become a fairly high volume commodity chemical and its homopolymer and copolymers could make a very large market impact (W.W. Kaeding, L.B. Young, and A.G. Propas, Chemtech, 556-562, September, 1982).

Bill M. Culbertson
Charles U. Pittman

CONTENTS

POLYIMIDINES - A NEW CLASS OF POLYMERS

Richard W. Thomas, Tejraj M. Aminabhavi
and Patrick E. Cassidy

Department of Chemistry
Southwest Texas State University
San Marcos, Texas 78666

INTRODUCTION

Intensive research has occurred in the last twenty years concerning the synthesis and processing of thermally stable polymers[1-3]. Early explorations in this area can be traced back to the late fifties impelled by the discovery of heterocyclic and aromatic amide polymers, which had the ability to withstand extreme temperatures.

One of the earliest and most common thermally stable polymers is polyimide synthesized originally by Bogart and Renshaw[4] in 1908. Since then, several useful polymers which contain imides or derivatives thereof have been synthesized[5,6]. Polyimides in various forms are marketed under the commerical names H-film, NR-150, RC-5081, Pyralin, Pyre-ML, Polymer SP, Vespel, PI-4701, and Kapton (DuPont); Kerimide, Nolimide and Kinel (Rhodia); Meldin (Dixon); P13N (TRW); P105A (Ciba-Geigy); PI-2080 (Upjohn); QX-13 (ICI)'; Skybond and Skygard (Monsanto); Thermid 600 (Gulf Oil Chemicals); and PMR (Riggs Engineering, Gulf Oil Chemicals, Ferro Corp., Fiberite, U.S. Polymeric and Hexcel).

In recent years, research efforts in these laboratories[7-13] and those of Imai, Ueda, and Takahashi[14-19] have been concentrated on the synthesis and characterization of a new class of polymers which are similar in structure to aromatic polyimides. Two advantages over the polyimides were recognized: (1) a one-step reaction, and (2) pendant phenyl groups which make polyimidines soluble.

The discovery of polyimidines in these laboratories was seren-
dipitous. 1,2-Dibenzoylbenzene was being sought for the synthesis
of benzylenebenzimidazole:

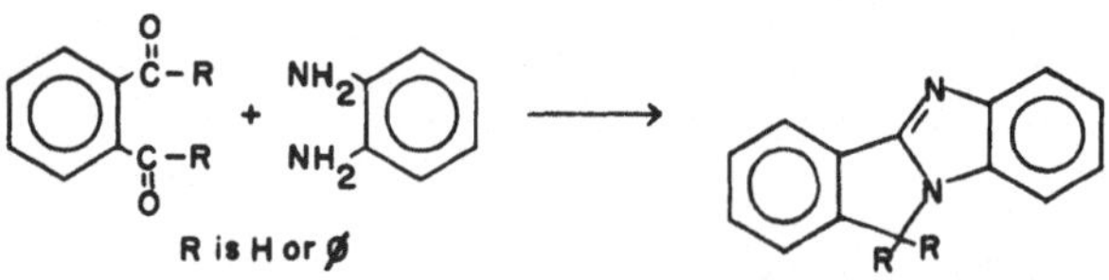

However, the dibenzoyl compound was isolated as the cyclic isomer
diphenylphthalide [1], reported by Friedel and Crafts[20]. By chance
during the same literature search the condensation of the phthalide
with aniline[30] to give a phthalimidine [5] was uncovered.

Interestingly, this research has now come full circle back to the
benzylenebenzimidazole. Because phthalaldehyde condenses with 1,2-
phenylene diamine and phthalides are isomers of phthalaldehyde, then
phthalides should do the same. So polymeric benzylenebenzimidazoles,
impossible to synthesize here-to-fore because of no route to pyromel-
litaldehyde, may be available through the tetraaldehyde isomers, bis-
phthalides.

MODEL COMPOUNDS

The model compounds used to explore the chemistry of imidines can
be divided into two basic groups, the diphenylphthalides and the
benzylidene phthalides. The first reported synthesis of 3,3-diphenyl-
phthalide [1] was by Friedel and Crafts in 1877[20]. Their method involved reacting phthaloyl chloride with benzene in the presence of
aluminum chloride.

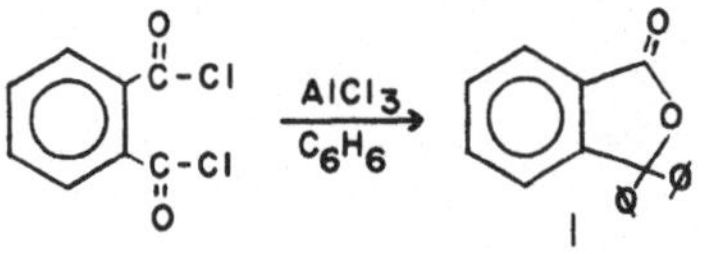

An alternate method of synthesizing this compound was reported in 1901
by the reaction of o-benzoyl benzoic acid chloride under similar conditions[21].

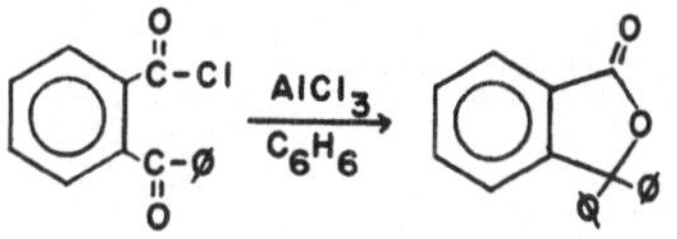

Both of these methods have been consistently used in this laboratory
and give overall yields from phthalic anhydride of 40% and 30% respectively.

Another phenylated lactone studied as a model compound is 3,3-
diphenylnaphthalide [2]. This compound was first reported by Mason[22]
by a two-step procedure similar to one used for 3,3-diphenylphthalide.

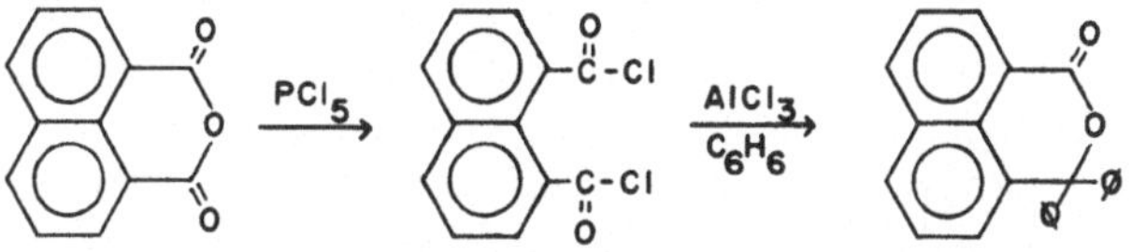

This was also duplicated in this laboratory and gave yields of about
35% overall[23]. A superior synthesis of 3,3-diphenylnaphthalide was
reported by Wittig and co-workers in 1931[24]. It was shown that naphthalic anhydride would readily react with phenyllithium at 0^0C to give
the product.

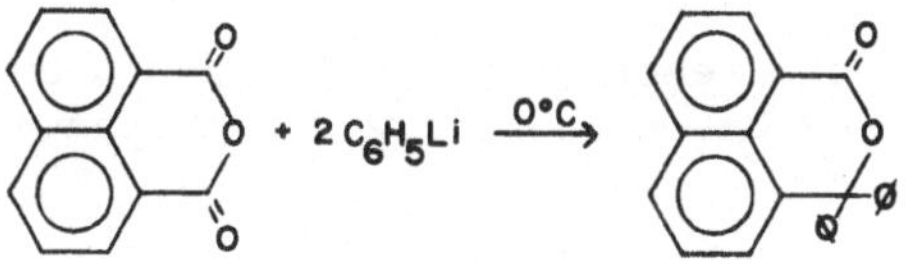

This method is less time consuming and can give yields up to 60%.
This method was also used to synthesize 3,3-diphenylphthalide in
yields of 25%[25].

A dithio derivative of 3,3-diphenylphthalide was reported by
Cassidy and co-workers in 1976[8] by the method of Prey and Kondler[26].

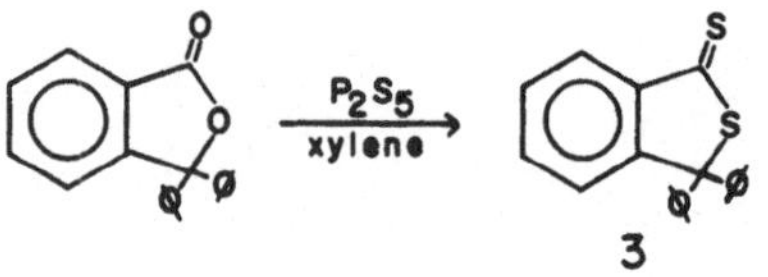

3

This procedure gives beautiful, bright red, rhombic crystals of 3,3-diphenyldithiophthalide [3] in yields up to 90%.

The other group of model compounds are phthalides substituted with a benzylidene group. As with the diphenylated phthalides, the chemistry of these models dates back to the nineteenth century, but the use of this chemistry to give polymers has been achieved primarily by Imai, Ueda and Takahashi. The model for this type of polyimidine was reported by Gabriel in 1885[27]. Phthalic anhydride was reacted with phenylacetic acid in the presence of sodium acetate to give 3-benzylidene phthalide [4].

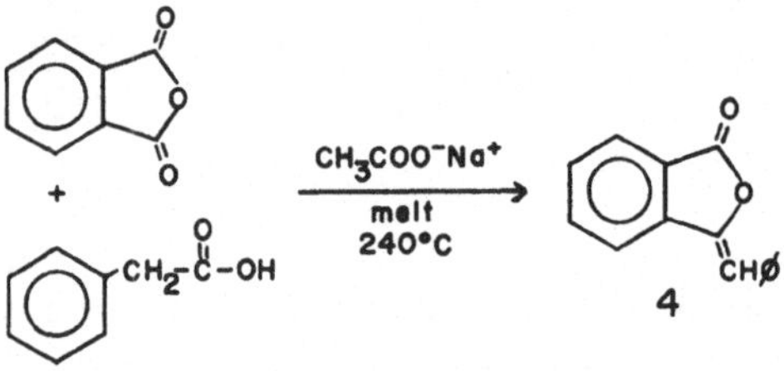

4

The synthesis was confirmed by Weiss[28] and yields are around 70%. In 1956, the product from this reaction was further characterized as trans-3-benzylidenephthalide[29].

As early as 1894, the reaction between 3,3-diphenylphthalide and aniline hydrochloride was known to give 2,3,3-triphenylphthalimidine [5] in good yields[30].

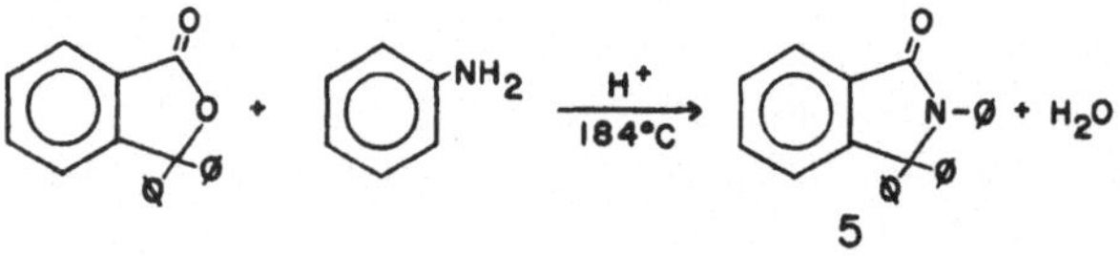

5

The reaction is believed to proceed through two steps:

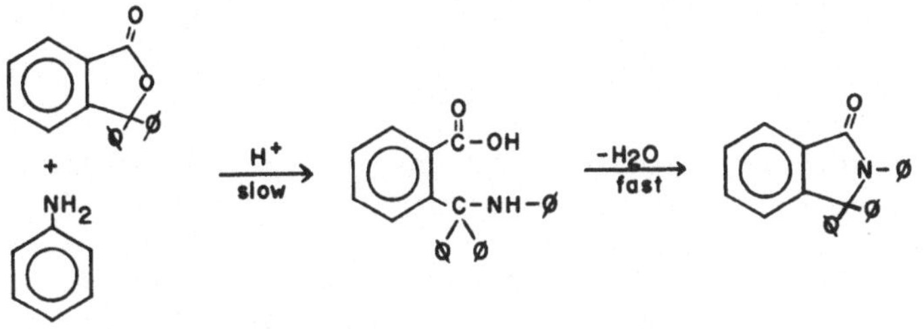

The structure of the proposed intermediate has not been confirmed but the fact that the reaction does not occur without an acid catalyst supports this belief. Upon protonation the strained ring can easily rearrange to give a very stable triphenyl carbocation which combines with the nucleophilic amine to give the unstable intermediate. This compound then easily undergoes cyclo-dehydration to give 2,3,3-triphenylphthalimidine.

In an effort to increase the facility of this model reaction, the synthesis was attempted with the use of isocyanates in place of amines [31]. Condensation was indicated at high temperature in sealed tubes by the evolution of carbon dioxide but the desired product was not obtained. Some Lewis acids were used as catalysts but were not effective.

The dithio derivative of 3,3-diphenylphthalide was found to react easily in boiling aniline and gave very good yields (83%).

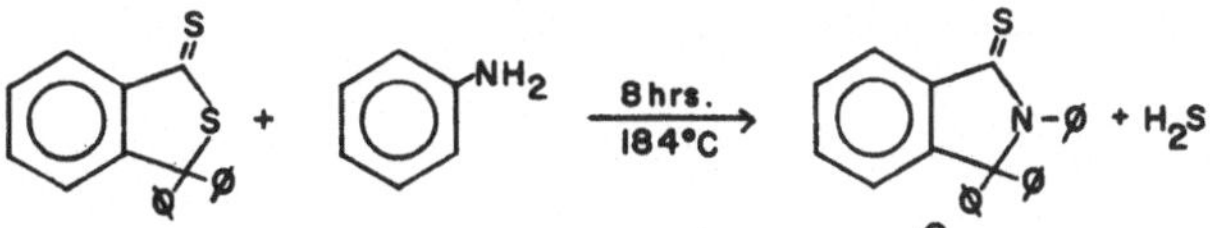

In solution, the reaction can be easily followed by the change in color from the bright red of the diphenyldithiophthalide to the light yellow of 2,3,3-triphenylthiophthalimidine [6]. There are not enough data available to postulate a mechanism for this reaction; but the relative reactivities between oxo and thio under a variety of conditions indicate that the two analogues proceed through different mechanisms.

The dithio compound was also successfully condensed with aromatic isocyanates and isothiocyanates[23,31].

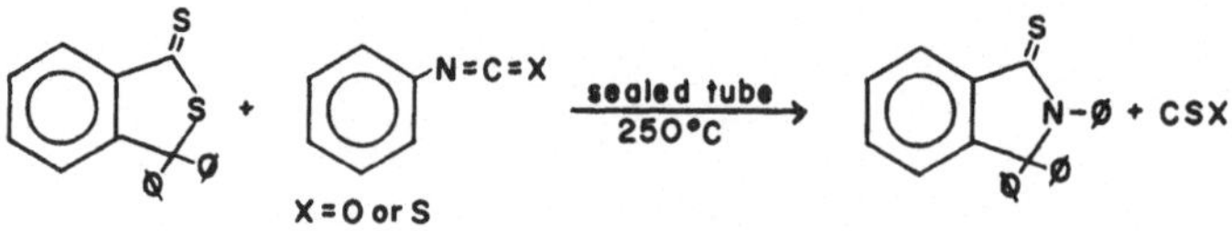

Although the reaction occurred, the use of isocyanates did not improve the facility of the relevant reactions as was anticipated.

With the intent of finding a more stable backbone structure for polyimidines, the reaction of 3,3-diphenylnaphthalide [2] and amines was studied[23]. Reactions were performed with aliphatic and aromatic amines, with and without catalysts, in solutions and in sealed tubes, and under a variety of reaction conditions. In no case was 2,3,3-triphenylnaphthalimidine isolated. The failure of this reaction to occur has been attributed to the stability of the 6-membered lactone ring.

The final model compound, 3-benzylidenephthalide [4], was found to be the most reactive towards amines[14]. 2-phenyl-3-benzylidene phthalimidine [7] is formed in good yields when 3-benzylidene phthalide is heated with aniline and a little boric acid in N-methylpyrrolidone[25].

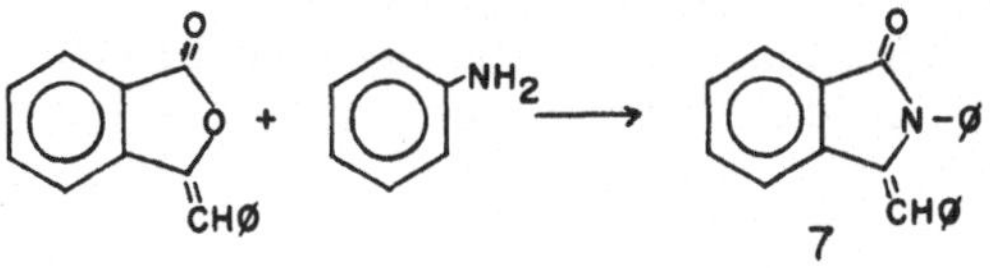

The reaction also proceeds through two steps, but unlike the diphenylphthalides, forms a stable intermediate.

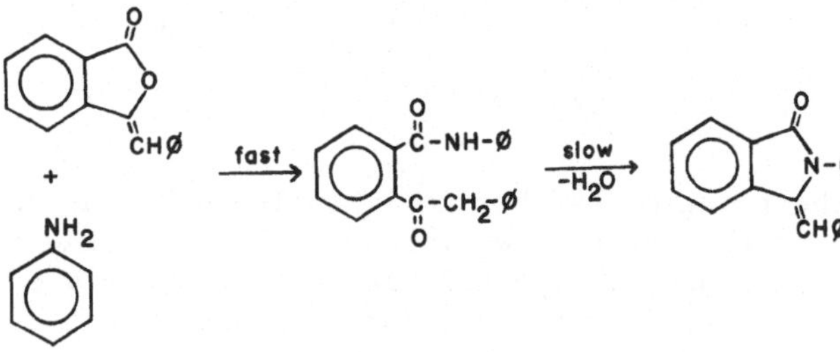

The formation of the keto-amide intermediate can occur in one day at room temperature which indicates that the cyclo-dehydration of the intermediate is the slow step in the synthesis of 2-phenyl-3-benzyl-idinephthalimidine.

MONOMERS

Tetraphenylpyromellitides

Two isomers namely, 3,3,5,5-tetraphenylpyromellitide (cis form; structure [8]) and 3,3,7,7-tetraphenylpyromellitide (trans form; structure [9]) were studied. In the first synthetic approach[7] pyromellitoyl chloride was reacted under Friedel-Crafts acylation conditions to give the cis isomer with only a 6% yield.

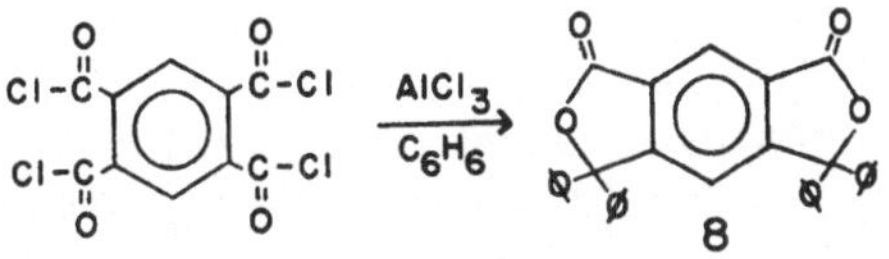

The low yield was believed to be caused by the multiple recrystallizations needed to remove the trans isomer.

An improved method of synthesis was later developed[9] by using pyromellitic dianhydride as a starting material to yield 23% and 27% of the cis [8] and trans [9] isomers respectively.

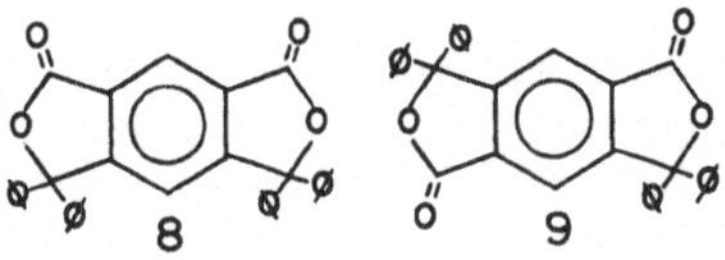

Pyromellitic dianhydride under Friedel-Crafts conditions gave 4,6-dibenzoylisophthalic acid [10] and 2,5-dibenzoylterephthalic acid [11].

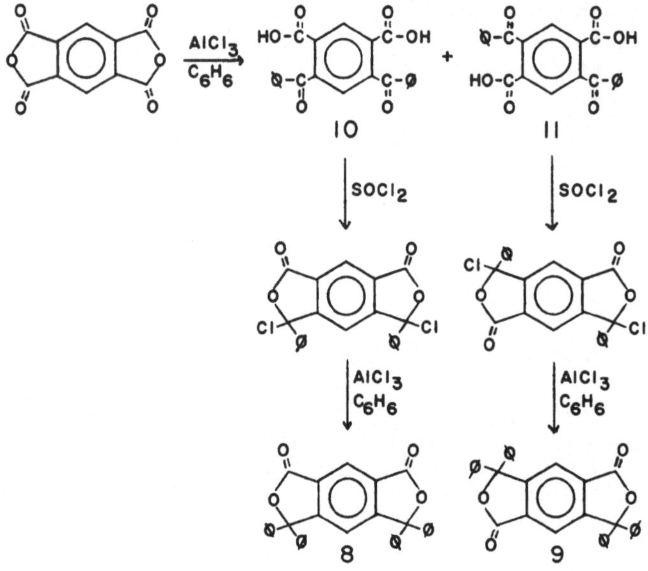

The two isomers were separated based on the relative solubilities of their potassium salts, the salt of the trans precursor [11] being far less soluble in aqueous solution. The free acid was then converted to the cyclized (pseudo) acid chloride with thionyl chloride followed by another Friedel-Crafts reaction to obtain the desired products, [8] and [9].

NMR spectra[32] serve to distinguish between the cis and trans structures, [8] and [9]. The central aromatic hydrogens in the trans structure are in identical environments and therefore give rise to just one singlet at $\delta 8.08$, whereas two singlets are observed for the non-identical, central hydrogens in the cis structure: one at $\delta 7.58$, the other at $\delta 8.29$.

A project was undertaken to apply Wittig's approach[24] to the synthesis of tetraphenylpyromellitides but the yields were poor[23]. However, it is feasible that the reaction be carried out in one step and with suitable reaction conditions one could get better yields also.

Dibenzylidenepyromellitides[14,15]

Two isomers, namely, 3,5-dibenzylidenepyromellitide [12] and 3,7-dibenzylidenepyromellitide [13] were synthesized by the Perkin reaction of pyromellitic dianhydride and phenylacetic acid in the presence of sodium acetate at 240-250 °C. The isomers were successfully separated by fractional recrystallization in dimethylformamide.

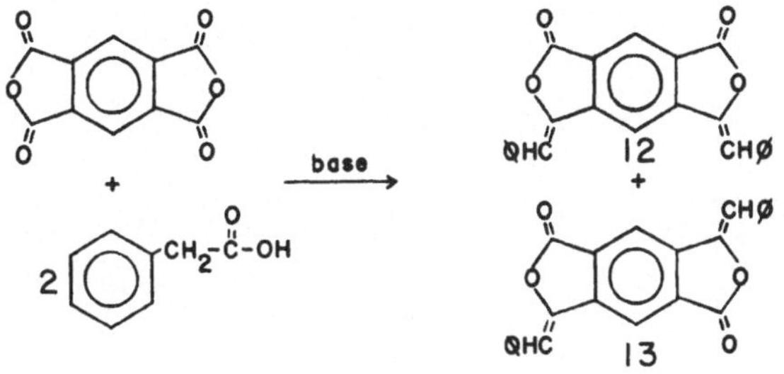

Tetraphenyltetrathiopyromellitides

Analogous to tetraphenylpyromellitides, two isomers namely, 3,3,5,5-tetraphenyltetrathiopyromellitide [14] and 3,3,7,7-tetraphenyltetrathiopyromellitide [15] were obtained in 90% yields from [8] and [9] respectively. The reactions of [8] and [9] were carried out with phosphorus pentasulfide in xylene or decalin to give the brilliant, dark-red, shiny crystals of their respective tetrathio derivatives.

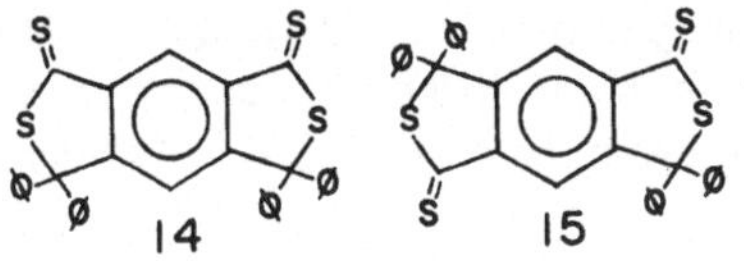

Bis(3,3-diphenyl-6-phthalidyl) Ketone

Three isomers [16], [17] & [18] of benzophenonedibenzoylcarboxylic acid were obtained[33] when 3,3',4,4'-benzophenonetetracarboxylic acid dianhydride was treated as described above.

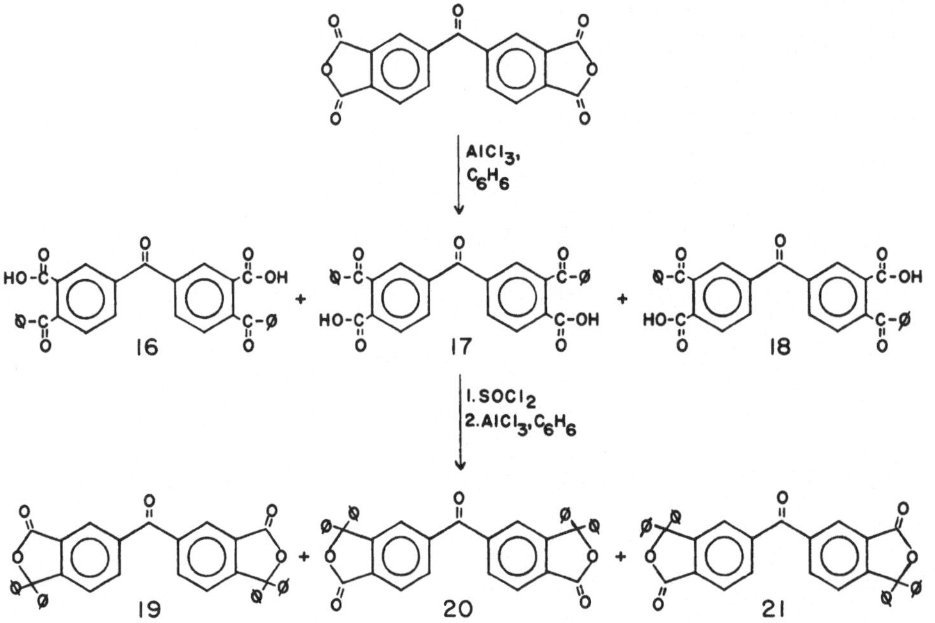

One of these [16] was separated due to its solubility in chloroform. Further treatment of [16] with thionyl chloride followed by another Friedel-Crafts reaction yielded the desired product namely bis(3,3-diphenyl-6-phthalidyl) ketone [19]. Similar reactions with other isomers, [17] and [18] gave [20] and [21] respectively.

The NMR spectrum (100 MHz) of [19] showed[33] only four types of aromatic protons, which indicated that the molecule is symmetrical about the carbonyl. The four pendant phenyl groups appeared to be a singlet at 7.37 ppm. Upon expansion, the six downfield aromatic protons appeared as three quartets.

Additionally, the three isomers namely, [19], [20] and [21] were also identified by the method of cleavage of aromatic ketones[34,35]. Cleavage of [19] would yield the 6-carboxyl; [20] would give 5-carboxyl; and [21] would result in a mixture of 5- and 6-substituted compounds. When Swan's method of cleaving nonenolizable ketones was applied to [19], 3,3-diphenylphthalide and 3,3-diphenyl-6-carboxy-phthalide were obtained in 75% and 73% yields respectively, thus strengthening its chemical proof.

Diphenylaminophthalides

Two diphenylaminophthalides have been synthesized[10,13]. The 3,3-diphenyl-6-aminophthalide [24] was synthesized in 20% yield by the following sequence of reactions which was patterned after early work[36-38]. This series involved nitration of phthalimide, hydrolysis and dehydration to 4-nitrophthalic anhydride, Friedel-Crafts reaction with benzene to 2-benzoyl-5-nitrobenzoic acid, cyclization with thionyl chloride to the pseudoacid chloride, Friedel-Crafts reactions with benzene, and, finally reduction of the nitro group to the amino function with Adams' catalyst.

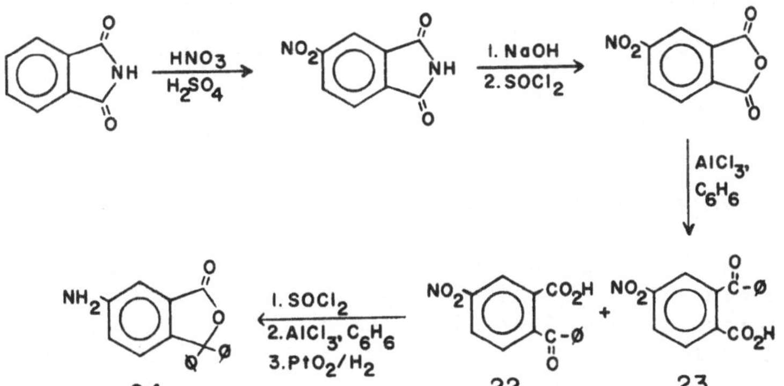

In the fourth step of the synthesis, two isomers are possible: 2-benzoyl-5-nitrobenzoic acid [22] and 2-benzoyl-4-nitrobenzoic acid [23]. The yield of isomer [22], however, is about five to ten times that of isomer [23].

Another aminophthalide, 3-(p-aminophenyl)-3-phenylphthalide [25], was synthesized[13] from 3,3-diphenylphthalide. In this case, the starting material was mononitrated on one of the pendant groups by the use of acetyl nitrate or nitronium tetrafluoroborate. The resulting nitrophthalide was then reduced to the amine with hydrogen over PtO_2 or with ammonium sulfide with overall yields of 85-90%.

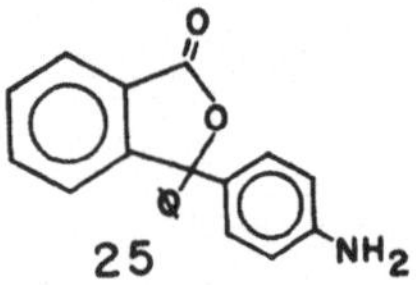

25

Two other monomers, 6-amino-3-benzylidenephthalide [26] and
3-(p-aminobenzylidene)phthalide [27], were reproduced from procedures
published elsewhere[39,40].

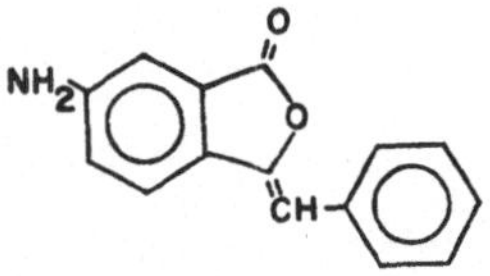

26 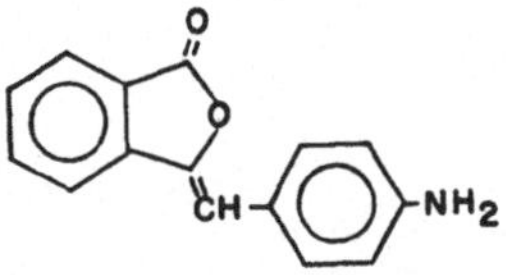   27

Chloroformylbenzylidenephthalides

Very recently, Imai and co-workers successfully synthesized two
monomers, 3-benzylidene-5-chloroformylphthalide [30] and 3-benzyl-
idene-6-chloroformylphthalide [31], from a Perkin reaction of trimellitic anhydride with phenylacetic acid, followed by chlorination.

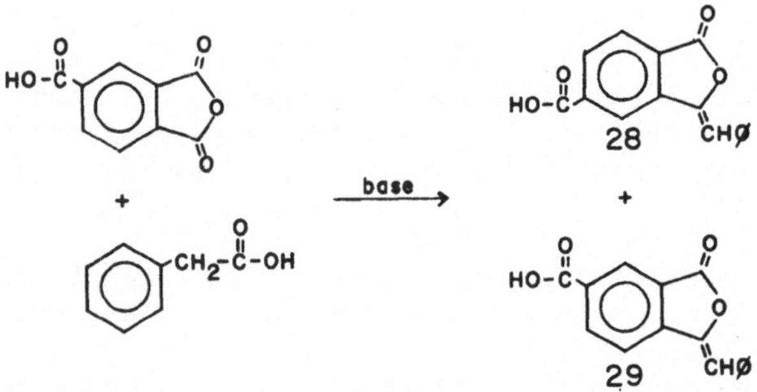

Intermediates, [28] and [29], were isolated by fractional crystallization with acetic acid. Subsequent chlorination of [28] and [29]
with thionyl chloride yielded the following chloroformylbenzylidene-
phthalides.

POLYMERS

A new class of polymers called polyimidines were reported for
the first time in 1976 almost simultaneously by two independent
research groups[7-13,14-19]. Polybenzodipyrrolediones, now
called polypyromellitimidines, were obtained[14-19] by reacting the
difurandione monomers [12 and 13] with aromatic diamines.

The reactants were heated at reflux in m-cresol (200 °C), with
boric acid as a dehydrating agent, to obtain aromatic polymers which
showed a 10% weight loss at 460-500 °C in nitrogen. It has been suggested that the exocyclic carbon-carbon double bond enhances the reactivity of the lactone with diamines. This is supported by the use of
relatively low reaction temperatures, and by the high yields and
molecular weights obtained.

Polyimidines with pendant phenyls were obtained by incorporating
tetraphenylated phthalides into a polymeric backbone[7-13]. The existence of pendant phenyl groups enhances the solubility of the polymers
to the extent that they show appreciable solubility in chloroform and
N,N-dimethylformamide.

Sixteen different phenylated polyimidines[7,9-13] and six of their
thio derivatives[8,9,32] have been discovered in these laboratories.
Table 1 lists these polymers along with some of their physical data.

All pendant phenyl polyimidines reported so far have been formed
by solution polymerization or by the reaction carried out neat (no
solvent) in a sealed reaction tube. The tetraphenylated-bis-phtha-
lides were condensed with a variety of diamines in one step, although
some required a postcure to complete cyclization.

In all cases the thio-monomers reacted faster, showed higher
yields and usually resulted in polymers which were completely cyclized
and more stable than the oxo-systems. Generally, the trans isomers
were found to be more reactive than the cis for both the thio- and
oxo- monomers. In general cis analogues were somewhat more soluble,
however.

Table 1.   Reaction Conditions and Properties of Polyimidines (Cassidy, et al.)

Backbone Structure	R	Reaction Temp.(°C)	Viscosity (dl/g)	Molecular Wt. ($\overline{M}n$)	TGA* Air/Nitrogen	Reference
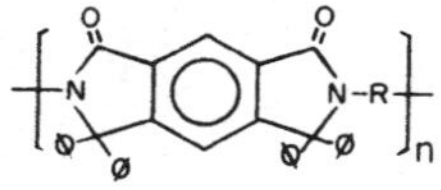	$-(CH_2)_6-$	250	0.13	14,300	300/340	7
	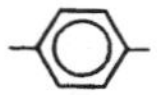 (phenyl)	250	0.07	5,000	420/450	7
	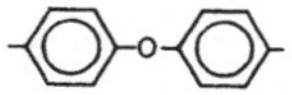 (diphenyl ether)	280	0.07	6,000	460/510	7
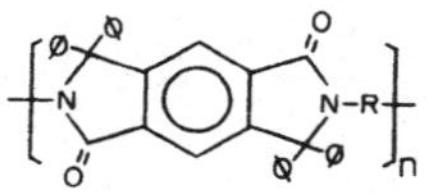	$-(CH_2)_6-$	280	0.08	2,900	340/380	9
	 (phenyl)	350	0.14	5,075	370/390	9
	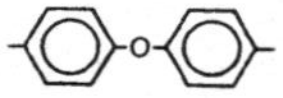 (diphenyl ether)	350	0.15	6,450	400/570	9
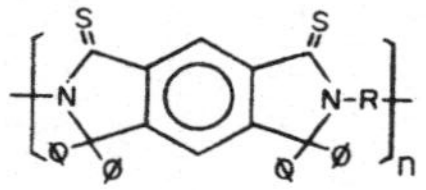	$-(CH_2)_6-$	76	0.56	6,840	350/380	8
	 (phenyl)	370	0.16	5,800	430/450	9
	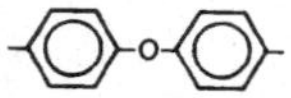 (diphenyl ether)	360	0.20	8,590	560/650	9
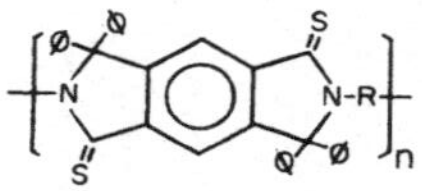	$-(CH_2)_6-$	350	0.45	5,500	370/400	9
	 (phenyl)	370	0.29	10,500	400/530	9
	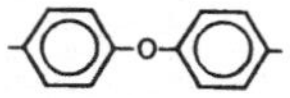 (diphenyl ether)	370	0.37	15,900	470/520	9

R					
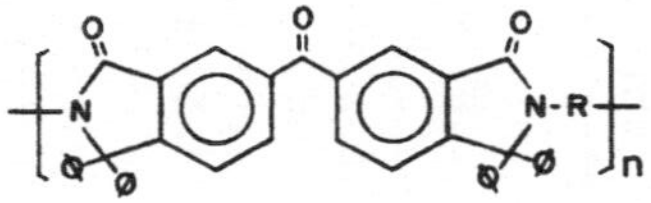 $-(CH_2)_6-$	250	0.10	1,220	365/430	11
 (phenyl)	300	0.13	--	510/535	11
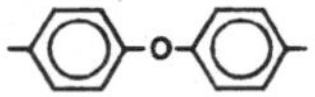 (diphenyl ether)	300	0.10	--	445/475	11
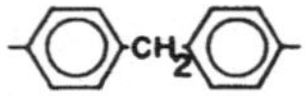 ($-CH_2-$ diphenyl)	300	0.08	--	445/540	11
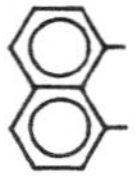 (naphthyl)	250	0.08	--	420/460	12
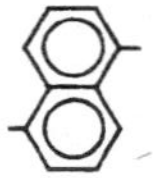 (naphthyl)	275	0.17	--	494/540	12
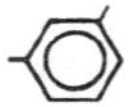 (phenyl)	250	0.08	--	470/505	12
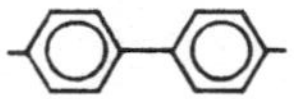 (biphenyl)	300	0.24	--	490/535	12
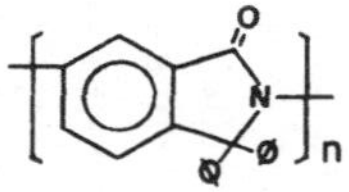	180	0.68	--	420/460	10
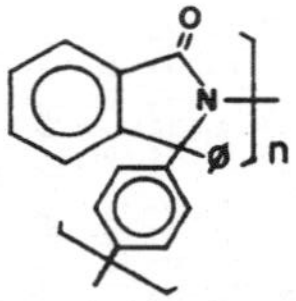	180	0.19	--	320/370	13

*TGA - temperature of 10% weight loss in air/N_2

Most polymers studied here were found to be thermally stable and fairly soluble. The highest number average molecular weight achieved so far is 15,900 for a backbone containing sulfur in the trans positions. The difficulty in obtaining high molecular weights has been attributed to a number of factors. One major problem was the purity of monomers, due primarily to the rapid oxidation of the diamines, even in solution. Problems encountered during the sealed tube reactions were thermal decomposition resulting from the high temperatures necessary to facilitate reaction and the removal of the off-gas, especially water, which interferes with polymerization.

Poly(benzylidenepyromellitimidines) as obtained by Imai and co-workers are listed in Table 2 together with polymerization conditions and other physical properties. The pendant benzylidene groups enabled a more facile polymerization than did pendant phenyl groups.

Thermal behavior and the cross-linking reaction of the aromatic polypyromellitimidines with pendant benzylidene groups have been recently explored[18]. In that study, thermally polymerizable ethylenic double bonds on the heterocycles were used as cross-linking sites to produce high temperature polymers with no volatile material being produced during the reaction. Two polymers were prepared according to the procedure described earlier[15] by the cyclopolycondensation of bis(4-aminophenyl) ether with dibenzylidene pyromellitides. Inherent viscosities as measured in m-cresol at 30 °C ranged from 0.54 to 0.58 dL/g. The cured polymers had better thermal stability than the corresponding uncured ones.

EXPERIMENTAL PROCEDURES

3,3-Diphenylphthalide [1][20,21,44]

A 500-mL round-bottom flask with condenser, mechanical stirrer, tube, and addition funnel attachments was used as the reaction vessel. Approximately 250 mL of benzene and 39.3 g (0.295 moles) of aluminum chloride were placed in the reaction flask. The reaction flask was cooled in an ice bath and then 30.0 g (0.148 moles) of phthaloyl chloride, which had been previously dissolved in 25 mL of benzene, was added slowly with stirring to the reaction flask. The red mixture was warmed until evolution of hydrogen chloride ceased. The reaction mixture was cooled to room temperature and then poured into 250 mL of crushed ice. The benzene layer was removed, washed with three-100 mL portions of six normal hydrochloric acid, and washed again with three-100 mL portions of distilled water. The washed benzene solution was dried over anhydrous magnesium sulfate and concentrated to 75 mL by rotary evaporation. When the benzene solution was cooled to room temperature crystals were produced. The crystals were dissolved in boiling 95% ethanol and the solution decolorized with charcoal. The

Table 2. Reaction Conditions and Properties of Polyimidines (Imai, et al.)

Backbone Structure	R	Polyimerization Conditions Temp (°C)	Time (days)	Solvent	Inherent Viscosity (dl/g)	TGA[1] Air/Nitrogen	Reference
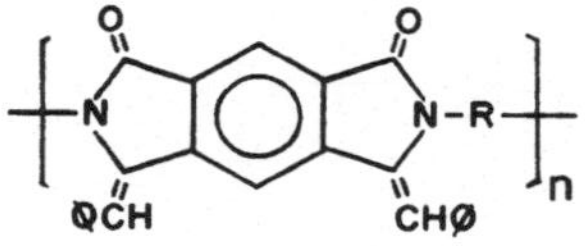	-(CH$_2$)$_6$-	200	1	NMP[2]	0.51	405/410	15
	 -CH$_2$-⬡-CH$_2$-	200	1	m-cresol[2]	0.20	420/430	15
	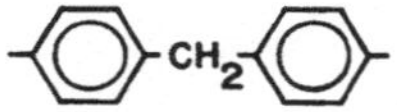 ⬡-CH$_2$-⬡	270	1	o-phenylphenol[2]	1.02	405/460	15
	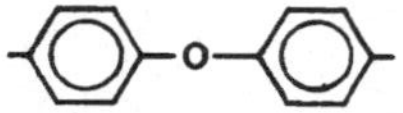 ⬡-O-⬡	270	1	o-phenylphenol[2]	0.58	425/500	15
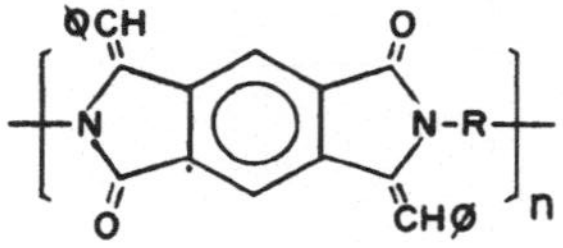	-(CH$_2$)$_6$-	200	1	m-cresol[2]	0.30	400/405	15
	 -CH$_2$-⬡-CH$_2$-	200	1	m-cresol[2]	0.36	415/430	15
	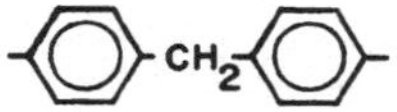 ⬡-CH$_2$-⬡	270	1	o-phenylphenol[2]	0.45	385/485	15
	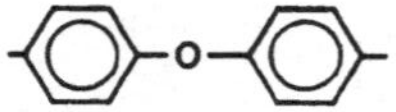 ⬡-O-⬡	270	1	o-phenylphenol[2]	0.72	400/490	15

(continued)

Table 2. Continued

Polymer structure	R	Temp.		Solvent		λ	Ref.
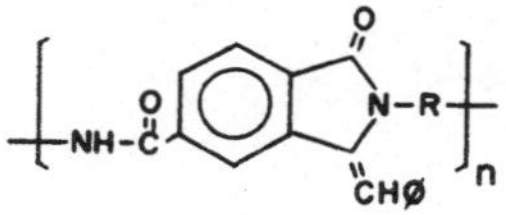 (structure with NH–C=O, N–R, CHØ, O)	 phenyl	200	1	NMP	0.20	410/460	19
	 phenyl (methyl-substituted)	200	1	NMP	0.35	450/470	19
	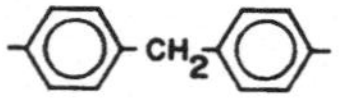 CH_2-bridged biphenyl	200	1	NMP	0.49	380/460	19
	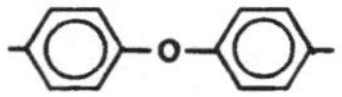 O-bridged biphenyl	200	1	NMP	0.46	365/435	19
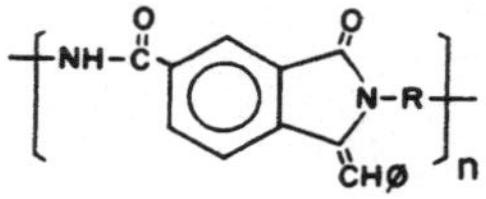 (structure with NH–C=O, N–R, CHØ)	 phenyl	200	1	NMP	0.22	360/475	19
	 phenyl	200	1	NMP	0.32	365/465	19
	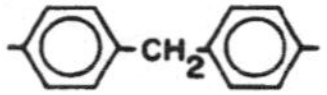 CH_2-bridged biphenyl	200	1	NMP	0.36	370/450	19
	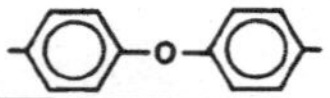 O-bridged biphenyl	200	1	NMP	0.48	370/460	19
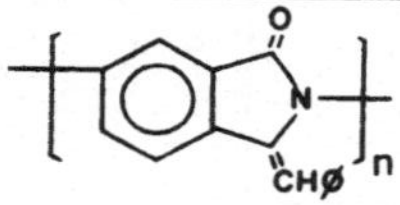 (structure with N, CHØ, O)		250	3	o-phenylphenol	0.51	420/460	17
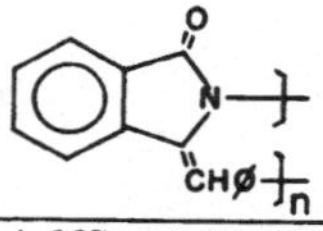 (structure with N, CHØ, O)		250	1	o-phenylphenol	0.12	455/490	17

1. A 10% weight-loss temperature by TGA
2. Boric acid catalyst

cooled ethanol solution produced 17.1 g (0.06 mol, 40% yield) of white crystalline, 3,3-diphenylphthalide, m.p. 115-116 °C.

3,3-Diphenylnaphthalide [2][22,23,24]

(Method A) A 500-mL round-bottom flask, equipped with a nitrogen inlet, a mechanical stirrer and a condenser attached to a gas trap, was charged with 47.0 g (0.22 mol) of nephthalic anhydride, 94.4 g (0.47 mol) of phosphoruspentachloride and 150 mL of phosphorusoxy-chloride. After a 60 hr reflux period the phosphorusoxychloride was removed by vacuum distillation. To the remaining yellow naphthaloyl chloride was added 58 g (0.44 mol) of anhydrous aluminum chloride in 300 mL of partially frozen benzene. The mixture was stirred at room temperature for 4 hours, heated at 50 °C for 2 hours and then heated at reflux for 3 hours. The deep red solution was hydrolyzed after it was carefully poured into a mixture of 500 mL of ice in 200 mL of dilute HCl. The layers were separated and the benzene layer was washed with 4 x 100 mL volumes of water. The aqueous layer was extracted repeatedly with benzene and all the benzene layers were combined. The combined benzene layers were dried and rotary evaporated to give a deep red oil. The oil was then dissolved in absolute ethanol, a volume of 10% KOH equal to 1/3 the volume of ethanol was added, and the solution was heated at boiling for 10 min., cooled and filtered to give yellow crystals. These were recrystallized in glacial acetic acid to give white needles, m.p. 198-200 °C (lit.[22] 202-203 °C); yield 26.6 g (36%).

(Method B) A 250 mL round-bottom flask equipped with a magnetic stirrer, a condenser, a drying tube and a gas by-pass dropping funnel was used as a reaction vessel. A volume of 100 mL of anhydrous diethyl ether was added to the flask and the system was purged with nitrogen through a gas bubbling tube for 30 min. The remainder of the reaction was performed under a flowing nitrogen blanket. Next, 3.47 g (0.50 mol) of lithium wire was added by cutting the wire into tiny pieces and allowing them to fall into the reaction vessel against a positive nitrogen pressure. The dropping funnel was charged with 30 mL (0.28 mol) of bromobenzene. The reaction was started by first stirring the lithium in ether at high speed followed by the addition of about 2 mL of bromobenzene. The exothermic reaction was controlled by adding bromobenzene dropwise at a rate which maintained a steady reflux. The reaction was determined to be complete when the lithium ceased to bubble. After standing in an ice bath for 15 min, a 2-mL aliquot was removed from the dark brown supernatent with a 5-mL syringe through a rubber serum cap and slowly added to 25 mL of water, forming LiOH, which was titrated against a standard solution of sulfuric acid.

The phenyllithium was then reacted with naphthalic anhydride by the method described by Wittig, Leo and Weimer[24]. To a 500-mL round-bottom flask equipped with a mechanical stirrer, a drying tube, and a

rubber serum cap, was added a stoichiometric quantity of naphthalic
anhydride as determined by the titration above. About 250 mL of
anhydrous diethyl ether was added as solvent and the reaction mixture
was cooled to 0 °C in an ice-acetone bath. The appropriate amount of
standard phenyllithium solution was slowly added dropwise to the
stirred mixture by pushing it through a 2-headed needle (cannula)
with nitrogen pressure. The red-orange suspension was allowed to
warm to room temperature with stirring and then left to stand overnight. The reaction mixture was hydrolyzed by being poured into
400 mL of water after which the mixture was light yellow. The ether
solvent was removed and the brown precipitate collected by suction
filtration. The crude product was then recrystallized in absolute
ethanol to produce long colorless needles of 3,3-diphenylnaphthalide,
m.p. 198-200 °C. The filtrate was acidified with dil HCl to yield a
red oil which was extracted into ether. After removal of the ether
the oil was treated with acetone , liberating crude crystals of naphthalic anhydride which were collected by suction filtration. The
filtrate was rotary evaporated and the solid recrystallized in
absolute ethanol to give colorless crystals of 3,3-diphenylnaphtha-
lide. Overall yields ranged from 30% to 60%.

3,3-Diphenyldithiophthalide [3][8,43]

A solution of 10 g (0.035 mol) of 3,3-diphenylphthalide, m.p. 115-
116 °C and 7.7 g (0.035 mol) of phosphorus pentasulfide in 150 mL of
xylene was heated under reflux for 12 hrs. The deep red solution was
filtered hot and steam-distilled until the residue precipitated. The
residue was collected and recrystallized twice from absolute alcohol
with a small amount of decolorizing charcoal to yield 8.0 g (72%) of
brightly shining orange leaflets which melted at 162-164 °C.

3-Benzylidenephthalide [4][25,27,28]

Phthalic anhydride (100g, 0.7 mol), phenylacetic acid (110 g,
0.8 mol) and sodium acetate (2.6 g) were placed in a still pot and
melt-fused at 200 °C until no more carbon dioxide evolved from the
mixture. Next, the temperature was raised to 250 °C and held until no
more water would distill. The resulting mass was recrystallized in
95% Et-OH to give 3-benzylidenephthalide (m.p. 100-1 °C) in yields up
to 60%.

2,3,3-Triphenylphthalimidine [5][30,25,42]

Aniline (3.2 mL, 0.035 mol),3,3-diphenylphthalide (10 g, 0.035
mol), and 0.25 g of aniline hydrochloride were refluxed in 50 mL of
NMP for 8 hrs. The solvent was removed and the oil recrystallized
from Et-OH/water to give 2,3,3-triphenylphthalimidine (m.p. 185-6 °C).
Yield 14%.

<u>2,3,3-Triphenylthiophthalimidine</u> [6][8,43]

To 1.0 g (0.003 mol) of 3,3-diphenyldithiophthalide [5] m.p. 162-
164 °C, 2.0 g (0.022 mol) of freshly distilled aniline was added.
The mixture was heated under reflux for 8 hrs. The solution was
cooled, and a dark residue formed. The excess aniline was removed by
suction filtration. The crude product was washed by multiple applications of warm water and was recrystallized from absolute alcohol
with a small amount of decolorizing charcoal. The slightly yellow
powder weighed 0.98 g (83% yield) and melted at 157-158 °C.

<u>2-Phenyl-3-benzylidenephthalimidine</u> [7][25]

A solution of 3-benzylidenephthalide (5 g, 0.02 mol), aniline
(2.1 mL, 0.02 mol) and NMP (25 mL) was heated at reflux for 8 hrs.
The solvent was removed and the resulting oil recrystallized from
Et-OH/water to give 2-phenyl-3-benzylidenephthalimidine (m.p. 197-
98 °C).Yield 50%.

<u>3,3,5,5-Tetraphenylpyromellitide</u> [8][7,9,32,44]

A solution of 24.0 g (0.07 mol) of 4,6-dibenzoylisophthalic
acid and 120 mL of thionyl chloride was heated at reflux for 3 hrs.
to give a clear yellow solution. A vacuum line was attached to the
flask, and the excess thionyl chloride was removed until the pseudoacid chloride was left as an off-white paste. Radiant heat from an
infrared lamp was useful in this last step. To the paste of pseudoacid chloride was added 1 L of dry benzene followed by slow addition
with stirring of 70.09 g (0.52 mol) of aluminum chloride. The resulting heterogeneous mixture was refluxed for 16 hrs, and then the benzene layer was concentrated to about 50 mL. After cooling, the aluminum chloride complex was destroyed by the slow addition of 500 g
of ice and 200 mL of 6 M HCl. The benzene layer was separated from
the resulting mixture, filtered, concentrated to 200 mL and 800 mL
of absolute alcohol was added. Upon cooling 3,3,5,5-tetraphenylpy-
romellitide precipitated. The precipitate was collected, dried, and
dissolved in a minimum amount of boiling benzene. To this was added
an equal amount of absolute alcohol and the solution was cooled to
yield white platelets: m.p. 280-282°C (47.0%).

<u>3,3,7,7-Tetraphenylpyromellitide</u> [9][7,9,32,44]

The process described above was repeated except that 2,5-diben-
zoylterephthalic acid was used. The reaction mixture workup was
changed as described below.

The final reaction mixture was steam distilled to remove the
benzene and the solid residue was isolated by filtration and dried.
The filtrate was extracted twice with 100-mL portions of benzene and
twice with 100-mL portions of chloroform. The extracts were evaporated on a steam bath and the residue was added to that solid removed

directly from the steam distillation residue. The combined solids
were subjected to Soxhlet extraction with benzene for 3 days. The
benzene solution was reduced in volume to 200-300 mL and an equal
amount of absolute alcohol was added to precipitate the tetraphenyl-
pyromellitide. The solid was recrystallized from benzene or benzene-
ethanol to yield 15.4 g (48.6%): m.p. 354-356°C.

3,5- and 3,7-Dibenzylidenepyromellitide [12,13][14],[15]

A mixture of pyromellitic dianhydride (39.3 g, 0.18 mol), phenylacetic acid (54.5 g, 0.4 mol), and freshly fused sodium acetate
(2.9 g) was heated at 240 °C for 10 hrs. The solid mass was fractionally recrystallized in DMF to give 3,5-dibenzylidenepyromellitide
(more soluble) and 3,7-dibenzylidenepyromellitide (less soluble).
After 2 recrystallizations the yields were 13.4 g (20%) and 9.0 g (14%)
respectively.

3,3,5,5-Tetraphenyltetrathiopyromellitide [14][9],[32],[44]

A solution of 3.0 g (0.006 mol) of 3,3,5,5-tetraphenylpyromelli-
tide and 0.80 g (0.012 mol) of phosphorus pentasulfide in 50 mL of
xylene (or decalin) was heated at reflux for 36 hrs. The resulting
clear red solution was filtered hot and steam distilled until a red
residue precipitated. After the residue was cooled to room temperature it was collected by vacuum filtration and dissolved in a minimum
amount of hot chloroform, filtered, and cooled. The yellow precipitate, probably the dithio derivative, was removed by filtration and
the filtrate was concentrated to one-half of its original volume. An
equal amount of alcohol was then added and the solution was cooled to
facilitate precipitation of the tetrathio compound. The fine, red
precipitate that formed was collected by vacuum filtration and dried
in a vacuum desiccator overnight to yield 3.1 g (90%) of 3,3,5,5-
tetraphenyltetrathiopyromellitide, m.p. 338-340 °C.

3,3,7,7-Tetraphenyltetrathiopyromellitide [15][9],[32],[44]

The reaction as described above was carried out on the trans
isomer except that the reaction was complete in less than 12 hrs. The
crude product was obtained in 90% yield and was extracted with boiling
chloroform. The solution was concentrated and cooled to yield beautiful, deep-maroon crystals, mp 357 ± 2 °C. (Slow heating results in
polymerization and no melting point. The value reported was obtained
by inserting capillaries into a preheated block).

Benzophenone-4,4'-dibenzoyl-3-3'-dicarboxylic Acid [16][33],[35]

To a mixture of 100 g (0.31 mol) of 3,3',4,4'-benzophenonetetra-
carboxylic dianhydride (Gulf Oil Chemical Co.) in 850 mL of dry
benzene was added 190 g (1.43 mol) of anhydrous aluminum chloride.
The mixture was heated at reflux for 4 hrs.   After cooling, the

aluminum chloride complex was carefully destroyed by the addition
of 500 g of ice and 100 mL of 12 M hydrochloric acid. The benzene
layer was extracted three times with 500 mL of 3 N sodium hydroxide.
The sodium hydroxide extract was acidified with hydrochloric acid, and
the isomeric acids were filtered, washed with water, and dried in a
vacuum desiccator. This yield was 87 g (57%).

This mixture of isomers was washed with chloroform to give 35 g
of diacid [16], m.p. 255-257 ^{0}C.

Bis(3,3-diphenyl-6-phthalidyl) Ketone [19][33,35]

A mixture of 48 g (0.10 mol) of benzophenone-4,4'-dibenzoyl-3,3'-
dicarboxylic acid and 140 mL of thionyl chloride was heated at reflux
for 4 hrs. The excess thionyl chloride was removed by distillation,
and the residue was dried under vacuum to give 51 g of the pseudoacid
chloride.

Dry benzene (260 mL) was added to the pseudoacid chloride followed by 56 g (0.42 mol) of aluminum chloride. The mixture was heated
at reflux until the evolution of hydrogen chloride ceased. After
being cooled, the aluminum chloride complex was destroyed with 6 N
hydrochloric acid. The benzene layer was extracted with 5% sodium
bicarbonate and dried over sodium sulfate. The benzene was removed
under vacuum, and the residue was recrystallized from acetone/water
to give 32 g (13% based on tetracarboxylic dianhydride) of bis(3,3-
diphenyl-6-phthalidyl) ketone, m.p. 238-39 °C.

3,3-Diphenyl-6-aminophthalide [24][10,36]

A sample of 4-nitrophthalic anhydride (64 g, 0.33 mol) was added
to 500 mL of dry benzene, followed by the slow addition with stirring
of 107 g (0.80 mol) of aluminum chloride. The mixture was then heated
at reflux until the evolution of hydrogen chloride ceased, then
cooled. The aluminum complex was destroyed by the addition of 500 g
of ice and 200 mL of 12 M HCl. The benzene was extracted with water
and with saturated NaCl and dried over sodium sulfate. The benzene
was then removed, and the crude benzoylbenzoic acids were placed in a
Soxhlet and extracted with methanol for 24 hrs. The extract was
allowed to cool and after filtration gave 2-benzoyl-4-nitrobenzoic
acid. Yield 10%, m.p. 158-160 °C (lit.[36] 163-5 °C). The other
isomer, 2-benzoyl-5-nitrobenzoic acid, was isolated by reducing the
filtrate to one-third its volume and cooling. The yield was 50.0 g
(55.6%); m.p. 207-211 °C (lit.[36] 212 °C). A solution of 50.0 g (0.18
mol) of 2-benzoyl-5-nitrobenzoic acid and 50 mL of thionyl chloride
was heated at reflux for 3 hrs. After the excess thionyl chloride was
removed under vacuum, 500 mL of dry benzene was added to the pseudoacid chloride. Aluminum chloride (46.0 g, 0.40 mol) was slowly added
and the mixture was heated at reflux until the evolution of hydrogen
chloride ceased. After the aluminum chloride complex was destroyed

with 400 g of ice and 150 mL of 12 M HCl, the benzene was extracted
with 6N NaOH, water, and saturated sodium chloride. The benzene was
then dried over sodium sulfate, filtered, and removed under vacuum to
give crude 3,3-diphenyl-6-nitrophthalide. After recrystallization
from 95% ethanol, the yield was 46 g (75%); m.p. 173-174.5 °C.

A mixture of 10.0 g (0.03 mol) of 3,3-diphenyl-6-nitrophthalide,
0.2 g of platinum oxide (Adams Catalyst-Alfa Inorganics, Inc.) and
0.03 mol of hydrochloric acid in 250 mL of 95% ethanol was hydrog-
nated in a Parr Pressure Reaction Apparatus model 3911. The initial
pressure was set at 55 psi and the mixture was shaken until the hydrogen uptake was completed. The mixture was filtered, and the ethanol
was removed under vacuum. The crude amine was recrystallized from
95% ethanol. The yield was 8.6 g (92%); m.p. 153.5-155.0 °C.

3-(p-Aminophenyl)-3-phenylphthalide [25][13]

Method one. A solution of 2.0 g (0.015 mol) of nitronium tetrafluoroborate [45] in 30mL of tetrathylene sulfone was added to a
solution of 4.32 g (0.015 mol) of 3,3-diphenylphthalide in 40 mL of
tetramethylene sulfone under a flowing nitrogen blanket. This mixture
was added to water which precipitated the crude nitro compound. One
gram of the crude product was dissolved in benzene and placed on a
2.5 x 30 cm silica gel (60-200 mesh) column which was slurry packed
with hexane. Elution successively with 100 mL of hexane, 100 mL of
25% (v/v) benzene in hexane, 200 mL of 50% (v/v) benzene in hexane
and 150 mL of benzene gave, upon work-up, 0.92 g of the pure 3-(p-
nitrophenyl)-3-phenylphthalide, m.p. 65-68 °C, with softening at
54 °C. The remaining crude nitro compound was reduced to 3-(p-aminophenyl)-3-phenylphthalide by using ammonium sulfide[30], and was isolated as the amine hydrochloride. The yield was 3.46 g (85.0%) after
compensating for the 1 g aliquot of the nitro compound removed for
purification.

Method two. Acetyl nitrate was prepared by the slow addition of
2.5 mL of cold, 90%, fuming nitric acid to 22.5 mL of iced acetic
anhydride. To this was added one drop of sulfuric acid. The nitra-
ting solution was then added dropwise to a solution of 12.87 g (0.045
mol) of 3,3-diphenylphthalide in 50 mL of acetic anhydride, at 0 °C.
After the addition was completed, the mixture was stirred for 30 min
at 0 °C and then added to 250 mL of 3 M sodium hydroxide maintained
at 0 °C. The crude nitro compound was collected by filtration, dissolved in ether, washed with 5% sodium bicarbonate and dried over
sodium sulfate. Yield 13.2 g (92%). After the ether was removed,
the nitro compound was catalytically reduced by the use of platinum
oxide, 95% ethanol, and an equivalent amount of hydrogen chloride.
The crude product was recrystallized from 95% ethanol to give 10.85 g
(90.0%) of 3-(p-aminophenyl)-3-phenylphthalide; m.p. 222-225 °C.

3-Benzylidene-6-aminophthalide [26][17]

By the method of Borsche et al.[39], phthalide was nitrated to give
6-nitrophthalide in 86% yield; m.p. 143 °C. This was then condensed
with benzaldehyde in the presence of piperidine to give 3-benzylidene-
6-nitrophthalide in 36% yield; m.p. 233-35 °C. Finally the nitro
group was reduced with stannous chloride to give 3-benzylidene-6-
aminophthalide in 52% yield; m.p. 213 °C from aqueous acetone.

3-(p-Aminobenzylidene)phthalide [27][17]

The method of Leupold[40] was used to prepare 3-(p-nitrobenzyli-
dene)phthalide in a 42% yield by the condensation of phthalic anhydride with p-nitrophenylacetic acid in the presence of potassium
acetate; m.p. 227-229 °C (lit[40] m.p. 222 °C). The nitro compound was
reduced by the conventional stannous chloride reduction to give 3-(p-
aminobenzylidene)phthalide in a 64% yield. Recrystallization from
aqueous acetone gave pale yellow needles; m.p. 233 °C.

3-Benzylidene-5-carboxyphthalide [28] and 3-Benzylidene-6-carboxy-phthalide [29][19]

In a flask were placed 96.0 g (0.5 mol) of trimellitic anhydride,
102 g (0.75 mol) of phenylacetic acid, and 5.0 g (0.05 mol) of freshly
fused potassium acetate. The mixture was heated with stirring at
230 °C for 5 hrs and the water produced was distilled off during the
reaction. The resulting solid mass was fractionally recrystallized
with acetic acid to give 3-benzylidene-6-carboxyphthalide (more soluble) and 3-benzylidene-5-carboxyphthalide (less soluble). These were
then separately recrystallized from glacial acetic acid to give yields
of 64% and 15% respectively. 3-benzylidene-6-carboxyphthalide melts
at 291 °C and 3-benzylidene-5-carboxyphthalide melts at 349 °C.

3-Benzylidene-5-chloroformylphthalide [30][19]

A mixture of 13.3 g (0.05 mol) of 3-benzylidene-5-carboxyphtha-
lide, 150 mL of thionyl chloride, and 2 drops of dimethylformamide
(DMF) was heated with stirring at 90 °C until a yellow solution was
obtained (1 hr). The excess thionyl chloride was removed under
reduced pressure. The yellow solid formed was recrystallized from
benzene to afford yellow plates; m.p. 205 °C. It weighed 11.1 g (78%).

3-Benzylidene-6-chloroformylphthalide [31][19]

A mixture of 26.6 g (0.1 mol) of 3-benzylidene-6-carboxyphtha-
lide, 300 mL of thionyl chloride, and 0.5 mL of DMF was heated at
reflux with stirring for 1 hr. The excess thionyl chloride was distilled under reduced pressure and the residual solid was recrystallized from benzene to give yellow plates; m.p. 171 °C (by DTA). The
yield was 20.1 g (70%).

CONCLUSIONS AND PRESENT RESEARCH

Several monomers were synthesized and polymers made from them. In general, it was found that the thio-monomers were more soluble and reactive than the oxo. They also produced more thermally stable polymers. The cis isomers of the monomers were more soluble than the trans, but the trans were more reactive. The polymers were stable in hot dilute hydrochloric acid and sodium hydroxide solutions. All polymers were soluble in chloroform, dimethylformamide, and sulfuric acid. Polymers which contained sulfur were also soluble in carbon tetrachloride, benzene, xylene and toluene. Brittle films could be cast from solution or melt-pressed.

Attempts at synthesizing the dimer model compound, 2,3,3-triphenylphthalimidine, with lactone and phenylisocyanate and phenylisothiocyanate were unsuccessful. Similar attempts with 3,3-diphenyldithiophthalide did produce the desired model compound, 2,3,3-triphenylthiophthalimidine, but only in sealed glass ampules.

This polyimidine research has led to another new backbone possibility, polybenzylenebenzimidazoles:[41]

Model compounds have been prepared by reaction of o-diaminobenzene with 3,3-diphenylphthalide and with α-bromo-o-cyanotoluene. Synthesis of bidentate analogues is proceeding.

The primary objectives of the research have become quite obvious. Initially, a new polymerization technique and/or ultrapurification of monomers must be employed in order to acheive high molecular weights. Secondly, the possibility of synthesizing monomers by a shorter and more facile approach should be investigated. If these goals are accomplished, then there is a possibility of producing a series of new polymers with commercial value. Some of these approaches are under active investigation in the authors' laboratory.

ACKNOWLEDGEMENTS

The results reported from the authors' laboratory were possible only because of the diligent work performed over the last decade by postdoctoral fellows, graduate students, and undergraduate students. Special contributions were made by Dr. Newton C. Fawcett and Dr.

Raymond A. Lohr. Gratitude is also due to The Robert A. Welch Foundation for financial support. Special thanks go to Ms. Roxie Smeal for typing the final manuscript. T. M. Aminabhavi thanks the administrators of Karnatak University, Dharwad, India for the opportunity of a leave of absence.

REFERENCES

1. P. E. Cassidy, "Thermally Stable Polymers, Syntheses and Properties," Marcel Dekker, Inc., New York (1980).
2. V. V. Korshak, "Heat Resistant Polymers," Moscow (1969) in Israli translation (1971).
3. A. H. Frazer, "High Temperature Resistant Polymers," Wiley Interscience, New York (1968).
4. T. M. Bogart and R. R. Renshaw, J. Am. Chem. Soc., 30:1140 (1908).
5. F. Darmory, Org. Coat. Plast. Preprints 34(1):181 (1974).
6. S. S. Hirsch and S. L. Kaplan, Org. Coat. Plast. Preprints 34(1):162 (1974).
7. P. E. Cassidy and A. Syrinek, J. Polym. Sci., Polym. Chem. Ed. 14:1485 (1976).
8. P. E. Cassidy and F. W. C. Lee, J. Polym. Sci., Polym. Chem. Ed. 14:1519 (1976).
9. P. E. Cassidy, J. C. Lin and N. C. Fawcett, J. Polym. Sci., Polym. Chem. Ed. 17:1309 (1979).
10. N. C. Fawcett, R. A. Lohr and P. E. Cassidy, J. Polym. Sci., Polym. Chem. Ed. 17:3009 (1979).
11. R. A. Lohr, P. E. Cassidy and A. Kutac, J. Polym. Sci., Polym. Chem. Ed. 18:1719 (1980).
12. P. E. Cassidy and S. V. Doctor, J. Polym. Sci., Polym. Chem. Ed. 18:69 (1980).
13. R. A. Lohr and P. E. Cassidy, Makromol. Chem. 181:1375 (1980).
14. M. Ueda, T. Takahashi and Y. Imai, J. Polym. Sci., Polym. Chem. Ed. 14:591 (1976).
15. Y. Imai, M. Ueda and T. Takahashi, J. Polym. Sci., Polym. Chem. Ed. 14:2391 (1976).
16. M. Ueda, T. Takahashi and Y. Imai, J. Polym. Sci., Polym. Chem. Ed. 16:1735 (1978).
17. Y. Imai, T. Takahashi and M. Ueda, J. Polym. Sci., Polym. Chem. Ed. 19:2841 (1981).
18. Y. Imai, T. Takahashi and M. Ueda, J. Polym. Sci., Polym. Chem. Ed., 20:249 (1982).
19. Y. Imai, T. Takahashi and M. Ueda, J. Polym. Sci., Polym. Chem. Ed. 20:1497 (1982).
20. G. Friedel and J. M. Crafts, Compt. Rend. 84:1453 (1877).
21. A. Haller and A. Guyot, Bull. Soc. Chim. 25:54 (1901).
22. F. A. Mason, Chem. Soc., London 125:2116 (1924).
23. R. W. Thomas, M.S. Thesis, Southwest Texas State University, San Marcos, Texas, Dec. (1981).

24. G. Wittig, M. Leo and W. Weimer, Ber. 64B:2405 (1931).
25. Unpublished results.
26. V. Prey and P. Kondler, Monatsh Chem. 89:505 (1958).
27. Gabriel, Ber. 18:2433 (1885).
28. Weiss, Org. Syn. Coll. Vol. II, A. H. Blatt, Ed., Wiley,
        N.Y. (1943) p 61.
29. G. Berti, Gazz Chim. Ital. 86:655 (1956).
30. O. Fischer and E. Hepps, Ber. 27:2791 (1894).
31. H. B. Taylor, M.S. Thesis, Southwest Texas State University,
        San Marcos, Texas, May (1980).
32. N. C. Fawcett, P. E. Cassidy and J. C. Lin, J. Org. Chem.
        42:2929 (1977).
33. R. Lohr, Jr., P. E. Cassidy, N. C. Fawcett and A. Kutac,
        J. Chem. Eng. Data 24(2):156 (1979).
34. G. A. Swan, J. Chem. Soc., 1408 (1948).
35. A. Kutac, M.S. Thesis, Southwest Texas State University,
        San Marcos, Texas, Dec. (1977).
36. W. A. Lawrance, J. Am. Chem. Soc. 42:1871 (1920).
37. E. H. Huntress, E. L. Shloss, Jr., and P. Ehrich, Org. Syn.
        Col. Vol. 2:457 (1943).
38. E. H. Huntress and R. L. Shriner, Org. Syn. Col. Vol. 2:459
        (1943).
39. W. Borsche, K. Diacont and H. Hanau, Ber. 67:675 (1934).
40. E. Leupold, Ber. 34:2837 (1901).
41. R. S. Wallace, M.S. Thesis, Southwest Texas State University,
        San Marcos, Texas, Dec (1982).
42. A. R. Syrinek, M.S. Thesis, Southwest Texas State University,
        San Marcos, Texas, Aug (1975).
43. F. Lee, M.S. Thesis, Southwest Texas State University, San
        Marcos, Texas, Aug (1975).
44. J. Lin, M.S. Thesis, Southwest Texas State University, San
        Marcos, Texas, Dec (1975).
45. S. J. Kuhn, G. A. Olah, J. Am. Chem. Soc. 83:4564 (1961).

SYNTHESIS AND PROPERTIES OF ACETYLENE

TERMINATED ARYL-ETHER OLIGOMERS

B.A. Reinhardt, G.A. Loughran and F.E. Arnold

Air Force Wright Aeronautical Laboratories
Materials Laboratory
Wright-Patterson AFB, OH 45433

and E. J. Soloski

University of Dayton Research Institute
Dayton, OH 45469

INTRODUCTION

A substantial effort in this laboratory has been directed toward the synthesis and characterization of acetylene terminated oligomers for use as addition-curable, moisture-resistant, thermoset systems. Early work on the acetylene-terminated phenylquinoxalines[1,2] demonstrated the moisture insensitivity of the product generated from the thermal cure. Studies of various difunctional acetylene-terminated monomers has demonstrated[3] that the polymerization is a free radical propogation of the acetylene moiety to a linear conjugated polyene. The kinetic chain length of this reaction is unusually short (6-8 acetylene units) and termination is first order.

More recent studies[4] have been concerned with the utilization of acetylene functionality and designing a system which would have all the processing criteria of an epoxide system. Materials which process analogously to the state-of-the-art epoxides require a very flexible backbone which will exhibit a low Tg before cure. The study provided a flexible aryl-ether system which incorporates a phenylsulfone backbone and has been referred to as ATS. The initial synthesis of ATS involved the nucleophilic displacement reaction of various leaving groups in the 4,4' positions of diphenylsulfone with the metallic salt of m-hydroxyphenylacetylene. Research[5] in our laboratory for lower cost precursors to ATS has led to the synthesis of bromo end-capped phenylsulfone oligomers via the Ullmann ether synthesis.

29

The bromo-oligomers can be converted to the acetylene functionality by the catalytically-induced reaction of 2-methyl-3-butyn-2-ol followed by base-catalyzed hydrolytic displacement of acetone.

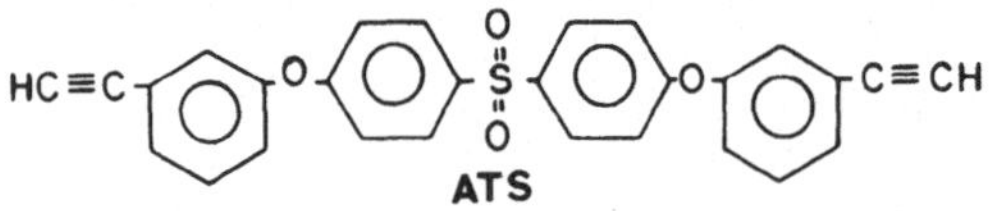

ATS

The objective of this work is to synthesize a series of acetylene terminated aryl-ether systems for the purpose of correlating their thermal and thermal mechanical properties with molecular structure.

EXPERIMENTAL

The general procedure for separation of monomer from oligomer was column chromatography. Separations were normally carried out after the Ullmann reaction, at the dibromo stage of the reaction sequence. The dibromo aryl-ethers were then converted as monomer or as a combination of monomer and oligomer to the acetylenes. The following are representative of the procedures used:

1,3-Bis(3-bromophenoxy)benzene. A mixture of m-dibromobenzene (47g, 0.20 mol), resorcinol (10g, 0.009 mol), anhydrous potassium carbonate (34.5g, 0.25 mol) and cuprous iodide (0.6g) in pyridine (100mL) was heated at reflux under nitrogen for 24 h. The reaction mixture was cooled and poured into 10% HCL (1L). Methylene chloride (200mL) was then added and the organic layer separated. The organic layer was washed with two 100mL portions of water, dried over anhydrous magnesium sulfate, and reduced in volume under reduced pressure to produce a dark viscous oil. The oil was dissolved in n-hexane and chromatographed on a 5cm x 50cm dry silica gel column (quartz) using hexane as the eluent. The second fluorescent band was collected and the solvent removed at 60°C under reduced pressure to give 10.4g of monomer as a colorless viscous oil. Elution with chloroform provided 10.3g of oligomer.
    Analysis Calcd for $C_{18}H_{12}O_2Br_2$: C,51.46; H,2.88; Br,38.04
                          Found: C,51.23; H,2.50; Br,37.73

1,3-Bis(3-ethynylphenoxy)benzene (1) A mixture of 1,3-bis(3-bromophenoxy)benzene (22.6g, 0.03 mol), 2-methyl-3-butyn-2-ol (6.03g, 0.072 mol), and triethylamine (100mL) was degassed by passing nitrogen through the solution for 20 min. To the reaction mixture was added bis-triphenylphosphine palladium chloride (.03g), cuprous iodide (0.13g) and triphenylphosphine (0.30g). The temperature of the reaction mixture was raised to 80°C and maintained there for 24 h. The reaction mixture was then cooled to room temperature and

the triethylamine removed under reduced pressure. The resulting
yellow-red oil was chromatographed on a 5cm x 50cm dry silica gel
column (quartz) using 1:1 hexane-ether as the eluent. The second
fluorescent band was collected (appears yellow on the column). The
solvent was removed under reduced pressure to yield 10.6g (83%) of a
dark viscous oil. The product was used in the next step of the
reaction sequence without further purification.

A mixture of the bis-butynol adduct (10.6g, 0.024 mol) and potassium hydroxide (0.75g) in (20mL) of anhydrous methanol were added
to toluene (100mL) and heated to reflux under nitrogen. The methanol and 40mL of the toluene were then removed by distillation over a
period of 2 h. The reaction mixture was cooled, and the toluene removed at 35°C under reduced pressure. The red, viscous residue was
chromatographed on a dry 5cm x 60cm column (quartz) of silica gel
using 3:1 hexane-methylene chloride. The first large fluorescent
band was collected and the solvent removed at 50°C under high
vacuum. The yield of pure monomeric product was 6.1g (79%).
Analysis Calcd for $C_{22}H_{14}O_2$: C,85.07; H,4.54
Found: C,84.72; H,4.23

4,4'-Bis(3-ethynylphenoxy)benzophenone (4) A solution containing
3-hydroxyphenylacetylene (5.3g, 0.04 mol), 4,4'-difluorobenzophenone
(3.2g, 0.15 mol), anhydrous potassium carbonate (12.5g) in dimethyl-
sulfoxide (250mL) was stirred under a nitrogen atmosphere at 80°C
for 24 h. The reaction mixture was cooled to room temperature and
poured into distilled water (250mL) and extracted with benzene. The
benzene extract was filtered through a bed of silica gel and the
benzene was removed under reduced pressure to give 6g (99%) of a
white crystalline solid. The product on crystallization from cyclohexane exhibited a mp of 83-84°C.
Analysis Calcd for $C_{29}H_{18}O_3$: C,84.04; H,4.38
Found: C,83.65; H,4.09

4,4'-Bis(3-bromophenoxy)biphenyl. A mixture of m-dibromobenzene
(58.9g, 0.25 mol), 4,4'-dihydroxybiphenyl (16.7g, 0.09 mol),
anhydrous potassium carbonate (69g, 0.05 mol) and cuprous iodide
(0.80g) in pyridine (100mL) was heated at reflux under nitrogen for
24 h. The reaction mixture was allowed to cool to room temperature
and poured into 1L of 10% HCL. Methylene chloride (200mL) was then
added and the organic layer separated. The organic layer was washed
with water, dried over anhydrous magnesium sulfate, and the solvent
removed under reduced pressure. The dark, viscous oil was dissolved
in chloroform and placed on silica gel. The impregnated silica gel
was placed at the top of a 5cm x 50cm dry silica gel column and
eluted with 4:1 hexane:chloroform to remove unreacted m-dibromobenzene. The solvent was then changed to 2:1 hexane:chloroform and the
monomeric dibromo-compound removed. The solvent was changed to pure
chloroform to elute the oligomeric dibromo-compound. The yield of
monomeric species was 8.76g (mp 89-92°C). The yield of oligomeric

species was 13.86g, weight ratio of monomer/oligomer 1/1.56.
Analysis Calcd for $C_{24}H_{16}O_2Br_2$:   C,58.09; H,3.25
Found: C,58.32; H,3.16

4,4'-Bis(3-ethynylphenoxy)biphenyl and Oligomer (5) A mixture of oligomeric dibromo-compound ℓ 2.62g, 0.045 mol), 2-methyl-3-butyn-2-ol
(1.15g, 0.08 mol), bis-triphenylphosphine palladium chloride (.03g),
triphenylphosphine (0.30g), cuprous iodide (0.12g), and triethylamine
(130mL) was degassed for 20 min with nitrogen. The above mixture was
heated to reflux for 24 h, cooled and the triethylamine was removed
under reduced pressure. The resulting oil was placed on silica gel
using methylene chloride. The impregnated silica gel was dried and
placed at the top of a dry silica gel column and eluted with 3:1
ether:hexane to remove any dibromo starting material and then with
ethyl acetate to remove the bis-butynol products (24.6g). The dark
viscous oil was used in the next step of the reaction sequence without further purification.

A mixture of the above bis-butynol adduct (24.6g), potassium
hydroxide (1.4g), and anhydrous methanol (20mL) was added to toluene
(200mL) and heated to reflux under a nitrogen atmosphere. The methanol and toluene (120mL) were then removed by distillation over a
period of 2 h. The reaction mixture was allowed to cool to room temperature, filtered and the toluene was removed at 35°C under reduced
pressure. The red, viscous residue was chromatographed on a dry
5cm x 50cm column (quartz) of silica gel using chloroform. The first
large fluorescent band was collected as product (18.8g) and the solvent was removed at 50°C under high vacuum.
Analysis Calcd for $(C_{28}H_{18}O_2)$:   C,87.02; H,4.69
Found: C,86.90; H,4.24

Characterization

Differential scanning calorimetry and thermal mechanical analysis data were obtained on a DuPont 990 thermal analyzer coupled with
a DuPont DSC or TMA cell. Isothermal aging studies were carried out
with an automatic multisample apparatus, which allowed weight loss
to be obtained in the aging environment. Thermal mechanical data for
all the systems prepared are listed in Table I.

RESULTS AND DISCUSSION

The synthesis of the oligomers was carried out by the reaction
of various aromatic bis-diols with m-dibromobenzene leading to a
series of bromo end-capped, aryl-ether systems. Pyridine was used as
the solvent for the reactions, and anhydrous potassium carbonate was
utilized to generate the metallic salts of the bis-diols. In an
effort to promote low molecular weight oligomers, the molar ratio
of m- dibromobenzene to aromatic bis-diol used in the synthesis

Table I. Thermomechanical Properties

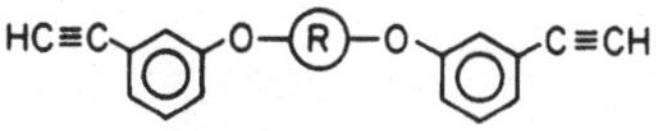

HC≡C-⟨ring⟩-O-(R)-O-⟨ring⟩-C≡CH

#	STRUCTURE (R)	RATIO mono/olig	DSC UNCURED T_m	T_g	CURED* DSC	TMA
1	⟨phenyl⟩ ; (⟨phenyl⟩-O-)$_n$	50/50		-39 / -46	255 exo / 210 exo	249 E / 108 E
2	-⟨ring⟩-$\overset{O}{\underset{O}{S}}$-⟨ring⟩- ; (-⟨ring⟩-$\overset{O}{\underset{O}{S}}$-⟨ring⟩-O-)$_n$	50/50		0 / -1	210 exo / 215 exo	252 E / 266 E
3	-⟨ring⟩-S-⟨ring⟩- ; (-⟨ring⟩-S-⟨ring⟩-O-)$_n$	65/35		-35 / -32	215 exo / 210 exo	232 E / 155 E
4	-⟨ring⟩-$\overset{O}{C}$-⟨ring⟩- ; (-⟨ring⟩-$\overset{O}{C}$-⟨ring⟩-O-)$_n$	40/60	85	3 / -10	225 exo / ~210 exo	263 E / 253 E
5	-⟨ring⟩-⟨ring⟩- ; (-⟨ring⟩-⟨ring⟩-O-)$_n$	40/60	109	-12	215 exo / 230 exo	160 E / 215 E
6	-⟨ring⟩-$\overset{CF_3}{\underset{CF_3}{C}}$-⟨ring⟩- ; (-⟨ring⟩-$\overset{CF_3}{\underset{CF_3}{C}}$-⟨ring⟩-O-)$_n$	40/60		-15 / -5	220 exo / 225 exo	155 E / 154 E

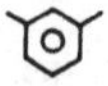

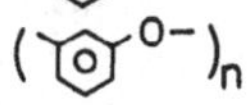

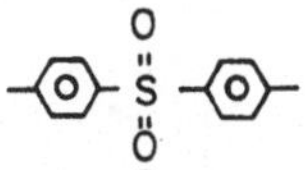

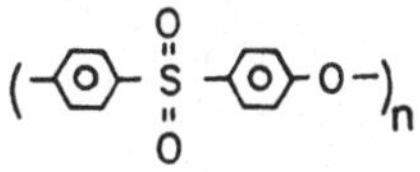

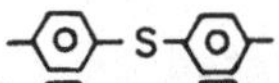

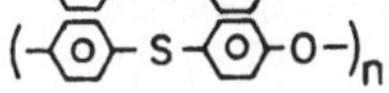

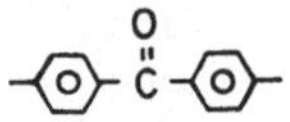

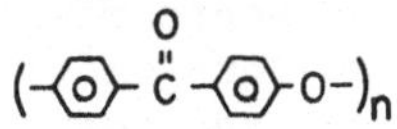

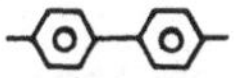

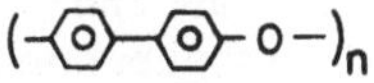

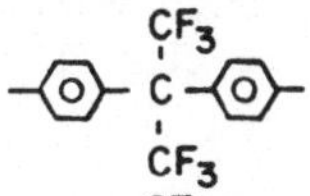

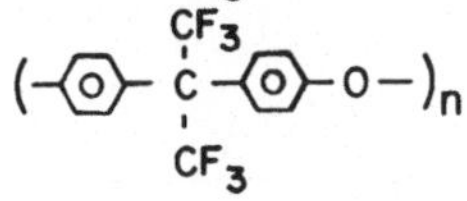

* CURED: 4 HRS @ 350°F
        1 HRS @ 425°F

4 to 1. In general, the amount of monomer obtained from the 4/1 ratio was from 30–50 weight percent (Table I) with the dimer being the second most prevalent oligomer.

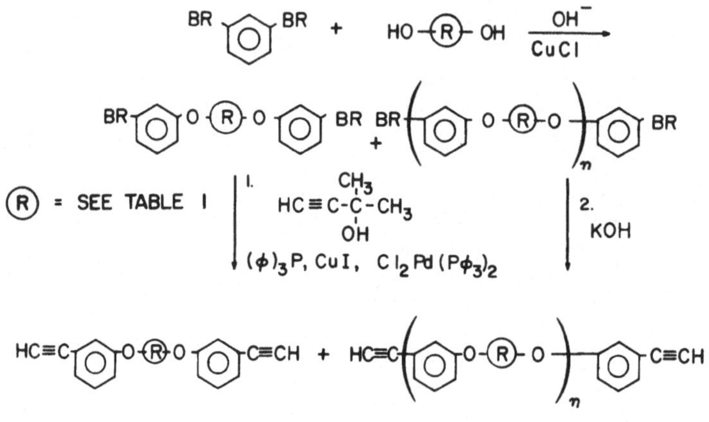

The bromo end-capped oligomers were converted to the acetylene terminated systems by the reaction with 2-methyl-3-butyn-2-ol, utilizing the catalyst system composed of triphenylphosphine, bis-triphenylphosphine palladium chloride, and cuprous iodide. Conversion of the generated butyn-adducts to the acetylene functionality was carried out by the hydrolytic displacement of acetone with potassium hydroxide in toluene. The displacement of acetone under basic conditions is an equilibrium reaction; therefore, the acetone must be removed by codistillation with toluene.

The benzophenone system was also prepared by the reaction of m-hydroxyphenylacetylene with 4,4'-difluorobenzophenone. The synthesis allowed an independent method of obtaining pure monomer which could be compared with the monomer obtained from the Ullmann ether synthesis.

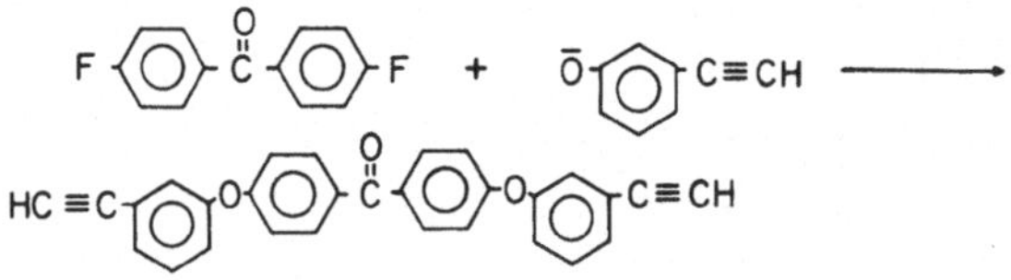

Thermal Mechanical Properties

The synthesis provided the acetylene-terminated, aryl-ether systems as a mixture of monomer and oligomer. The monomer was separated from the oligomer by means of column chromatography preferentially after the Ullmann reaction, at the dibromo stage of the reaction sequence. Samples of the monomer and monomer/oligomer were cured four hours at 350°F followed by one hour at 425°F under a nitrogen atmosphere. Their glass transition temperatures before and after cure were determined using differential scanning colorimetry (DSC) plus thermomechanical analysis (TMA). Both DSC and TMA were obtained at a heating rate of 20°C/min under a nitrogen atmosphere. In all cases, the extrapolated onset of the DSC baseline shift was taken as the Tg. For the determination of the Tg after cure, the TMA expansion (change of rate of expansion) and for penetration modes (point of highest rate of penetration) were used. The results of the thermo-analytical measurements are summarized in Table I.

Analysis of the aryl-ether systems by DSC showed an initial strong baseline shift attributed to the Tg of the material and a strong exotherm initiating at 140°C and maximizing at approximately 240°C, for the polymerization of the terminal acetylene groups. The Tg's of the pure monomers are all at 0°C or below with the exception of the benzophenone and biphenyl systems which exhibit crystalline transitions. The crystallinity of the benzophenone system could be removed by heating the material beyond its melting point (90°C) and cooling. The resulting material was amorphous with a Tg of 3°C. The biphenyl system with a Tm of 109°C retains a high degree of its crystallinity on cooling, after thermal treatment. Cooling and heating showed a polymorphous crystalline form with a Tm of 90°C, which appeared more stable, persisting through several heating and cooling cycles. The DSC studies of the uncured monomer/oligomer samples showed the presence of the oligomer generally to be beneficial with only small changes noted in Tg's. The greatest change was noted with the biphenyl system which was no longer crystalline having a Tg of -14°C.

The materials after cure showed Tg's at or above the cure temperature except for the biphenyl and perfluoroisopropyl systems. DSC studies on all the cured samples indicate incomplete cure with substantial exotherms still evident. Samples of the benzophenone and biphenyl monomeric moieties were cured at 500°F for 16 hours and their DSC's showed no evidence of exotherm in the 250°C region. An exotherm at 350°C (Figure 1) was related to the thermal decomposition of the materials. Thermogravimetric-mass spectral analysis was performed in vacuum on the benzophenone system at the same heating rate as the DSC scan. The ion intensity as a function of temperature is shown in Figure 2 and indicates that thermal decomposition starts at approximately 350°C with the evolution of phenol, biphenol, and water.

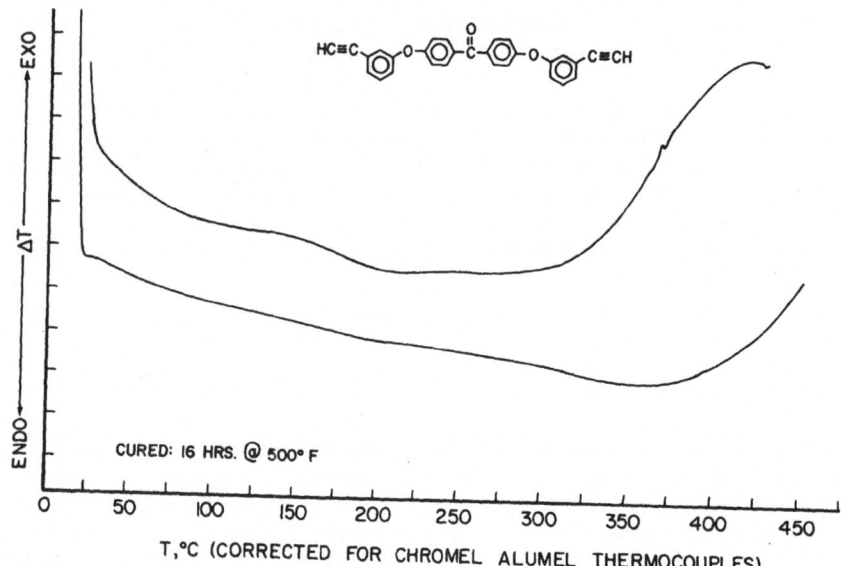

Figure 1. DSC Scan of the Benzophenone System Cured at 500°F for 16 hours.

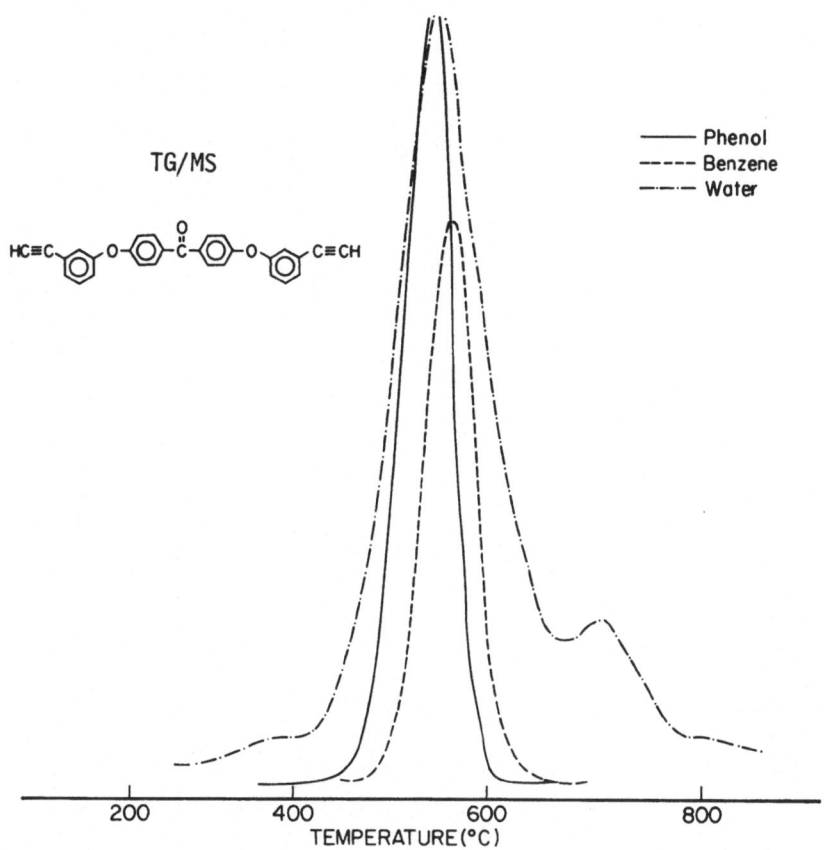

Figure 2.   Thermogravimetric-Mass spectral analysis showing ion
             intensity as a function of temperature.

Thermal Properties

The thermooxidative properties of the polymers by isothermal
aging in air, at 600°F (Figure 3). Very little to no weight loss
was noted when the aging studies were carried out at temperatures
below 500°F. As can be seen, the most stable systems were the
biphenyl and diphenylperfluoroisopropyl. All the materials, except
for the diphenylsulfide system, undergo initial air oxidation as
indicated by early weight gain in the isothermal aging studies.
This is most likely due to the air oxidation of residual acetylene
groups, since the materials were not fully cured at 425°F. Isothermal aging of the samples which were cured at 500°F for 16 hours
did not show an early gain in weight (Figure 4).

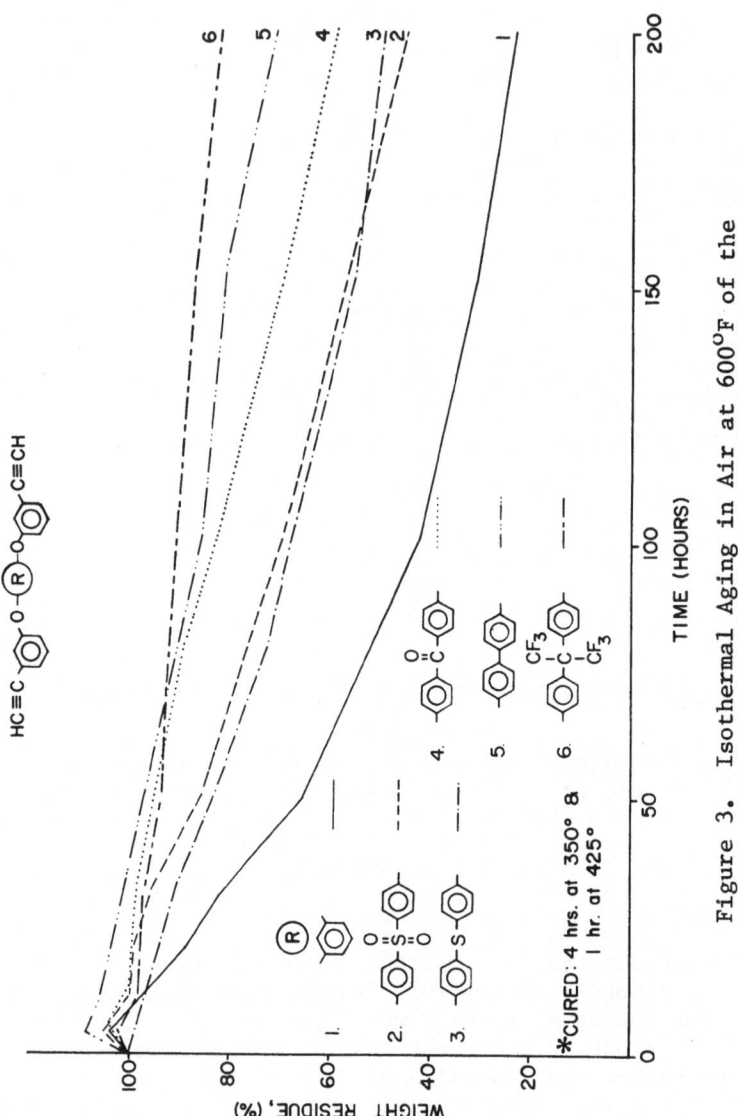

Figure 3. Isothermal Aging in Air at 600°F of the Cured (425°F maximum) Aryl-Ether Systems.

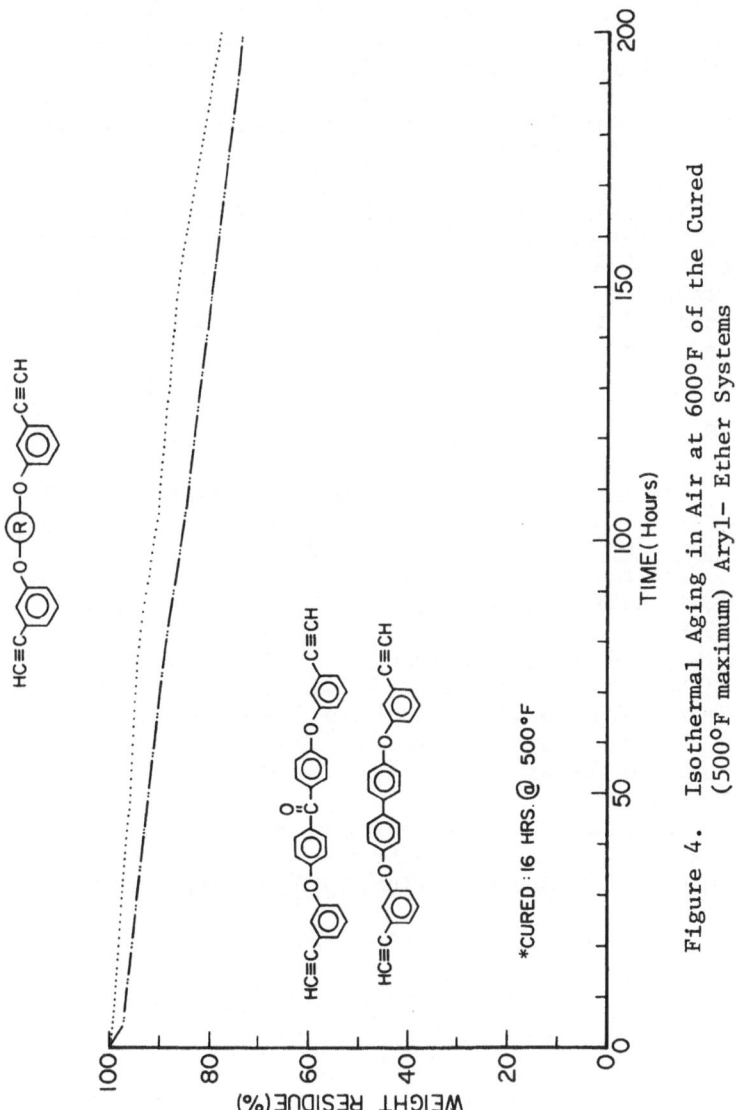

Figure 4.   Isothermal Aging in Air at 600°F of the Cured
(500°F maximum) Aryl- Ether Systems

SUMMARY AND CONCLUSIONS

    A series of acetylene-terminated, aryl-ether thermoset systems
were prepared by an Ullmann ether synthesis involving the condensation of various salts of aromatic bis-diols with m-dibromobenzene.
The bromo end-capped oligomers were converted to the acetylene-
terminated systems by the catalytically-induced, bromo-displacement
reaction with 2-methyl-3-butyn-2-ol, followed by base hydrolysis.
The synthesis provided the materials as mixtures of monomer and
oligomers which were separated by means of column chromatography
and characterized. All of the materials in the amorphous state exhibited Tg's below room temperature, which would provide excellent
tack and drape properties as a graphite prepreg. The thermalooxi-
dative properties of the aryl-ether systems were also excellent, far
exceeding the 300-350^{0}F requirement of an epoxide. The results are
positive and encouraging, although it is recognized that future work
will be required to define more clearly the completeness of cure and
the mechanical properties after cure.

REFERENCES

1. R. F. Kovar, G. F. L. Ehlers, and F. E. Arnold, "Thermosetting
    Acetylene Terminated Polyphenylquinoxalines," J. Polym. Sci.,
    Polym. Chem. Ed., 15:1081 (1977).
2. F. L. Hedberg and F. E. Arnold, "New Acetylene Terminated Phen-
    ylquinoxaline Oligomers," J. of Appl. Polym. Sci., 24:763
    (1979).
3. J. M. Pickard, E. G. Jones and I. J. Goldfarb, "The Kinetics
    and Mechanism of the Bulk Thermal Polymerization of Bis-4-
    (3-ethynylphenoxy)phenylsulfone," Macromol., 12:895 (1979).
4. G. A. Loughran and F. E. Arnold, "Synthesis and Properties of an
    Acetylene Containing Aryl-Sulfone Resin," Polym. Prepr.,
    21:199 (1980).
5. J. J. Harrision, E. T. Sabourin, C. M. Selwitz, W. A. Feld,
    M. R. Unroe and F. L. Hedberg, "A Novel Low-Cost Route to
    Acetylene Terminated Resins," Polym. Prepr., 23:189 (1982).

PHENYLATED AROMATIC HETEROCYCLIC POLYPHENYLENES CONTAINING

PENDANT DIPHENYLETHER AND DIPHENYLSULFIDE GROUPS

B. A. Reinhardt, T. T. Tsai, and F. E. Arnold

Air Force Wright Aeronautical Laboratories
Materials Laboratory
Wright-Patterson AFB, OH 45433

INTRODUCTION

The utilization of pendant groups to promote solubility in common organic solvents was first demonstrated[1] with the aromatic polyphenylenes. Polymers[2] prepared by the oxidation of benzene or dehydrogenation of polycyclohexadiene are crystalline materials and insoluble. Phenylated polyphenylenes obtained from the Diels-Alder polymerization of bistetracyclones with m- or p- diethynylbenzene are amorphous materials and are soluble in toluene. The pendant phenyl groups serve to decrease crystallinity and promote solubility.

Aromatic heterocyclic polymer systems which contain pendant phenyl groups also exhibit unusual solubility properties. The polyphenylquinoxalines[3] exhibit good solubility in chloroform and sym-tetrachloroethane, whereas the polyquinoxalines[4] are soluble only in high boiling aprotic solvents such as hexamethylphosphoramide, dimethylforamide and n-methylpyrolidone. Other heterocyclic systems which contain pendant phenyl groups and are soluble in low boiling chlorinated hydrocarbon solvents, include the polyquinolines[5], polyimides[6], poly-as-triazines[7], and polyimidines[8].

Longer pendant groups which encompass diphenylether moieties have been incorporated into the quinoxaline and as-triazine polymer systems[9]. No increase in solubility was described; however, a substantial decrease in glass transition temperatures was obtained from the diphenylether pendant when compared to the phenyl pendant polymer systems.

This work is concerned with the synthesis and characterization

of several new polymers where diphenylether and diphenylsulfide
pendants are in conjunction with pendant phenyl groups along heterocyclic polymer backbones. The objective of the work was to determine how the substituents affected the properties of the polymers.

EXPERIMENTAL

Biscyclopentadienone Monomers

3,3'-(1,3-Phenylene)bis[2,5-diphenyl-4-p-phenyloxyphenylcyclo-
pentadienone] (1) A stirred suspension of m-bis(p-phenoxyphenyl-
glyoxyl)benzene (10g, 19 mmole) and 1,3-diphenylacetone (8.4g,
40 mmole) in 250mL of 95% ethanol was heated to reflux. To the
mixture was added 20mL of a potassium hydroxide solution (0.6g KOH
in 20mL H_2O) in two portions over a period of five minutes. The
reaction mixture immediately turned purple with the first addition
of base. Heating was continued for forty minutes during which time
a purple solid began to precipitate. The reaction mixture was then
cooled in ice, filtered, and air-dried. The crude product was purified by chromatography on a dry silica gel column using a (9/1)
toluene/ether mixture as the eluent to give 15.2g (82.7%) of purified product, mp 205-206°C.
    Analysis Calcd for $C_{64}H_{42}O_4$: C,87.85; H,4.84
                          Found: C,87.50; H,5.10

3,3'-(1,3-Phenylene)bis[2,5-diphenyl-4-p-phenylthiophenylcyclo-
pentadienone] (2) A solution of m-bis(p-thiophenoxyphenylglyoxyl)-
benzene (10g, 17.9 mmole) and 1,3-diphenylacetone (10.5g, 50 mmole)
in 225mL of 95% ethanol was heated to reflux. To the reaction
mixture at reflux was added a solution of potassium hydroxide
(0.6g, in 25mL of H_2O). Upon addition of base the solution turned
brown in color. Reflux was continued for a total of thirty minutes
at which time the reaction mixture was allowed to cool to room
temperature, filtered, and air-dried to give 16g (98.8%) of crude
product. The material was purified by column chromatography on dry
silica gel using toluene as the eluent, mp 202-203°C.
    Analysis Calcd for $C_{64}H_{42}S_2O_2$: C,84.73; H,4.66
                          Found: C,84.25; H,4.58

Heterocyclic Acetylene Monomers

2-(3-Bromophenyl)-5-bromobenzothiazole. A mixture of 2-amino-4-
bromomercaptobenzene hydrochloride (20g, 83 mmole) and polyphosphoric acid (274g) was heated at 70°C until dehydrochlorination was
complete. To the mixture was then added sulfolane (109g) and
m-bromobenzoic acid (22g, 110 mmole). The mixture was heated at
90°C for 3 h, at 145°C for 18 h, at 190°C for 1 h, and finally, at
200°C for 1 h. The reaction mixture was allowed to cool to room

temperature and precipitated into a 50/50 methanol/water solution. The product was collected by filtration, washed with dilute ammonium hydroxide, water, and air-dried. Recrystallization from toluene gave 24.5g (80%) of the expected compound, mp 162–163°C.

Analysis Calcd for $C_{13}H_7NSBR_2$: C,42.30; H,1.91; N,3.80; S,8.69
Found: C,42.21; H,1.56; N,3.81; S,9.15

2-[3-(3-Methyl-3-hydroxy-1-butynyl)phenyl]-5-(3-methyl-3-hydroxy-1-butynyl)benzothiazole. A mixture consisting of 2-(3-bromophenyl)-5-bromobenzothiazole (24g, 65 mmole), dichlorobis(triphenylphosphine)-palladium II (.09g), triphenylphosphine (0.7g), cuprous iodide (0.26g) and triethylamine 250mL was heated to reflux and cooled to room temperature under a nitrogen atmosphere. To the mixture was added 2-methyl-3-butyn-2-ol (8g, 0.33 mole), heated to 90°C and maintained at that temperature for 24 h. The cooled reaction mixture was then filtered through a Celite pad to remove the amine hydrobromide and the triethylamine was removed under reduced pressure. The resultant brown solid was dissolved in chloroform, washed several times with sulfuric acid (10%), water and dried over anhydrous magnesium sulfate. Removal of the chloroform gave 23g (94%) of the product which was used without further purification.

2-(3-Ethynylphenyl)-5-ethynylbenzothiazole (3) To a solution containing 2-[3-(3-methyl-3-hydroxy-1-butynyl)phenyl]-5-(3-methyl-3-hydroxy-1-butynyl)benzothiazole (23g, 0.06 mole) dissolved in toluene (250 mL) was added potassium hydroxide (2g). The mixture was heated to reflux, 100 mL of the toluene was distilled, 100 mL of fresh toluene was added and also distilled. The reaction mixture was filtered, washed with water, dried over anhydrous magnesium sulfate, and the toluene removed under reduced pressure. The residual solid was chromatographed on silica gel using methylene chloride as the eluent. Removal of the methylene chloride under reduced pressure provided 8.4g (54%) of product, mp 180–181°C.

Analysis Calcd for $C_{17}H_9NS$: C,78.42; H,2.99; N,5.22
Found: C,78.74; H,3.20; N,5.40

2-(3-Ethynylphenyl)-6-ethynylbenzothiazole (4) The monomer (mp 161–162°C) was prepared by analogous procedures starting with 2-amino-6-bromomercaptobenzene hydrochloride.

N-3-Ethynylphenyl-4-nitronaphthalimide. A mixture of 4-nitro-naphthalic anhydride (24.3g, 0.1 mole) and 3-aminophenylacetylene (11.7g, 0.1 mole) in glacial acetic acid (300mL) under a nitrogen atmosphere was refluxed for 1 h, cooled and filtered. The solid was washed with methanol and air-dried. Recrystallization from toluene gave 20g (58%) of product mp, 265–267°C.

Analysis Calcd for $C_{20}H_{10}N_2O_4$: C,70.17; H,2.94; N,8.18
Found: C,69.85; H,2.73; N,8.41

N-3-Ethynyl-4-(3-ethynylphenoxy)naphthalimide (5) To a solution of
3-ethynylphenol (5.9g, 0.05 mole) in dry N,N-dimethylacetamide was
added sodium methoxide (2.7g, 0.05 mole) and benzene (100mL). After
the benzene was distilled, N-3-ethynylphenyl-4-nitronaphthalimide
(17.1g, 0.05 mole) was added and the mixture was heated for 1 h.
The dark solution was cooled and poured into ice water (2.5L). The
precipitate was collected by filtration and air-dried. Chromatography of the solid on silica gel using methylene chloride as the
eluent gave 7.5g (36%) of product, mp 200-202°C.

Analysis Calcd for $C_{28}H_{15}NO_3$: C,81.34; H,3.65; N,3.38
Found: C,81.45; H,3.67; N,3.42

4-(3-Ethynylphenoxy)-N-(3-ethynylphenyl)phthalimide (6) The monomer
(mp 159-160°C) was prepared by analogous procedures starting with
4-nitrophthalic anhydride.

2,2-Bis[4-(3-Ethynylphenyl-N-phthalimide)]hexafluoropropane (7) The
bisimide was prepared from the reaction of 2,2-bis(3',4'-dicarboxy-
phenyl)hexafluoropropane dianhydride with two moles of 3-aminophenyl-
acetylene in glacial acetic acid. Purification by chromatographing
on silica gel provided the material with a mp 185-186°C, reported[10]
mp 180-186°C.

Polymers

The following are representative of the procedures used in the
preparation of polymers:
Poly[(1,3-dioxo-2,5-isoindolinediyl)oxy[4',6'''-bis(p-phenoxy-
phenyl)-2',2''',5',5'''-tetraphenyl-m-quinquephenyl-3,3'''-ylene]]. In
a 25mL polymerization tube were placed 3,3'-(1,3-phenylene)bis-
[2,5'-diphenyl-4-p-phenyloxyphenyl cyclopentadienone] (0.839g,
0.96 mmole), 4-(3-ethynylphenoxy)-N-(3-ethynylphenyl)pthalimide
(0.3448g, 0.96 mmole) and 1,2,4-trichlorobenzene (7mL). The contents of the tube were degassed by several freeze-thaw cycles at
liquid nitrogen temperatures and then sealed in vacuo. Approximately 50mL of 1,2,4-trichlorobenzene was added to a Parr pressure
reactor, the polymerization tube was placed inside, and the reactor
heated at 220°C for 18 h. The temperature was then increased to
240°C and heating continued for an additional 24 h. After the
reaction had cooled to room temperature, the tube was opened, the
contents diluted with chloroform (30mL) and precipitated into 1L of
rapidly stirring methanol. The fluffy white polymer was filtered,
reprecipitated from chloroform into absolute methanol, and freeze-
dried from benzene. After drying at 120°C (0.4 mm Hg) overnight,
the polymer had an intrinsic viscosity at 30°C of 0.61 in DMAC.

Analysis Calcd for $(C_{86}H_{55}NO_5)$: C,87.36; H,4.69; N,1.18
Found: C,87.16; H,4.49; N,1.31

Poly[(1,3-dioxo-1H-benz[de]isoquinoline-2,6(3H)-diyl)oxy[2',-
2''',5',5''-tetraphenyl-4',6'''-bis[p-(phenylthio)phenyl]-m-quinque-
phenyl-3,3''''-ylene]]. A solution of 3,3'-(1,3-phenylene)bis[2,5-
diphenyl-4-p-phenylthiophenyl cyclopentadienone] (0.493g, 0.54 mmole)
and 5-(3-ethynylphenoxy)-N-(3-ethynlphenyl)naphthalimide (0.224g,
0.54 mmole) in 7mL of 1,2,4-trichlorobenzene was degassed under
vacuum by several freeze-thaw cycles and sealed in a tube. The tube
containing the reaction mixture was heated at 220°C for 18 h and the
temperature then raised to 240°C for 24 additional h. The tube was
cooled to room temperature, frozen in liquid nitrogen, and opened.
When the contents had warmed to room temperature, the solution was
diluted with chloroform (30mL) and precipitated into methanol (1L).
The off-white polymer was reprecipitated from chloroform to give a
quantitative yield of polymer with an intrinsic viscosity in DMAC of
0.48.

Analysis Calcd for ($C_{90}H_{57}NO_3S_2$): C,85.48; N,4.54
Found: C,84.89; N,4.23

RESULTS AND DISCUSSION

A series of new polymers were prepared in an effort to determine
what effects a pendant diphenylether or diphenylsulfide group in
conjunction with pendant phenyl groups would have on the physical
properties of two different heterocyclic systems. The thiazole and
imide heterocyclic ring structures were selected for this study
because of their good thermal oxidative stability and poor solubility
properties. High molecular weight polybenzothiazoles are only soluble
in strong mineral or organic acids and aromatic polyimides exhibit
solubility only in high boiling aprotic solvents. The polymers were
conveniently prepared by using the Diels-Alder polymerization of
substituted biscyclopentadienone monomers with heterocyclic diacetylenes.

Monomer Synthesis

Introduction of the diphenylether and diphenylsulfide pendants
into polymers was accomplished by incorporation of such groups into
biscyclopentadienone monomers. The synthesis of the monomers was
carried out by the base catalyzed condensation of the appropriately
substituted bis-tetraketone with 1,3-diphenylacetone. The resulting
products were purified to monomer grade purity by column chromatography. The monomers 3,3'-(1,3-phenylene)bis[2,5-diphenyl-4-p-phenyl-
oxyphenylcyclopentadienone] (1) and 3,3'-(1,3-phenylene)bis[2,5-
diphenyl-4-p-phenylthiophenyl cyclopentadienone] (2) are highly
colored compounds with mp of 205-206°C and 202-203°C respectively.
This method provided a 2/1 ratio of phenyl to diphenyl ether or
diphenyl sulfide pendants per repeat unit.

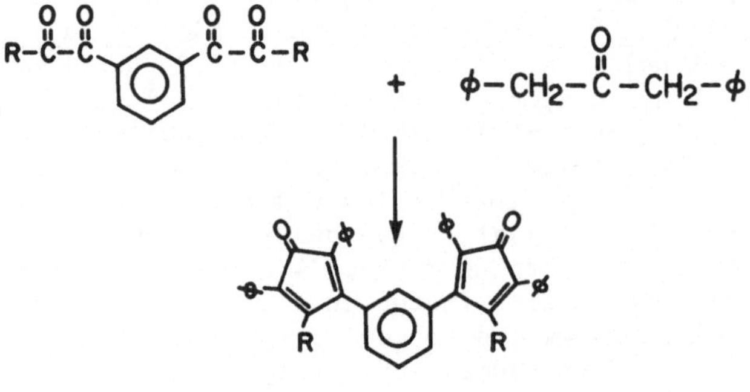

Five heterocyclic acetylenic monomers were prepared for this study and are shown in Table I along with their mp's.

Table I Aromatic Heterocyclic Acetylenes

Structure	No.	mp °C
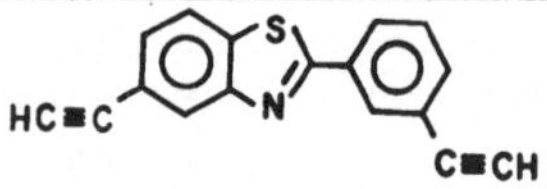	3	180–181
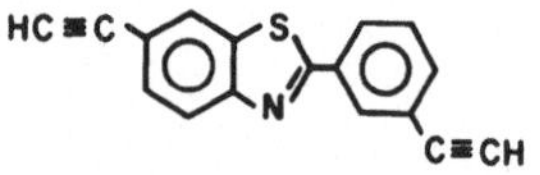	4	161–162
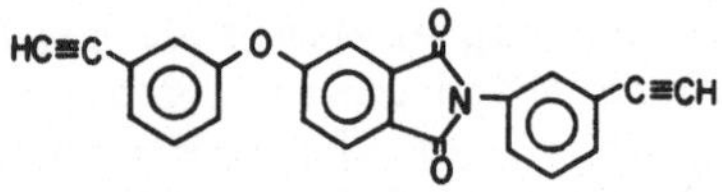	6	159–160
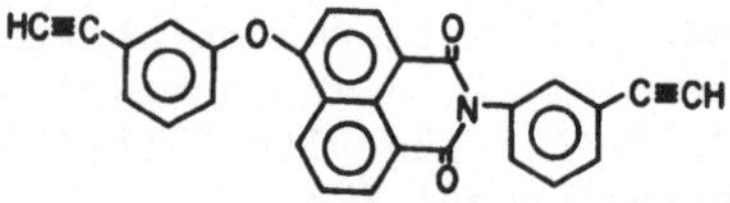	5	200–202
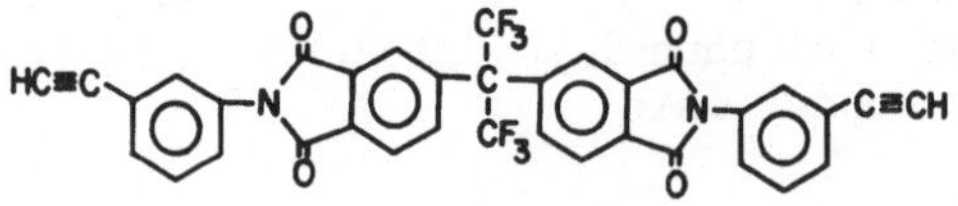	7	185–186

Two isomeric acetylenic benzothiazole monomers, 2-(3-ethynyl-phenyl)-5-ethynylbenzothiazole (3) and 2-(3-ethynylphenyl)-6-ethynyl-benzothiazole (4) were prepared according to the general reaction scheme shown below. The synthesis of the benzothiazole heterocyclic structure was carried out by the condensation of m-bromobenzoic acid with isomeric bromo-substituted o-aminomercaptobenzenes in poly-phosphoric acid (PPA). The bis-bromobenzothiazoles were converted to the acetylene systems by the reaction with 2-methyl-3-butyn-2-ol and subsequent displacement of acetone with base. The bromo displacement reaction utilized a catalyst composed of triphenylphosphine, (bis-triphenylphosphine)palladium dichloride and cuprous iodide.

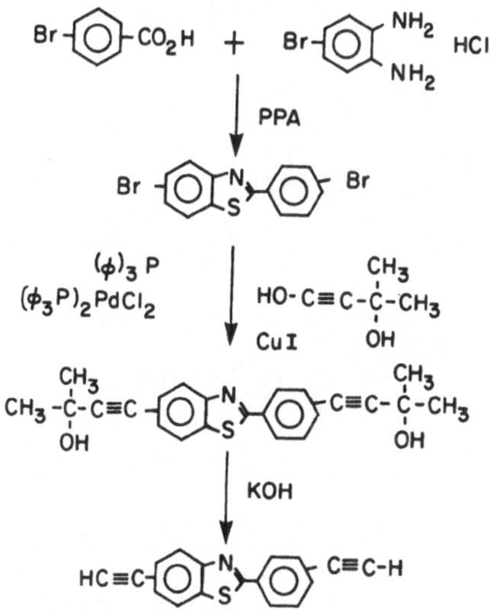

The second class of acetylenic monomers prepared for this study was those containing the imide heterocyclic ring system. Treatment of 4-nitronaphthalic anhydride with 3-aminophenylacetylene in glacial acetic acid gave 4-nitro-N-(3-ethynylphenyl)napthalimide in 58% yield. As the imide, the nitro group becomes activated for nucleophilic displacement. Subsequent treatment with the sodium salt of 3-ethynylphenol in DMAC gave 4-(3-ethynylphenoxy)-N-(3-ethynylphenyl)-napthalimide (5) in 36% yield after purification. In an analogous fashion, the five-membered 4-(3-ethynylphenoxy)-N-(3-ethynylphenyl)-phthalimide (6) was obtained in a 44% yield from 4-nitrophthalic anhydride.

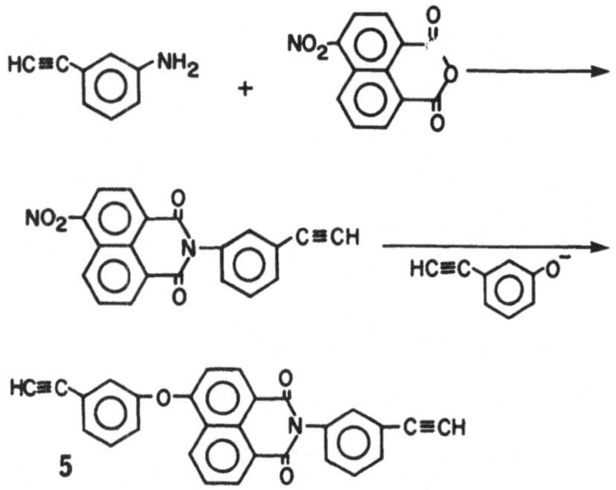

Another diacetylene containing two imide rings was synthesized in an effort to provide a higher imide content along the polymer backbone. The condensation of 2,2-bis(3,4-dicarboxyphenyl)hexafluoro-propane dianhydride with two moles of 3-aminophenylacetylene provided a 73% yield of 2,2-bis 4-(3-ethynylphenyl-N-phthalimide)hexafluoro-propane (7) after purification by column chromatography.

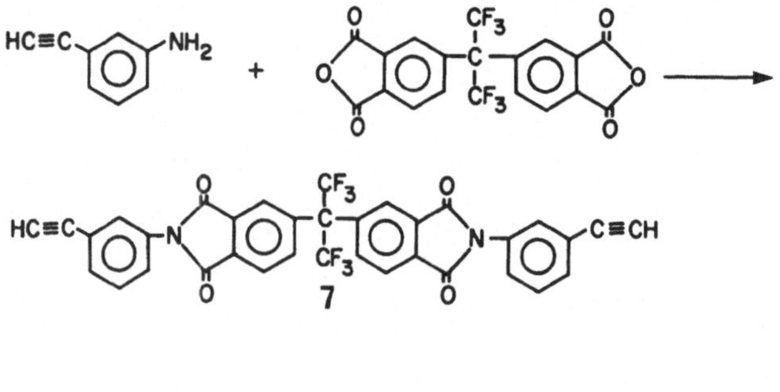

Polymers

The Diels-Alder polymerizations of the appropriately substituted biscyclopentadienone monomers with the various diacetylenic monomers were carried out in sealed tubes using 1,2,4-trichlorobenzene as the solvent. All polymerizations were conducted at 220°C for 18 h followed by heating at 240°C for an additional 24 h. The resulting polymers (Table II) upon precipitation into methanol, reprecipitation from chloroform-methanol, freeze-drying from benzene and drying at 150°C, gave intrinsic viscosities of 0.28 to 0.61 in DMAC or toluene. The general reaction scheme is shown below. Tough flexible films could be cast from 2% solutions of polymers in chloroform.

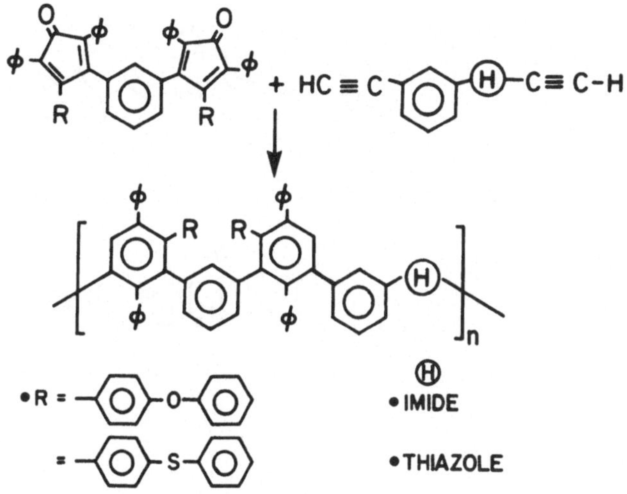

PROPERTIES

Polymer Solubility

Solubility properties of the new polymers were very interesting and quite different from any other heterocyclic polymer containing pendant phenyl groups. Most interesting, the polymers were found to be soluble in aromatic hydrocarbons such as benzene, toluene and xylene. Concentrations as high as 10% could be obtained in benzene. This offered a convenient method by which the polymers could be purified for combustion analysis. As part of the purification procedure, the polymers were freeze-dried from benzene which provided

a large surface area for combustion. The solubility increase is
presumed to result from a modification of chain packing, which allows
greater solvent-polymer interaction and increases the distance between
adjacent chains. The polymers were also soluble in chloroform, sym-
tetrachloroethane and chlorinated aromatic hydrocarbons, as well as,
aprotic solvents such as dimethylforamide, dimethylacetamide and
n-methylpyrolidone.

In an effort to obtain a direct comparison of the solubility
properties for an analogous phenylated system the bis-imide acetylene
monomer was polymerized with 3,3'-(1,3-phenylene)bis[2,4,5-triphenyl-
cyclopentadienone]. The polymer gelled from trichlorobenzene when
allowed to cool to room temperature and exhibited only partial
solubility in refluxing toluene.

Thermal Mechanical Behavior

The glass transition temperatures (Tg's) reported (Table II)
were determined by differential scanning colorimetry (DSC) at a
heating rate of 20°C/min in nitrogen. In all cases, the extrapolated
onset of the DSC baseline shift was taken as the Tg. Tg's for the
polymers were in the range of 230-280°C which are generally lower
than normal for aromatic imide and thiazole polymer backbones having
the same degree of flexibility. Lower Tg's for these polymers is
apparently due to both a decrease molecular symmetry and intermolecular associations. Molecular symmetry is decreased since various
isomeric structures are formed via the Diels-Alder polymerization.
The decrease in intermolecular association (induced dipole-dipole
interactions) is due to the lower heterocyclic content per repeat
unit. The polyimide containing only pendant phenyl groups exhibited
a Tg of 292°C, whereas substitution of two phenyl pendants with
diphenylether groups lowers the Tg to 240°C. It is interesting to
note that the same equivalent lowering of the Tg has been shown to
occur for the as-triazine and quinoxaline polymer systems.[9]

Table II Polymer Properties

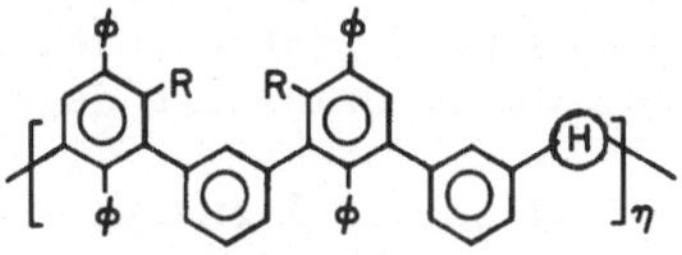

(H)	R	$[\eta]$	Tg^a	ISOTHERMAL AGING % b
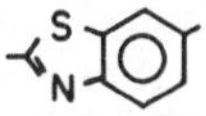 (benzothiazole–phenyl structure)	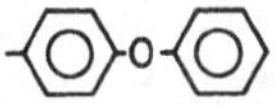 –⟨O⟩–O–⟨O⟩–	0.30	282	50
" "	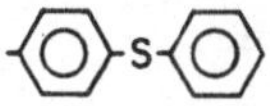 –⟨O⟩–S–⟨O⟩–	0.35	255	80
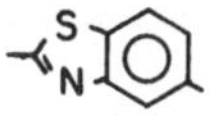 (benzothiazole–phenyl structure)	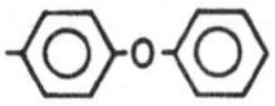 –⟨O⟩–O–⟨O⟩–	0.45	269	55
" "	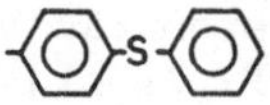 –⟨O⟩–S–⟨O⟩–	0.28	266	35
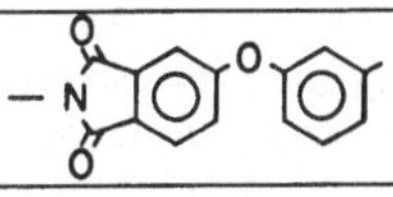 (phthalimide–O–phenyl structure)	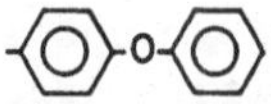 –⟨O⟩–O–⟨O⟩–	0.61	238	80
" "	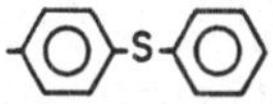 –⟨O⟩–S–⟨O⟩–	0.57	233	89
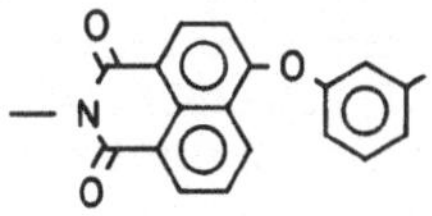 (naphthalimide–O–phenyl structure)	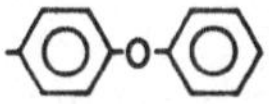 –⟨O⟩–O–⟨O⟩–	0.54	273	44
" "	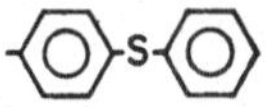 –⟨O⟩–S–⟨O⟩–	0.48	264	29
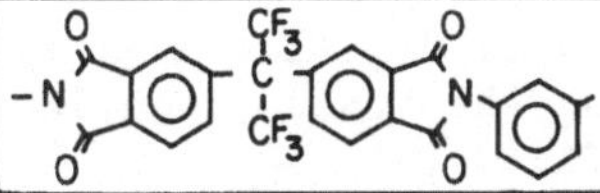 (bis-imide hexafluoroisopropylidene structure)	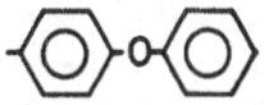 –⟨O⟩–O–⟨O⟩–	0.55	240	86

a = GLASS TRANSITION TEMP. AS DETERMINED BY DSC ▲ RATE 10^0/ MIN

b = % WEIGHT RETAINED (ISOTHERMAL AGING IN AIR 650F FOR 200 HR)

Thermal Evaluation

The thermal properties of the polymers were evaluated using isothermal aging and thermogravimetric–mass spectral analysis. Isothermal aging studies were carried out with an automatic multi-sample apparatus, which allowed weight loss to be obtained in the aging environment. Freeze–dried samples were examined in circulating air at 343°C (650°F). The weight retention data after 200 hours

exposure time are summarized in Table II. Of the two heterocyclic
systems studied, the more stable seems to be the five-membered polyimide. The instability of the six-membered imide polymers is likely
associated with the exposed peri proton which has been shown in our
laboratory to be susceptible to oxidative degradation.

Thermogravimetric-mass spectral analysis conducted under vacuum
on a phenylphenoxy pendant benzothiazole polyphenylene showed that
degradation commences at 500°C and continues to the highest temperature studied, 850°C. There are basically two regions of volatile
product evolution, 600° and 750°C. The major volatile products
released at 600°C include benzene, phenol, biphenyl, and phenylether
with the highest molecular weight product identified as triphenyl.
The higher temperature cracking processes (750°C) release methane,
carbon monoxide, and hydrogen.

SUMMARY AND CONCLUSIONS

High molecular weight aromatic heterocyclic polyphenylenes
have been prepared through Diels-Alder reactions of biscyclopenta-
dienones and heterocyclic diacetylenes. The imide and thiazole
polymers with pendant phenyl groups in conjunction with either
diphenylether or diphenylsulfide groups exhibit unusual solubility
in aromatic hydrocarbon solvents. Introduction of the diphenylether
or diphenyl sulfide pendants were made via the biscylopentadienones
monomers. The thermal and thermooxidative properties of the polymers
were very good as evidenced by thermogravimetric-mass spectral
analysis and isothermal aging studies. Glass transition temperatures for the polymers were in the range of 230–280°C which are
lower than normal for aromatic imide and thiazole polymer backbones.

REFERENCES

1. H. Mukamal, F. W. Harris, and J. K. Stille, "Diels-Alder
    Polymers III. Polymers Containing Phenylated Phenylene
    Units," J. Polym. Sci., 5:2721 (1967).
2. P. E. Cassidy, C. S. Marvel and S. Ray, "Preparation and
    Aromatization of Poly-1,3-cyclohexadiene and Subsequent
    Crosslinking III," J. Polym. Sci., 3:1553 (1965).
3. P. M. Hergenrother, "Linear Polyquinoxalines," J. Macromol. Sci.,
    Rev. Macromol. Chem., 6:1 (1971).
4. J. K. Stille and F. E. Arnold, "Polyquinoxaline III," J. Polym.
    Sci., 4:551 (1966).
5. J. K. Stille, J. E. Wolfe, S. O. Norris and W. Wrasidlo,
    "Polyquinolines A New Class of High Performance Materials,"
    Polym. Prepr., 17:41 (1976).

6. F. W. Harris, W. A. Feld and L. H. Lanier, "The Polymerization of Phenylated Bis(phthalic anhydrides) with Diamines," J. Polym. Sci., B13:283 (1975).

7. P. M. Hergenrother, "Poly-as-triazines," J. Polym. Sci., 7:945 (1969).

8. P. E. Cassidy and A. Syrinek, "Polyimidines, A New Class of Polymers I. Phenylated Polypyromellitimidenes," J. Polym. Sci. Polym. Chem. Ed., 14:1485 (1976).

9. P. M. Hergenrother, "Poly(phenyl-as-triazines) and Poly(phenyl-quinoxalines). New and Cross-Linked Polymers," Macromol., 7:575 (1974).

10. N. Bilow, "Acetylene-Substituted Polyimides as Potential High Temperature Coatings," in: "Resins for Aerospace," C. A. May, ed., Amer. Chem. Soc. Symposium Series #132 (1980).

SUBSTITUTED POLYAMIDES AS PRECURSORS FOR ALKYL AND ALKENYL

POLYBENZOXAZOLES

Lon J. Mathias

Dept. of Polymer Science
Univ. of Southern Mississippi
Hattiesburg, MS 39406

Sharf U. Ahmed and Peter D. Livant

Dept. of Chemistry
Auburn University
Auburn, AL 36830

INTRODUCTION

Polymers with heteroaromatic repeat units have been known
for many years to exhibit good thermal and oxidative stability.
Polybenzimidazoles (1),polybenzothiazoles (2), and polybenzoxazoles (3,4) were first reported in the late '50's and early '60's
and have since been under widespread investigation. Both AA-BB
and AB monomer systems have been employed in the synthesis of
high molecular weight polymers. (5) The thermal stability in
nitrogen atmosphere of the completely aromatic AB polymers
prepared from the various 3,4-disubstituted benzoic acids decreased in the order benzimidazoles>benzothiazole>benzoxazole.
In air, however, all three types of polymers possessed similar
stability.

In addition to the all-aromatic AB polymers, which are
infusible and generally insoluble (6), several polymers which
contain alkyl chains have been obtained. The earliest such
benzoxazole polymers were prepared from the ω-(3-amino-4-
hydroxyphenyl-derivatives of propionic acid, 3-methyl-butyric
acid and 5-methylhexanoic acid (7). Thermal polymerization of
the propionic acid monomer under nitrogen atmosphere with gradual
temperature increase from 195° to 285° gave a polymer which was

molded at 400° into films. The films swelled slightly in concentrated sulfuric acid but were unaffected by other solvents,
suggesting crosslinking.

    In this paper we report the synthesis of new AB monomers
3-6 which have potential for forming polybenzoxazoles with backbone alkane and alkene moities. Application to monomers 1-6
of the previously developed low-temperature, high-yield condensation procedure (8) employing triphenylphosphine, hexachloroethane
and pyridine is described. The soluble aryl-alkyl and aryl-
alkenyl polyamides thus obtained were characterized and converted
to polybenzoxazoles. Comparisons are presented between polymers
obtained by this two-step method and those prepared by traditional
methods.

COOH   CH₂COOH   CH₂CH₂COOH

HO   NH₂   NH₂
      NH₂   OH   OH

   1   2   3

CH₂CH₂COOH   CH=CHCOOH   CH=CHCOOH

HO   NH₂   HO
      NH₂   OH   NH₂

   4   5   6

RESULTS AND DISCUSSION

Monomer Synthesis and Characterization

    Literature procedures were employed in the synthesis of 4-
amino-3-hydroxybenzoic acid 1 (9) and 3-amino-4-hydroxyphenylace-
tic acid 2 (10). The propanoic and propenoic compounds 3-6 were
obtained as outlined in Figure 1. In both isomeric systems, the
intermediate hydroxynitrobenzaldehydes were cleanly converted to
the cinnamic acid derivatives by Knoevenagel condensation with
malonic acid followed by spontaneous decarboxylation. Reduction
of the nitro groups with sodium dithionate then gave the alkene

monomers while simultaneous reduction of the alkene and nitro
groups with hydrogen and palladium-on-carbon led to the alkane
compounds. Methyl easter derivatives were prepared in very high
yields for most of 1-6 by bubbling hydrochloric acid gas into
methanolic solutions of the appropriate acids.

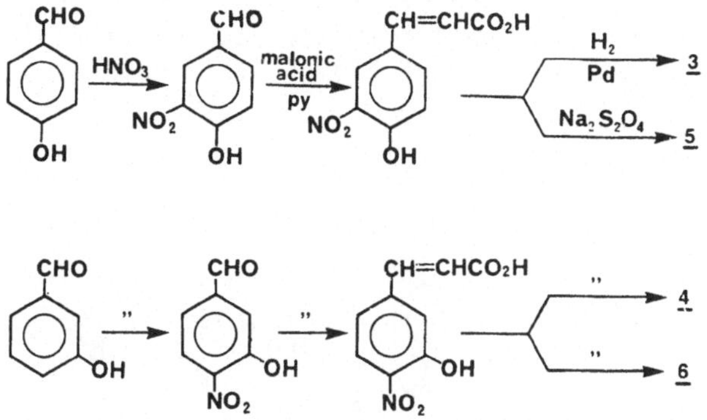

Figure 1

New compounds were characterized by elemental analysis,
melting point, IR, ^{1}H and ^{13}C NMR. Table 1 presents ^{13}C NMR
spectral data for monomers 3-6. Assignment of peaks was based on
reported assignments for analogs and from off-resonance decoupled
or fully coupled spectra. Some assignments are tentative and
may be in error.

Table 1. ^{13}C NMR chemical shifts in ppm from TMS of alkyl and
aklenyl monomers 3-6 in DMSO-d_6.

Cmpd.	Carbon Atoms*								
	1	2	3	4	5	6	7	8	9
3	142.4	136.2	114.6	131.8	114.4	116.1	30.1	35.9	174.0
4	134.5	144.0	114.4	129.0	119.7	114.5	29.9	36.0	174.0
5	146.9	137.0	114.4	125.8	114.4	114.4	145.2	118.7	168.0
6	140.7	143.7	113.5	122.7	113.0	110.7	145.9	122.0	167.8

*Carbons numbered as in Figure 3 with C-1 always para to alkyl or
alkenyl group.

Polymer Synthesis and Characterization

      Two main types of polymerization procedures were employed.
The traditional methods (Figure 2, lower path) use high temperatures (>200°C) for conversion in the melt and solid-state (7) or
in polyphosphoric acid (PPA) solution (2,6). The milder, two-
step procedure (Figure 2) involves polyamide formation and isolation followed by thermal cyclization to the polybenzoxazole.
We are aware of only one (4) previous application of this process
which involved use of thionyl chloride and phenol for in situ
generation of an active ester followed by immediate formation of
polyamide.

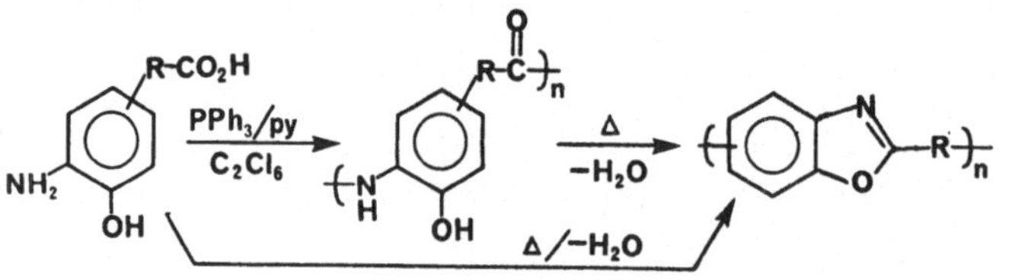

Figure 2

      Our procedure is based on recent reports of mild polyamide
formation from aryl and alkyl diamines, diacids and amino acids.
(8, 11-13) These related methods employ acid-activating reagents
which are phosphorous derivatives. The method we employed uses
triphenylphosphine with hexachloroethane as co-reagents in
pyridine solvent. (8) Diphenyl- and tripenyl-phosphite in
pyridine (11, 12) and even PCl_3 on $POCl_3$ in pyridine (13) gave
good yields and molecular weights of various polyamides and polyureas. We found that the careful purification of reagents and
solvent (especially hexachloroethane) was crucial for good
polymerizations. With almost all monomers employed here, the
room temperature conversion led to initial polyamide precipitation within minutes and to complete conversion in less than one
hour.

      Table 2 lists all polyamides obtained with the triphenyl-
phosphine method along with selected properties of the polymers.
Values for monomer conversion ranged from 70% to 100%. Extraction
and reprecipitation to remove reagent by-products probably led
to loss of low molecular weight polymers and thus to lower yields.
Figure 3 gives the ^{13}C spectra of representative alkyl and alkenyl
polyamides.

Table 2. Properties of polyamides obtained with triphenylphos-
          phine and hexachloroethane in pyridine.

Monomer	Color	η_{inh} (dl/g)*
1	brown	0.18
2	light brown	0.13
3	pale yellow	0.25 (0.13 g/dl)
4	brown	0.29 (0.1 g/dl)
5	bright-yellow	0.14
6	orange-yellow	0.49

*Obtained at 0.2 g/dl unless indicated otherwise

Table 3 contains reaction conditions and polymer properties
for polybenzoxazoles prepared by one-step thermal polymerizations
and by cyclization of the intermediate polyamides. The latter
dehydration process was readily followed by IR. The polyamides
have strong bands near 1655 cm^{-1} which gradually disappear during
the cyclization. Concomitant appearance of the characteristic
benzoxazole band at 1600-1620 cm^{-1} confirms the process and the
product structure. In addition, microanalysis data have been
obtained for the polybenzoxazoles from monomers 3 and 4.
Calculated and found values for C, H, and N were within 0.3% of
each other for the former. The values for the latter were consistant with either ½ equivalent of bound H_2O per repeat unit
and/or incomplete cyclization.

The two-step procedure offers little or no advantage over
direct polybenzoxazole formation for monomer 1. The polyamide
initially obtained is apparently chain-extended during the cyclization process as indicated by increasing viscosity from the
260°-280° conversions. This is consistant with the behavior of
the polyamides obtained with the thionyl chloride-phenol method.
(4) The final viscosity of this polymer is comparable to (but
not better than) those of polymers obtained by other methods.

Monomer 2 gave insoluble polybenzoxazole from the polyamide
and while direct thermal polymerization of 2 and its methyl
ester gave soluble polymers, their molecular weights were not
high as indicated by viscosities. The insolubility is attributed
to cross-linking since the polymer containing the flexible methylene unit in the backbone should be much more soluble than the all-
aromatic polymer from 1. The mechanism involved in this crosslinking reaction is not known.

Table 3. Polymerization conditions and polymer properties for
          polybenzoxazoles from 1-6 or monomer methyl ester
          1a-6a.

Monomer	Method[a]	Time/Temp	η_{inh}[b]	Color
1	A	260°/6h	0.56	brown
	A	280°/7h	1.2	"
	C	200°/4h	0.86	"
2	A	240°/4h	insol.	gray
	B	240°/3h	0.16	lt. brown
2a	B	250°/6h	0.24	dk. brown
3	A	240°/5h	2.25	lt. brown
	B	270°/2h	0.69	"
	C	<250°/1-3h	low	brown
3a	B	<250°/2-4h	low	dk. brown
	B	270°/2h	insol.	"
4	A	245°/5h	insol.	dk. brown
	A	190°-210°- 240°/3h ea.	0.95	yl.-orange
5	A	220o/12h	insol.	dk. yellow
	A	245°/20 min.	0.38	"
	B	220°/1h	insol.	yl.-orange
	C	160°/6h	0.18	black
5a	B	200°/2h	low	dk. yellow
	B	240°/5 min.	insol.	"
6	A	220°/12h	insol.	dk. yellow
	B	220°/1h	"	dk. brown
	C	160°/12h	"	black
6a	B	200°/45 min.	low	brown

a) Methods: A=neat thermal treatment of polyamides from
   Table 2; B=neat sealed tube of monomer; C=solution in
   PPM.
b) At 0.2 g/dl in H_2SO_4

    Some polymerizations of 3-6 also gave insoluble materials,
especially when the traditional one-step procedures were employed.
Monomer 3 is the best example of the ability of the two-step
synthesis to give high molecular weight polybenzoxazole. While the
intermediate polyamide had an inherent viscosity of 0.25 dl/g,
that of the final polybenzoxazole was 2.25 dl/g. Monomer 4 could

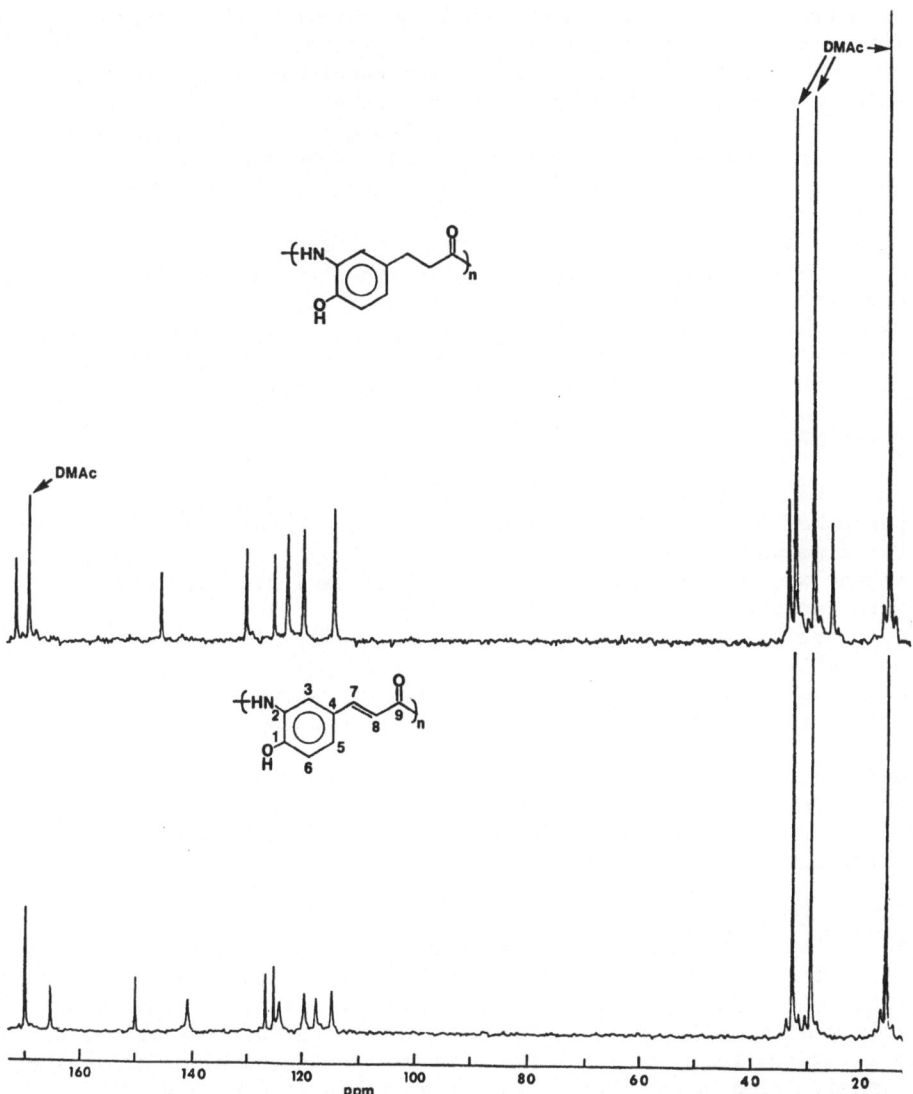

Figure 3. ^{13}C NMRs in DMAc of polyamides obtained from 3 and 5.

also be converted to polybenzoxazole with a reasonable viscosity
(0.95 dl/g) but only if slow and careful temperature increases
were used.

Obtaining soluble, high viscosity polybenzoxazoles from
monomers 5 and 6 was found to be very difficult with any of the
procedures described here. Well-characterized polyamides were
produced from 5 and 6 (see Figure 3), but cyclization conditions
which led to chain-extended products for other polyamides gave

only cross-linked materials. Milder conditions did result in
fairly complete cyclizations although product viscosities were
not high. The traditional PPA method generally gave only insoluble polymers with most monomers, probably due to side-
reactions involving the alkene and alkane groups. The products
obtained, however, display a shiny black appearance which may
indicate extended conjugation and potential conductive properties.

In summary, four new monomers, 3-6, have been synthesized
and converted to polyamides and polybenzoxazoles with varying
degrees of success. Further work is underway to extend the
synthetic and characterization efforts and, especially, to
further examine the potential of the polyamides described for
application in composites and conductive materials.

EXPERIMENTAL

The starting materials m- and p-hydroxybenzaldehyde were
purchased from Aldrich Chemical Company and used as received.
Pyridine was distilled from KOH under nitrogen and hexachloroethane was sublimed three times. Other chemicals and solvents
were reagent grade or purfied prior to use. IR spectra were
obtained as KBr pellets on Perkin-Elmer 583B. ^{1}H NMR spectra
were obtained on a Varian EM-390 and ^{13}C NMR spectra on a
Varien CFT-20 and a JEOL FX90Q. Microanalyses were performed
by Galbraith Laboratories, Nashville, Tennessee.

4-Hydroxy-3-nitrobenzaldehyde. This compound was prepared
according to the procedure described (14) and recrystallized
twice from ethanol to yield a yellow solid. m.p. 143-144°
(lit. m.p. (14) 143-144°).

4-Hydroxy-3-nitro-cinnamic acid. This compound was prepared
according to the procedure described (14) with slight modification.
After the acid was obtained by pouring the pyridine solution
into cold aqueous HCl solution it was filtered and dried. The
acid was then dissolved in NaHCO$_3$ solution and was precipitated
by treating with conc. HCl. The acid was collected after filtering,
washed, dried and recrystallized from ethanol; m.p. 224-225°
(lit. m.p. (14) 223°).

3-Hydroxy-4-nitro-cinnamic acid. This acid was prepared according
to the procedure used for the 4-hydroxy-3-nitro isomer. It was
recrystallized from ethanol with decolorizing carbon to yield a
yellow solid (≈70% yd); m.p. 256-257°C (lit. m.p. (15) 248°).

Methyl-4-hydroxy-3-nitro-cinnamate. The acid (5g) was dissolved
in 400 ml methanol. Into this solution was bubbled HCl gas until
saturated. The whole solution was stirred for 2 more hours
during which time a yellow precipitate formed. The solvent was

removed by rotary evaporation. The residue was dried in a vacuum and recrystallized from methanol to yield yellow needles (5g, 93%); m.p. 148-149° (lit. m.p. (16) 142-144°).

Methyl-3-hydroxy-4-nitro-cinnamate. This compound was prepared according to the above procedure and after recrystallization from methanol was obtained in 84% yeild; m.p. 164-165°.

3-Amino-4-hydroxy-cinnamic acid. The nitro compound (4g) was dissolved in 250 ml of acetone and cooled to 10-12°. To this cold solution a slurry of 20 g $Na_2S_2O_4$ in 60 ml H_2O was added rapidly with stirring. Immediate precipitation occured with a rise in temperature to ca. 33° while the color changed to orange-yellow. The mixture was stirred for 45 min. during which time the precipitate dissolved and the solution turned very light yellow. The solution was then extracted with 300 ml ether. The ether layer was separated, dried with anhydrous sodium sulfate and evaporated to yield a light yellow solid. This material was recrystallized from water to yeild 2.4 gm (70%) of a white solid; m.p. 158-159°.

4-Amino-3-hydroxy-cinnamic acid. This material was prepared according to the above procedure in 65% yield; m.p. 169-170°.

3-Amino-4-hydroxy-hydrocinnamic acid. The nitro alkene (4g) was dissolved in 150 ml of THF and 0.5 gm of 10% Pd/C was added. The mixture was hydrogenated in a Parr low-pressure apparatus for 4 h at 40 psi. The catalyst was removed by filtration and the solvent removed by rotary evaporation. The white residue was recrystallized from methanol to yield 2.9 gm (84%); m.p. 196-197° (lit. m.p. (17) 196°).

4-Amino-3-hydroxy-hydrocinnamic acid. This compound was prepared according to the above procedure. Product recrystallization from water gave a yellow solid (88%); m.p. 188-189°.

Methyl-3-amino-4-hydroxycinnamate. The nitro ester was reduced with $Na_2S_2O_4$ using the same procedure as for the corresponding acid. Chloroform instead of ether was used as extractant. Recrystallization employed a 1:4 mixture of CH_3OH in CCl_4 to give a white powder (64% yd); m.p. 146-147°.

Methyl-4-amino-3-hydroxycinnamate. Reduction of the nitro compound was carried out as described for the acid except that the reaction temperature was maintained at 55° for 45 min. Extraction with $CHCl_3$ gave a light yellow solid which was recrystallized from 1:4 CH_3OH in CCl_4 to give a light brown solid (60% yd); m.p. 145-146°.

Methyl-3-amino-4-hydroxyhydrocinnamate. Hydrogenation of the nitro alkene in THF, work-up and recrystallization from $CHCl_3$ gave a white solid (88% yd); m.p. 111-112°C.

4-Hydroxy-3-nitrophenylacetic acid. This material was prepared according to the procedure described (18) in 68% yd; m.p. 144-145° (lit. m.p. (18) 148°).

3-Amino-4-hydroxyphenylacetic acid. Reduction of the nitro compound according to the described procedure (10) gave this compound in 90% yd.; m.p. 228-230° (lit. m.p. (10) 231-232.5°).

Methyl-4-hydroxy-3-nitrophenylacetic acid. The acid (7.5g) was dissolved in 250 ml CH_3OH, the solution saturated with HCl and stirred for 2 h. Rotary evaporation gave a yellow solid which was recrystallized from CH_3OH to give yellow needles (6.4g, 80%); m.p. 67-68° (lit. m.p. (19) 68-69°).

Methyl-3-amino-4-hydroxyphenylacetate. The nitro compound (2g) was dissolved in 75 ml CH_3OH and 0.3 g 10% Pd/C added. Hydrogenation was carried out for 4 h at 30 psi. Filtration and solvent evaporation gave a tan solid which was recrystallized from $CHCl_3$ (70% yd); m.p. 102-103°.

General Polyamide Synthesis (8). Monomer (0.004 mol) and triphenylphosphine (0.0048 mol) were dissolved in 10 ml pyridine. Hexachloroethane (0.006 mol) was added in one portion to the rapidly stirring solution which immediately turned yellow to amber-red depending on the monomer. Mild heat evolution and polymer precipitation occurred within minutes. The mixture was stirred an additional 30 min and 100 ml CH_3OH added. The solid polymer was filtered, washed sequentially with 300 ml H_2O and 200 ml CH_3OH, and then dried in vacuo for 12 h.

Thermal Cyclization of Polyamides to Polybenzoxazoles. The polyamide (0.2-1.0g) was put into a 25 ml flask, put under vacuum and immersed in a preheated oil bath for times indicated in Table 3. IR spectra were used to monitor disappearance of the amide bands to insure complete reaction.

Solution Polymerization in Polyphosphoric acid. PPA (20g) was placed in a 50 ml flask equipped with an inlet and outlet for nitrogen. The flask was heated under nitrogen to 160°C for 3 h with stirring and then cooled to room temperature. The monomer (0.5g) was then added and the solution heated under nitrogen to the required temperature and kept there as indicated in Table 3. The solution was then poured into 300 ml CH_3OH or water with rapid stirring. The polymer was collected by filtration, washed with H_2O and CH_3OH and dried under vacuum at 100°C for 10 h.

Polymerization in a sealed tube. The monomer (0.5g) was placed
in a polymerization tube and the tube kept under vacuum for
several hours. It was then sealed, placed into a preheated oil
bath at the indicated temperature and kept at that temperature
for the time listed in Table 3. The tube was then taken out,
cooled and opened to collect the polymer. In case of incomplete
cyclization the polymer was washed with boiling methanol and
heated under vacuum again at higher temperature until the cyclization was complete.

REFERENCES

1. H.A. Vogel and C.S. Marvel, J. Polym. Sci., 50, 511 (1961).
2. P.M. Hergenrother, W. Wrasidlo and H.H. Levine, J. Polym.
    Sci., A3, 1665 (1965)
3. K.C. Brinker and I.M. Robinson, U.S. Pat. 2,895,948 (1958).
4. W.W. Moyer, C. Cole and T. Anyos, J. Polym. Sci., A3, 2107
    (1965).
5. C. Arnold, Jr., J. Polym. Sci. Macromol. Rev., 14, 265 (1979).
6. Y. Imai, K. Uno and Y. Iwakura, Macromol. Chem., 83, 179
    (1965).
7. Brit. pat. 811,758 (1959); Chem. Abstr., 53, 14582C (1959).
8. G. Wu, H. Tanaka, K. Sanui, and N. Ogata, Polym. J., 14, 571
    and 797 (1982).
9. L. Jannelli and P.G. Orsini, Gazz. Chim, Ital., 88 331 (1958);
    Chem. Abst. 53, 16623b (1959).
10. J.B. Wright, J. Heterocycl. Chem. 9, 681 (1972) and
    references therin.
11. N. Yamazaki, F. Higashi and J. Kawabata, J. Polym. Sci.
    Polym. Chem. Ed., 12, 2149 (1974).
12. N. Yamazaki, T. Iguchi and F. Higashi, Ibid., 13, 785 (1975).
13. N. Ogata and H. Tanaka, Polym. J., 7, 412 (1975).
14. W. Freund, J. Chem. Soc. 3072 (1952).
15. S.N. Chakravarti, K. Granapati and S. Aravamudhachari,
    J. Ind. Chem. Soc., 171 (1938).
16. T.B. Johnson and E.F. Kohmann, J. Amer. Chem. Soc.,37, 162 (1915).
17. Brit. Patent. No. 811, 758; Chem. Abst., 53, 14582 (1959).
18. J.R. Dimmock, J. Sci. Fd. Agric, 18, 368 (1967).
19. T. Kametani, et. al., Takugaku Zasshi, 84, 432 (1964).

THE SYNTHESIS OF AROMATIC POLYFORMALS

F.J. Williams, A.S. Hay, H.M. Relles, J.C. Carnahan,
G.R. Loucks, B.M. Boulette, P.E. Donahue and
D.S. Johnson

General Electric Co.
Corporate Research & Development
P.O. Box 8
Schenectady, New York 12301

INTRODUCTION

The synthesis of aliphatic polyformals (1) has been known for many years and is relatively straightforward[1]. As shown in equation 1, aliphatic diols can be reacted directly with formaldehyde or dialkyl formals (2) to produce these polymers. The principal complication of the reaction occurs when the diol used contains 2, 3 or 4 carbon atoms in which case the formation of the cyclic formal (3) is favored.[2] Musser and Jackson[3] have more

(1)

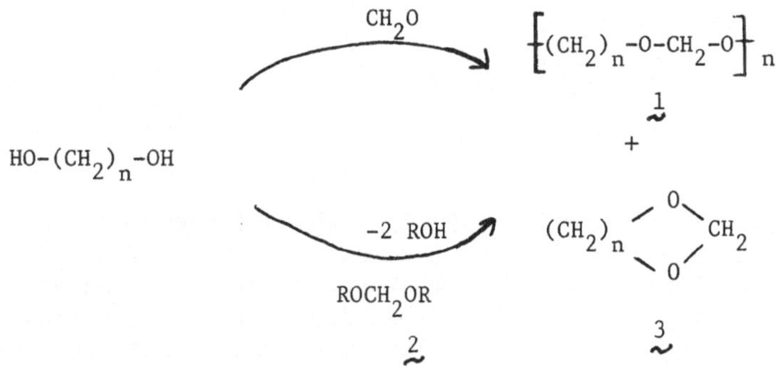

$$HO-(CH_2)_n-OH \xrightarrow{CH_2O} \left[(CH_2)_n-O-CH_2-O\right]_n \quad \mathbf{1}$$

$$+$$

$$\xrightarrow[ROCH_2OR]{-2\ ROH} (CH_2)_n \overset{O}{\underset{O}{\diagdown}} CH_2 \quad \mathbf{3}$$

$$\mathbf{2}$$

recently prepared some aliphatic polyformals based on tetrabromobisphenol A diethanol (4) (equation 2). These low molecular weight polymers were reported to be useful as flame retardant additives in polyester fibers.

67

(2)

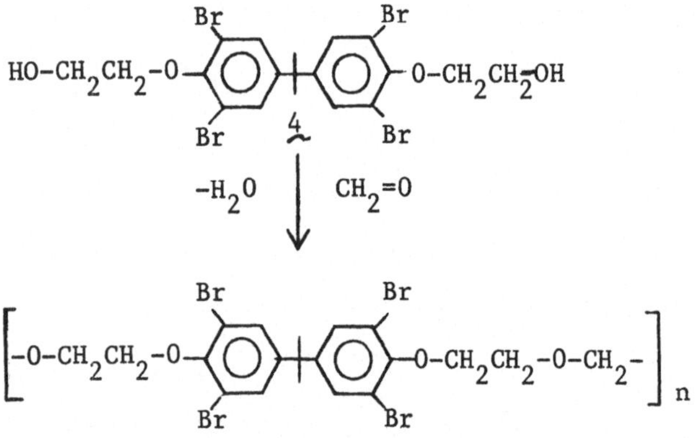

$$\text{HO-CH}_2\text{CH}_2\text{-O-}\underset{\underset{\text{Br}}{\overset{\text{Br}}{\bigcirc}}}{}\underset{\underset{4}{}}{}\underset{\underset{\text{Br}}{\overset{\text{Br}}{\bigcirc}}}{}\text{-O-CH}_2\text{CH}_2\text{OH}$$

$$-\text{H}_2\text{O} \quad \Big| \quad \text{CH}_2\text{=O}$$

$$\left[-\text{O-CH}_2\text{CH}_2\text{-O-}\underset{\underset{\text{Br}}{\overset{\text{Br}}{\bigcirc}}}{}\underset{\underset{\text{Br}}{\overset{\text{Br}}{\bigcirc}}}{}\text{-O-CH}_2\text{CH}_2\text{-O-CH}_2\text{-}\right]_n$$

$$[\eta] = 0.12 \text{ dl/g}$$

Aromatic polyformals cannot be synthesized by an extension of this chemistry. Reaction of aromatic phenols or bisphenols with formaldehyde does not give the formal derivatives derived from O-alkylation but instead produces the well known phenol-formaldehyde resins which result from C-alkylation of the phenol[4] (equation 3). As a result, an alternate to formaldehyde must be found to supply the methylene group of the formal linkage. Barclay first reported such an approach in 1962[5] when he reacted the

(3)

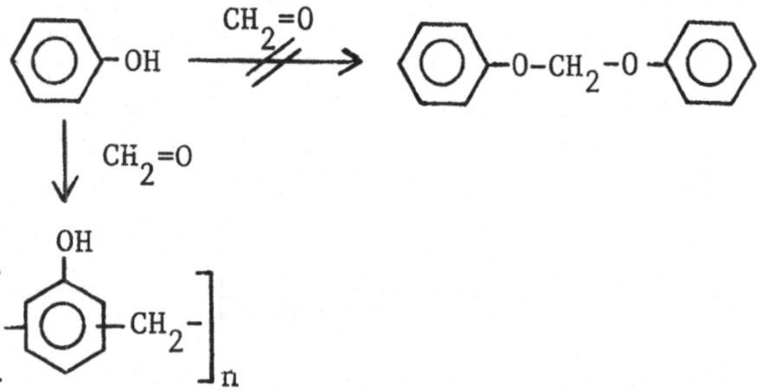

$$\bigcirc\text{-OH} \quad \xrightarrow[]{\text{CH}_2\text{=O}} \quad \bigcirc\text{-O-CH}_2\text{-O-}\bigcirc$$

$$\Big\downarrow \text{ CH}_2\text{=O}$$

$$\left[\underset{\text{OH}}{\bigcirc}\text{-CH}_2\text{-}\right]_n$$

anhydrous disodium salt of BPA (5) with one equivalent of bromo-chloromethane in DMSO. Low molecular weight formal (6, n≈17) was isolated and further reacted with phosgene to give a formal-carbonate copolymer (7). The initial reaction to form the formal likages required the use of DMSO, elevated temperatures and long reaction times (equation 4) and produced only low molecular weight polymer.

(4)

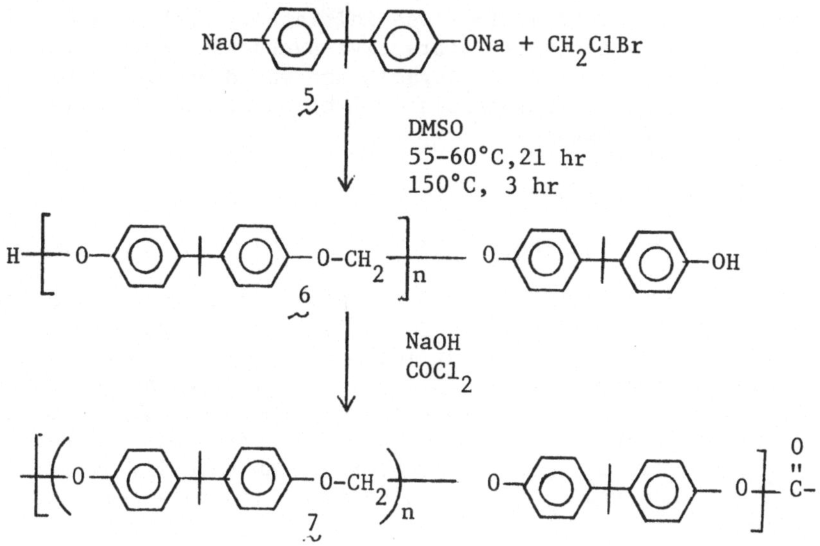

$$ \text{NaO}-\phi-\phi-\text{ONa} + \text{CH}_2\text{ClBr} $$

$$ \underset{\sim}{5} $$

DMSO
55-60°C, 21 hr
150°C, 3 hr

$$ \text{H}\left[\text{O}-\phi-\phi-\text{O-CH}_2\right]_n \text{O}-\phi-\phi-\text{OH} $$

$$ \underset{\sim}{6} $$

NaOH
COCl$_2$

$$ \left[\left(\text{O}-\phi-\phi-\text{O-CH}_2\right)_n \text{O}-\phi-\phi-\text{O}\overset{O}{\underset{}{\text{C}}}\right] $$

$$ \underset{\sim}{7} $$

Matzner, Noshay and McGrath[6] have also reported the use of low molecular weight aromatic linear formals (based upon BPA and Tetramethyl BPA) to form organosiloxane block copolymers. However, no details were provided on the synthesis of these low molecular weight formals.

In 1974, McKillop and co-workers[7] reported the formation of diphenoxymethane derivatives from the attempted alkylation of p-t-butylphenol with benzyl chloride and a catalytic amount of a phase transfer catalyst (PTC) in methylene chloride. More recently, Dehmlow and Schmidt [8] reported the synthesis of a series of dialkyl and diaryl formals ($\underset{\sim}{8}$) from the reaction of alkoxide and phenoxide anions with methylene chloride in the presence of a phase transfer catalyst (equation 5). For example, reaction of 2,3,5-trimethylphenol, powdered potassium hydroxide

(5) $\text{ArOH} + \text{CH}_2\text{Cl}_2 \xrightarrow[\text{KOH}]{\text{PTC}} (\text{ArO})_2\text{CH}_2$ (80-97% yields)

$$ \underset{\sim}{8} $$

and methylene chloride in the presence of Aliquat 336[9] as the phase transfer catalyst yielded the corresponding aromatic formal. In addition, the authors reported that Aliquat 336 was a more effective catalyst than triethylbenzylammonium chloride.

As a result of these more recent efforts, an investigation of the reaction of bisphenols and methylene halides under phase

transfer catalyzed conditions was initiated. Such a reaction
would provide an extremely attractive synthetic route for the
preparation of aromatic polyformals and would hopefully provide
material of sufficient molecular weight to allow for a preliminary
evaluation of properties.

DISCUSSION OF RESULTS

The Polyformal of Bisphenol A[10]

     Reaction of a mixture of BPA, potassium hydroxide pellets,
Aliquat 336 (0.1 mole per mole of BPA) in refluxing methylene
chloride for 15 hr gave, after methanol precipitation, a high
molecular weight polyformal (6), [η] = 0.59 dl/g in 70% isolated
yield. Proton and ^{13}C NMR spectra confirmed the structure of 6.
Interestingly, because of the apparent kinetics of this reaction,
$k_b >> k_a$ (equation 6), methylene chloride can actually be used as
the solvent and reactant in this condensation polymerization
reaction.

(6)

     The reaction was repeated in a mixed chlorobenzene/methylene
chloride (3/1) solvent system. With this mixture, the reaction
temperature could be maintained at 75°C and a 93% yield of 6,
[η] = 0.54 dl/g, was obtained after only four hours. Cloudiness
in injection molded pieces of several of these polyformal samples
suggested the presence of crystalline impurities. Analysis of
these samples by gel permeation chromatography indicated the
presence of a large amount of low molecular weight material.
Further examination by liquid-liquid chromatography (LLC), using
reverse phase gradient elution conditions, resolved this low
molecular weight material into an extended series of cyclic
oligomers 9 (n = 1,2... >20). See Figure 1. The structure of
the cyclic dimer and cyclic trimer was confirmed by a combination
of preparative LLC and mass spectrometry. In addition, if the
linear formal dimer of BPA(10) was used in place of BPA, only

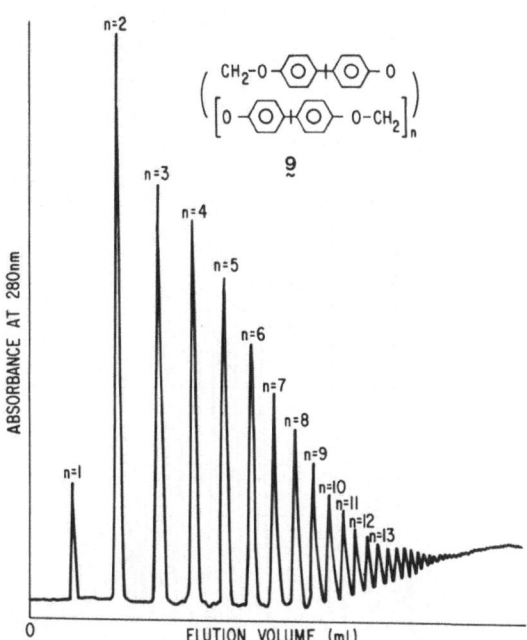

Fig. 1. Gradient elution liquid chromatogram of low MW
cyclic formals in a BPA polyformal sample.

cyclics containing even number of units were seen (9 n = 1,3,5,7,
etc.) If the reactions were run for shorter periods of time, then
LLC analysis also resolved the linear oligomers as shown in Figure
2. Analysis by ^{13}C NMR of the material containing a high proportion of linear oligomers showed the presence of protonated
carbon atoms ortho to the phenol end group (115 ppm) These same
peaks were absent in the spectra of material containing the cyclic
oligomers.

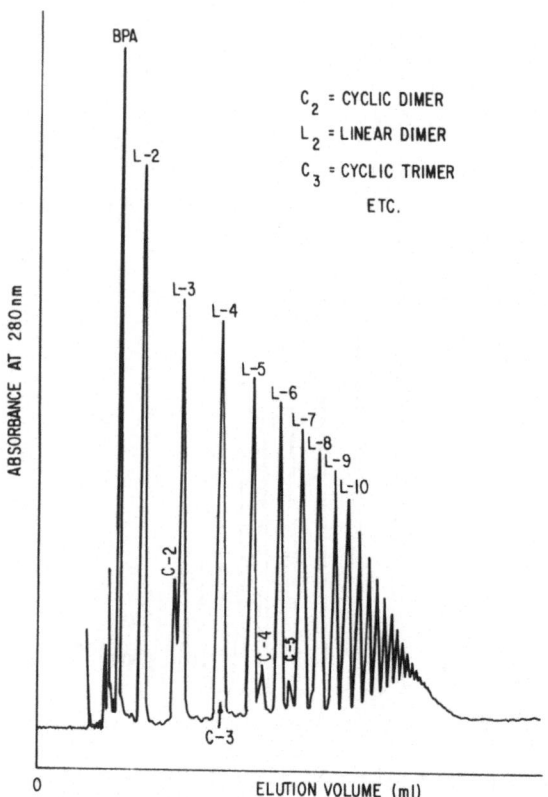

Fig. 2   Gradient elution liquid chromatogram of low MW
         oligomers in a BPA polyformal sample

        The cyclic content of the polymers initially synthesized
using phase transfer catalysis ranged from between 40 to 50%.   At
the present time the best rationale to explain this unusually
high formation of cyclic oligomers is the low solubility of the
BPA dianion in this system which essentially results in high
dilution conditions for the reaction.   In an attempt to increase
the efficiency of the reaction and hopefully reduce the amount of
cyclics formed, a detailed study of different phase transfer catalysts was made.   These studies, which involved the reaction of
4-methylphenol and sodium or potassium hydroxide in a 3/1 mixed
chlorobenzene/methylene chloride solvent system are summarized in

Table 1. Symmetrically substituted onium salts (e.g. tetrabutyl, tetraheptyl) are the best catalysts. Salts containing methyl or benzyl groups (such as Aliquat 336) are relatively poor catalysts in these reactions, presumably because of competing side reactions involving nucleophilic attack of these groups to give methyl or benzyl ethers or alcohols. Tetrabutylammonium bromide (TBAB) can be used with either sodium or potassium hydroxide whereas Aliquat 336 is best used only with potassium hydroxide. Finally, the onium salts are superior to the two crown ethers which were used and, as expected, 18-crown-6 gave better results with potassium hydroxide while 15-crown-5 gave better results with sodium hydroxide.

Based on these studies polymerization reactions were repeated using 0.1 moles of TBAB per mole of BPA in a mixture of chloro-benzene/methylene chloride (3/1) (see Table 2). When potassium hydroxide was used as the base, high molecular weight polymer was obtained but the cyclic content was extremely high. Increasing the amount of TBAB to 1.0 mole per mole of BPA gave a 93% yield of 6 ($[\eta]$= 0.58 dl/g) which contained 7% cyclics after precipi-tation into methanol. Analysis of this material before methanol precipitation showed a cyclic level of 11%.

An alternate approach to increasing the solubility of the BPA dianion and thus decreasing the amount of cyclics formed was to use a dipolar aprotic solvent. Reaction of BPA and sodium hydroxide in a 60/40 N-methylpyrrolidone (NMP)/methylene chloride mixture gave a high molecular weight polymer with a cyclics level between 7-10% before isolation of the polymer. Addition of the reaction mixture to methanol generally resulted in a polymer which, upon isolation, contained 4-7% cyclics. However, by diluting the re-action mixture with chlorobenzene, adding to it an equal volume of methanol/acetone (50/50) and slurrying the resulting precipitate with additional methanol/acetone, a polymer could be obtained which contained <1% cyclics. Similar reactions were also run substituting methylene bromide or chlorobromomethane for methylene chloride but there was no reduction in the amount of cyclics formed. Optimum concentration for these reactions was about 35% solids; higher concentration led to reaction mixtures which were extremely difficult to stir.

A sample of 6 containing <1% cyclics was injection molded at ca 410°F to give colorless, transparent parts whose preliminary properties were determined (see Table 3). For purpose of com-parison, properties are also reported for a sample of BPA poly-carbonate. The polyformal of BPA has a lower T_g and notched Izod impact than the corresponding polycarbonate.

Table 1.   Studies of Phase Transfer Catalysts using 4-Methylphenol
and Sodium Hydroxide/Potassium Hydroxide.[a]

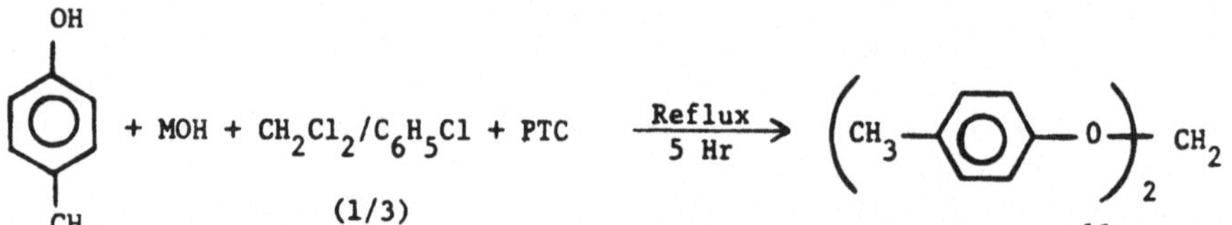

$$\text{OH–}\langle\text{C}_6\text{H}_4\rangle\text{–CH}_3 + MOH + CH_2Cl_2/C_6H_5Cl + PTC \xrightarrow[\text{5 Hr}]{\text{Reflux}} \left(CH_3-\langle\text{C}_6\text{H}_4\rangle-O\right)_2 CH_2$$

(1/3)   11

PTC[b]	NaOH % 11	KOH % 11
$Bu_4\overset{\oplus}{N}\ Br^{\ominus}$	99	99
$Bu_4\overset{\oplus}{N}\ Cl^{\ominus}$	99	
$Bu_4\overset{\oplus}{N}\ I^{\ominus}$	61	
$(n\text{-Propyl})_4\overset{\oplus}{N}\ Cl^{\ominus}$	73	
$(n\text{-Heptyl})_4\overset{\oplus}{N}\ Cl^{\ominus}$	99	
$Bu_4\overset{\oplus}{P}\ Cl^{\ominus}$	86	
Aliquat 336	3	56
Adogen 464	17	
$\emptyset CH_2\overset{\oplus}{N}Et_3\ Br^{\ominus}$	28	13
$Et_3\overset{\oplus}{N}CH_2Cl\ Cl^{\ominus}$	3	
$(CH_3)_3\overset{\oplus}{N}\emptyset\ Br^{\ominus}$	0	
$\emptyset_3\overset{\oplus}{P}CH_3\ Br^{\ominus}$	1	
$CH_3(CH_2)_{15}\overset{\oplus}{N}(CH_3)_3\ Br^{\ominus}$	1	
Dibenzo-18-crown-6	0	85
15-crown-5	83 (8 hr)	

a) Reactions were run with enough solvent to give a solution which would be
20% in final product (g of formal/ml of solvent x100=20%)

b) 0.05 moles of PTC per mole of 4-methylphenol.

Table 2. PTC Polymerizations[a]

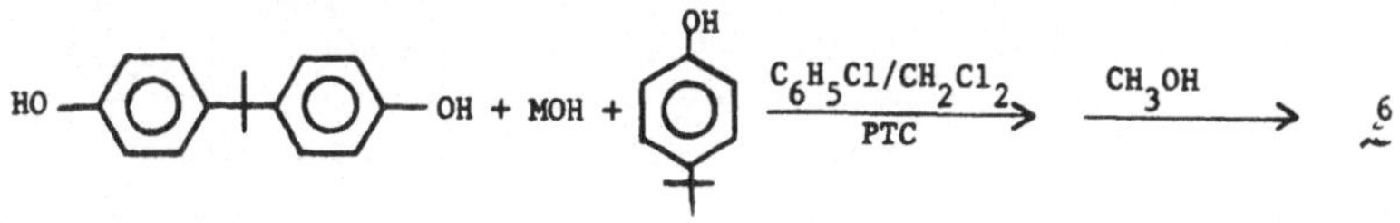

$$HO-\text{(ring)}-\text{C}-\text{(ring)}-OH + MOH + \text{(ring with OH and cyclohexadienone)} \xrightarrow[\text{PTC}]{C_6H_5Cl/CH_2Cl_2} \xrightarrow{CH_3OH} \underline{6}$$

PTC (amount)[c]	MOH	Yield	$[\eta](dl/g)$	% cyclics (before isolation)
Bu$_4$N$^{\oplus}$ Br$^{\ominus}$ (0.1 mole)	KOH	86	0.48	45
Bu$_4$N$^{\oplus}$ Br$^{\ominus}$ (0.1 mole)	NaOH		Methanol soluble	
Bu$_4$N$^{\oplus}$ Br$^{\ominus}$ (1.0 mole)	KOH	93	0.58	7 (11%)
Bu$_4$N$^{\oplus}$ Br$^{\ominus}$ (1.0 mole)	NaOH	86	0.56	7 (10%)
Bu$_4$N$^{\oplus}$ Br$^{\ominus}$ (1.0 mole)[b]	KOH	93	0.57	7

a) Reactions were run at a concentration of 35% for 5 hours using a 3/1 mixture of chlorobenzene/methylene chloride and 0.75% of chain stopper.

b) Run in just methylene chloride

c) Moles of catalyst per 1 mole of BPA

Samples of $\underline{6}$ were also prepared using DMSO in place of NMP as the dipolar aprotic solvent. In a DMSO solvent system, a homogeneous solution of the anhydrous disodium salt of bisphenol A can be prepared. Addition of methylene chloride to this homogeneous solution appeared to offer the opportunity of preparing $\underline{6}$ with decreased cyclic oligomers. A mixture of bisphenol A, aqueous sodium hydroxide, DMSO and toluene was heated at reflux and water was removed by azeotropic distillation. Following the removal of water, toluene was removed by distillation to give a homogeneous solution of the dianion in DMSO. The mixture was held at 110°C and an excess of methylene chloride was added as rapidly as possible. Isolation of the polymer from methanol gave a sample of 6 ($[\eta]$= 0.88 dl/g) which still contained 7% cyclics. Interestingly, if the reaction was repeated but the DMSO solution was cooled to 25°C before the methylene chloride addition (resulting in an extremely inhomogeneous reaction mixture), a sample of $\underline{6}$ was

Table 3.   Property Profiles

	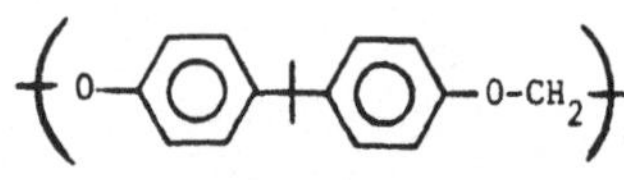 BPA-Polyformal	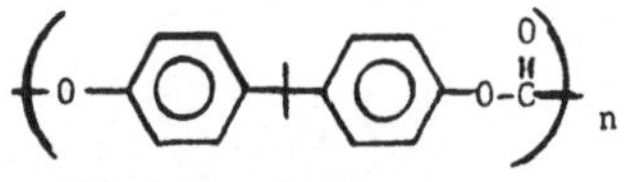 BPA-Polycarbonate
Tg (°C)	94	149
Oxygen Index	20	27
Yield Stress (psi)	7100	9400
Flexural Modulus (psi x 10^3)	370	335
Flexural Strength (psi x 10^3)	12	14.4
% Elongation	80	110
Gardner Impact (ft lbs)	> 320	> 320
Notched Izod Impact (ft lbs/in notch)	1.0	16

isolated containing 66% of the cyclic oligomers. Thus, although we were unable to prepare a cyclic-free polymer, we could prepare 6 with extremely high amounts of these cyclics. Further treatment of this sample with acetone in a soxhlet extractor allowed for isolation (in the acetone solution) of ca. 75% of the cyclics formed. Both the cyclic dimer and cyclic tetramer preferentially precipitated from this acetone solution and could be recovered by filtration.

It is interesting to speculate why Barclay[5] did not prepare high molecular weight polyformal. Repeating Barclay's reaction conditions with the exception of eliminating the final heating in DMSO at 150°C gave 6 ([η] = 0.42 dl/g) which contained ∿15% cyclics. If the reaction was repeated but the DMSO solution was heated at 150°C for 3 hr., a low molecular weight material ([η] = 0.09 dl/g) was obtained which was almost exclusively linear oligomers. Heating a previously prepared sample of 6 ([η] = 0.66 dl/g) in DMSO at 150°C totally degraded the polymer to linear oligomers. The decomposition of DMSO upon heating to give a strong acid has been reported.[11] Reaction of such a strong acid with 6 at 150°C could easily destroy the formal linkages. Confirmation of this hypo-

thesis was obtained by heating a sample of 6 in DMSO at 150°C in
the presence of potassium carbonate and recovering a polymer
which was essentially unchanged. It is possible that Barclay did
prepare higher molecular weight polyformals but that he decomposed these polymers by heating them in DMSO at 150°C.

Some very preliminary experiments indicate that the aromatic
formal linkage may have surprising stability to acidic, nucleophilic and oxidative conditions. As shown by the DMSO experiments
run at 150°C, the aromatic formal linkage is sensitive to strong
acid. However, the initial results of heating methanolic hydrochloric acid solutions of either the linear or cyclic dimer of
BPA (summarized in Table 4) indicate that the linkage certainly
has some stability in acidic environments. The decreased
stability of the cyclic dimer is likely related to the relief of
strain resulting from initial opening of the ring. Reaction of
diphenoxymethane ($\underset{\sim}{8}$, Ar = C_6H_5) with either sodium 4-methyl-
phenoxide or sodium 4-methylthiophenoxide in N-methylpyrrolidone
(NMP) at 60°C showed no products derived from the attack of these
nucleophiles at the methylene group of the model formal.

Dialkyl formals such as $\underset{\sim}{12}$ have been successfully oxidized
to the corresponding carbonate esters with ozone[12] (equation 7).
However, if ozone was bubbled through a 2% solution of diphen-

(7)

$\underset{\sim}{12}$

oxymethane in methylene chloride at -78°C for twenty minutes,
followed by warming of the solution to 20°C with continued ozone
flow for an additional 40 minutes, no reaction was observed. In
addition, if oxygen was bubbled through a molten solution of
diphenoxymethane in the presence of several oxidizing reagents,
no formation of diphenyl carbonate was observed (see Table 5).
TGA (heating rate = 10°C/min) of the polymer $\underset{\sim}{6}$ showed less than
5% weight loss in air up to 418°C and in nitrogen up to 438°C.

These preliminary results suggest that aromatic polyformals
such as $\underset{\sim}{6}$ may have better oxidative and chemical stability than
would have been predicted based on the reactivity of aliphatic
formals. However, the effect of long term aging (either to
oxidative or chemical environments) on the properties of $\underset{\sim}{6}$ have
not been determined.

Table 4. Acid Hydrolysis of Aromatic Formals

Starting Material	Reaction Conditions[a]	Half-Life (hr)
Linear Dimer (10)	1.0N HCl, 55°C	∿ 4.7
	0.5N HCl, 55°C	∿ 14.1
Cyclic Dimer (9, n=1)	1.0N HCl, 55°C	∿ 2.7
	0.5N HCl, 55°C	∿ 7.8

a) Reactions were run in methanol using enough 12N HCl to give the final normalities shown. Disappearance of starting material was followed by liquid chromatography vs an internal standard.

Table 5. Attempted Oxidation of Diphenoxymethane[a]

Reagent	Temperature	Time
Oxygen	285°C	3 days
$H_2O_2/FeSO_4$	25°C	30 min.
Dicumyl Peroxide	145-170°C	6 hr
t-butylhydroperoxide	285°C	4 days

a) Oxygen was bubbled through the mixture of diphenoxymethane and oxidizing agent under the reaction conditions shown. Reaction was monitored by a combination of IR and VPC analysis.

Polyformals Based on Other Bisphenols

The reaction of 1,1-dichloro-2,2-bis(4-hydroxyphenyl)ethylene (BPC, 13)[13] with methylene chloride to form the corresponding formal 14 has also been extensively studied. (equation 8). Reaction of 13 with a 60/40 mixture of methylene chloride/NMP and

(8)

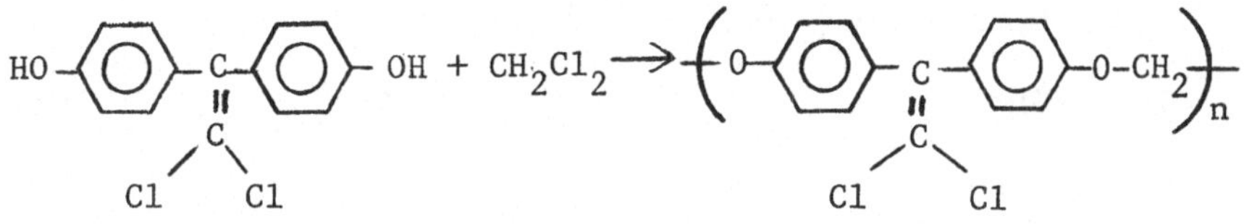

13

14

sodium hydroxide gave the polymer 14 which contained ca. 10-15% cyclic oligomers. The distribution of these cyclic oligomers was almost identical to that obtained when bisphenol A was used (see Figure 1). The T_g of 14 was about 30°C higher than 6 and it had a greatly improved oxygen index. Other properties of 14 are presented in Table 6.

Preliminary information on new aromatic polyformals based on other bisphenols is contained in Table 7. Most of these materials have been made only once and no extensive characterization of the polymers has been made. None of these polyformals were examined for cyclic oligomers. Where possible, the T_g for the corresponding polycarbonate is also given. All polymerizations were run in NMP/CH_2Cl_2 mixtures except when the bisphenol contained an electron withdrawing group (15, X = $-\overset{O}{\underset{\|}{C}}-$, $-SO_2-$). In these instances a

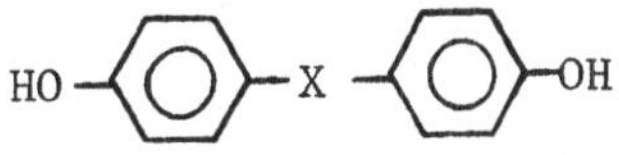

15

NMP/CH_2Br_2 mixture gave the best results.

Copolyformals Based on BPA and Other Bisphenols

Using the NMP/CH_2Cl_2 system, the reaction of a mixture of two bisphenols to form copolyformals was relatively straightforward. The majority of our efforts were focused on mixtures of BPA with other bisphenols in an attempt to raise the T_g of the resulting polymer relative to the homopolymer based on BPA (6). Typical results are presented in Table 8. No further properties were obtained for these copolymers. Incorporation of the bisphenols

Table 6.   Property Profiles

Properties	BPA Polyformal (6)	BPC Polyformal (14)
Specific Gravity (g/CC)	1.10	
Tg (°C)	94	120
Oxygen Index	20	43
Yield Stress (psi)	7100	10700
Flexural Modulus (psi x 10^3)	370	380
Flexural Strength (psi x 10^3)	12	16
% Elongation	80	85
Gardner Impact (ft lbs)	>320	>320
Notched Izod Impact (ft lbs/in notch)	1	1

into the copolymers was verified by ^{13}C NMR analysis of the isolated polymer. Figures 3 and 4 contain examples of ^{13}C NMR spectra obtained for two mixtures of bisphenols and clearly indicate the three different types of formal linkages which are formed as well as the proportion of each bisphenol present in the isolated polymers.

Block Copolyformalcarbonates

Various oligomeric mixtures of BPA polyformal (6) were pre-pared with the NMP/CH_2Cl_2 system, but using less sodium hydroxide than was required to make high molecular weight polymer. No phenolic chain stopper was used since it was desired that the oligomeric mixtures be hydroxy terminated. The reaction mixtures were worked up using the reverse precipitation method described previously to minimize the level of cyclics obtained in the isolated products. Typical data for these oligomeric mixtures is given in Table 9. The molecular weights of the isolated materials were less than the theoretical value (calculated assuming complete utilization of the sodium hydroxide for formation of the formal linkage). There are, however, at least three hydrolytic side

Table 7. Aromatic Polyformals[a,d]

Bisphenol	$T_g°C(T_m°C)$ Polyformal	$T_g°C$ Polycarbonate
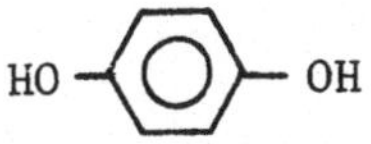 HO—⬡—OH (b)	41 (200)	
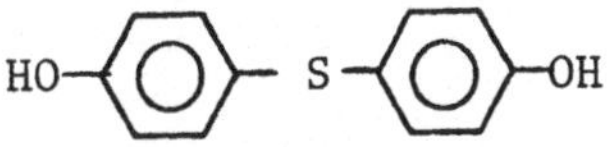 HO—⬡—S—⬡—OH (c)	63 (164)	
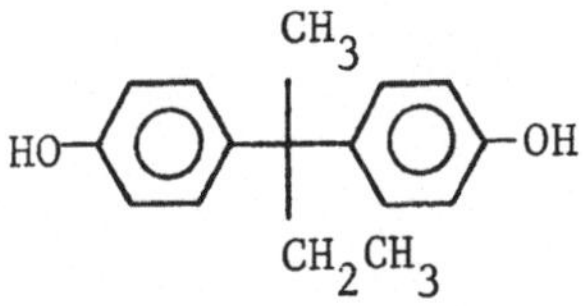 HO—⬡—C(CH₃)(CH₂CH₃)—⬡—OH	87	134
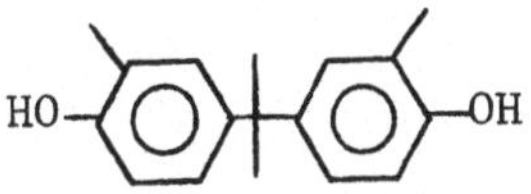 HO—⬡—C—⬡—OH	90	
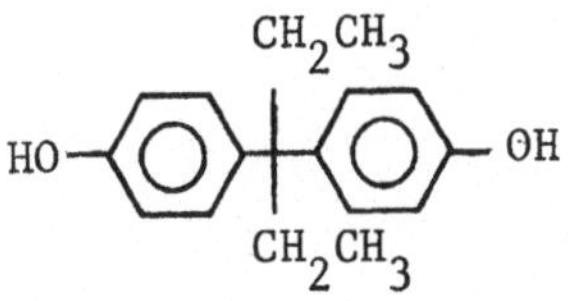 HO—⬡—C(CH₂CH₃)(CH₂CH₃)—⬡—OH	95	
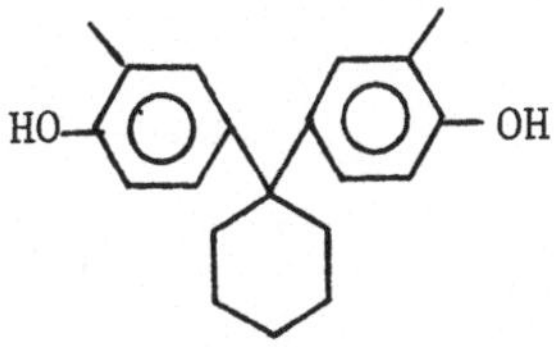 HO—⬡—(cyclohexyl)—⬡—OH	105	
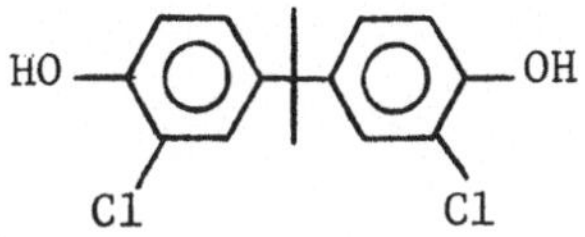 HO—⬡(Cl)—C—⬡(Cl)—OH	106	146

(continued)

Table 7.   Aromatic Polyformals[a,d]

Bisphenol	$T_g°C(T_m°C)$ Polyformal	$T_g°C$ Polycarbonate
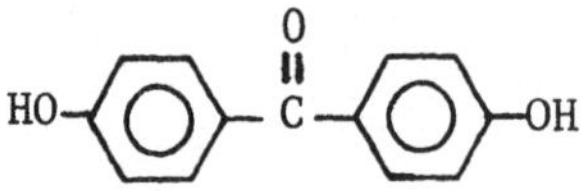 (e)	116	
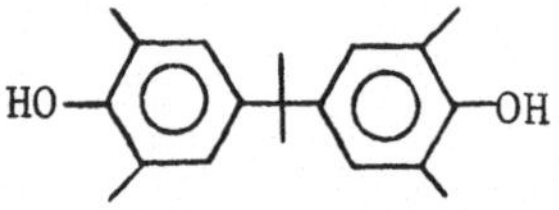	121	203
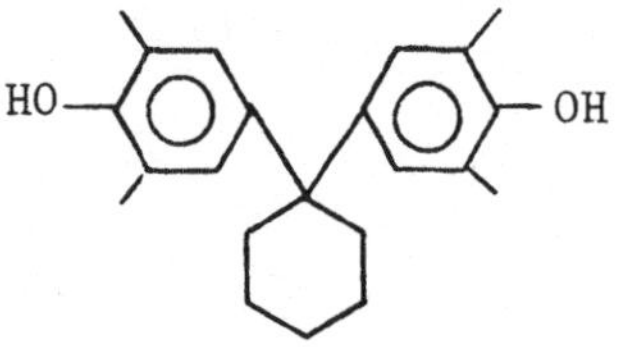 (b)	136	
 (b)	136 (250)	
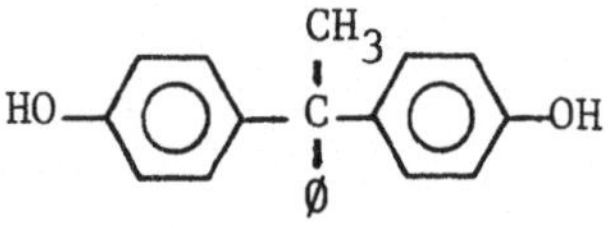	137	180
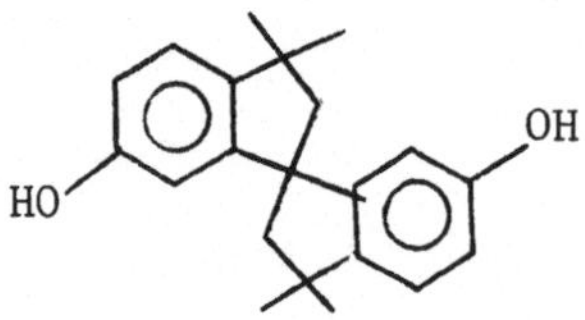	152	
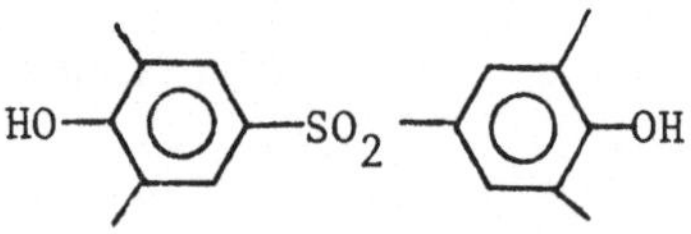 (e)	169	

Bisphenol	T_g°C(T_m°C) Polyformal	T_g°C Polycarbonate
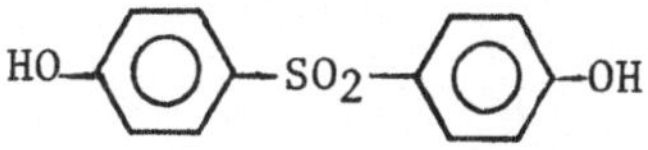 HO—⬡—SO₂—⬡—OH (c)(e)	179	
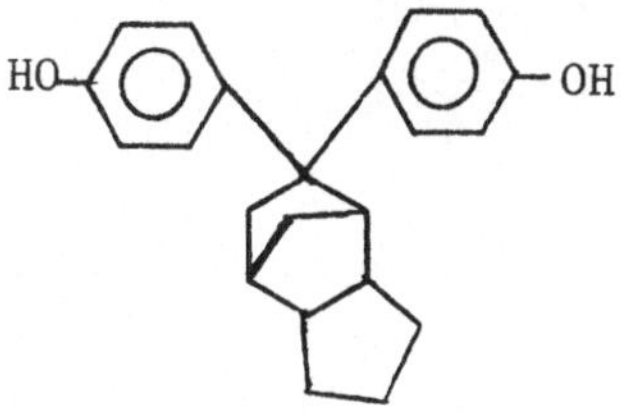 HO—⬡—C(—⬡—OH)(norbornane/cyclopentane)	202	
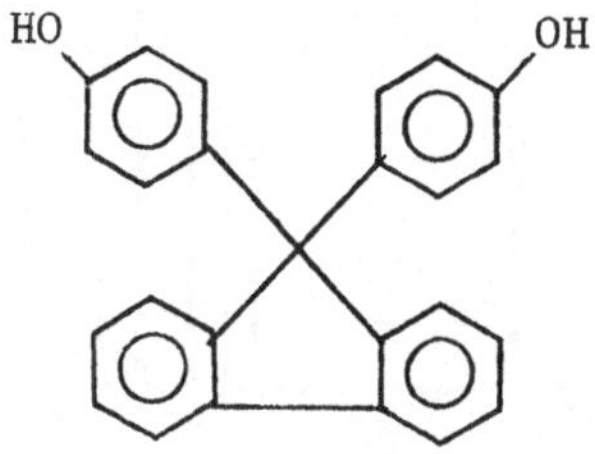 HO—⬡ ⬡—OH (spiro fluorene type)	218	
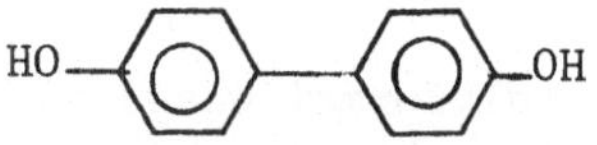 HO—⬡—⬡—OH (b)	(324)	

a) All polymers had intrinsic viscosities ≥0.35 dl/g in chloroform unless indicated.
b) Polymer insoluble, no intrinsic viscosity was obtained.
c) [η] determined in NMP
d) All reactions were run in NMP/CH_2Cl_2 unless indicated
e) No high molecular weight polymer could be obtained in NMP/CH_2Cl_2. A NMP/CH_2Br_2 mixture had to be used.

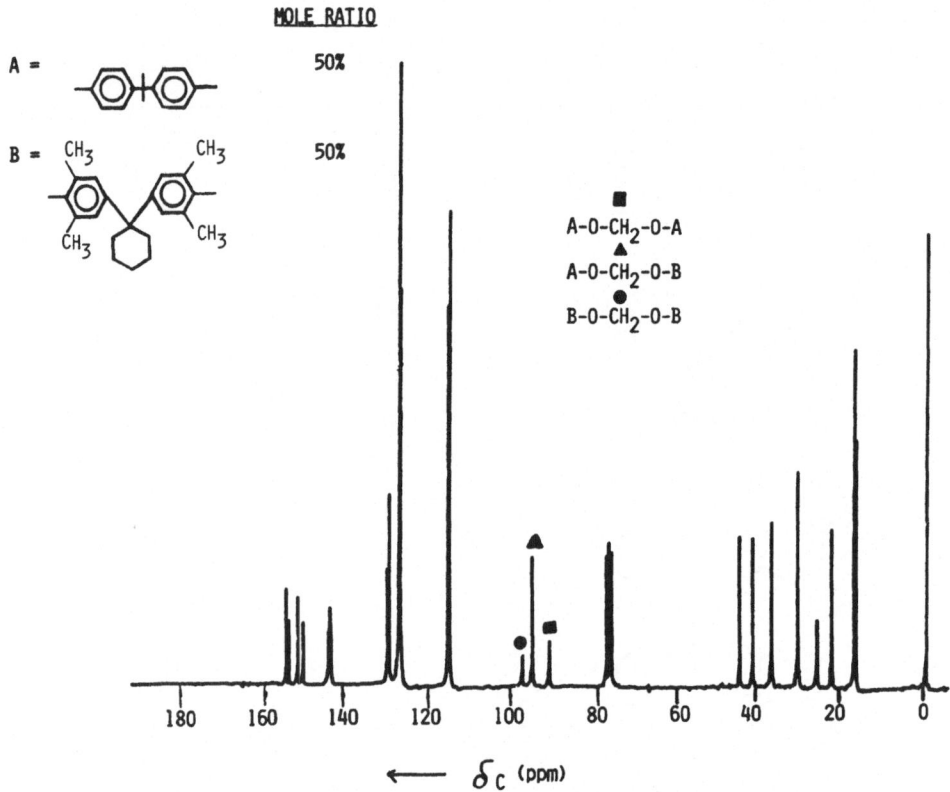

Fig. 3 ^{13}C NMR spectra for a 50/50 copolyformal based on
         BPA and the bisphenol derived from B

reactions which could be responsible for the consumption of the
sodium hydroxide.

Polycarbonate formation, using these oligomeric polyformal
mixtures in combination with BPA (and phenol as a chain stopper)
was carried out using a two-phased phosgenation procedure to give
the block copolyformal-carbonates 16.

16

Table 8. Copolyformals with BPA

Bisphenol	Mole Ratio of BPA/Bisphenol Initial	^{13}C-NMR	$[\eta]$ (dl/g)	T_g °C
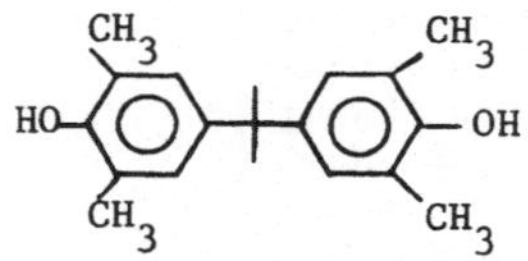 (tetramethyl bisphenol, isopropylidene)	75/25	71/29	0.58	102
	50/50	47/53	0.50	110
	25/75	31/69	0.62	122
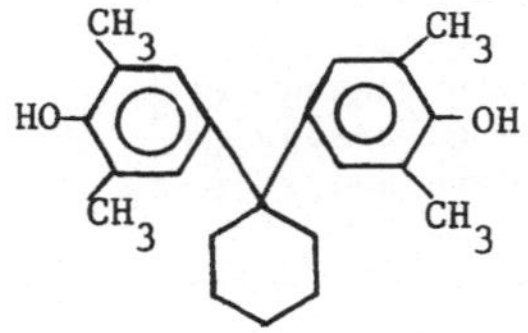 (tetramethyl bisphenol, cyclohexylidene)	75/25	75/25	0.32	106
	50/50	51/49	0.59	124
	25/75	35/65	0.58	133
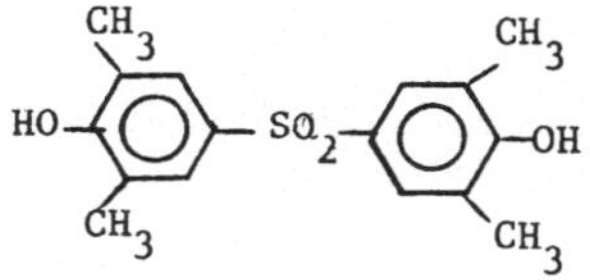 (tetramethyl bisphenol S, SO_2)	75/25	80/20	0.42	114
	50/50	53/47	0.52	135
	25/75	30/70	0.25	152
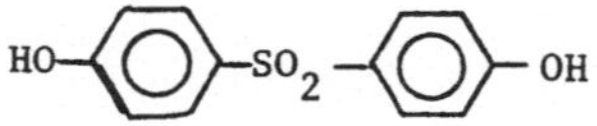 (bisphenol S, SO_2)	80/20	81/19	0.52	106
	65/35	68/32	0.37	119
	40/60	39/61	0.26	138
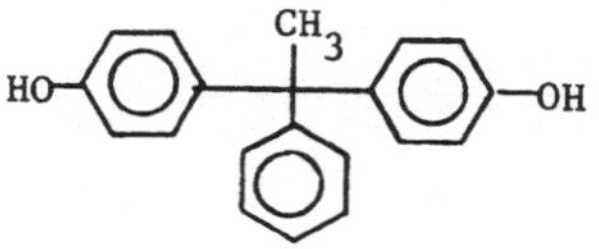 (triphenyl methane type)	50/50	56/44	0.37	114

Table 9.   Preparation of Oligomeric BPA-Polyformals

Moles NaOH / Moles BPA	% Yield	[η] (HCCl₃)	Tg(°C)[a]	Weight % C-2 through C-6 Cyclics (H.P.L.C.)	$\bar{M}_n$ [b] (G.P.C.)	$\bar{M}_n$ [c] (¹³C-NMR)	$\bar{M}_n$ (Calc.)[d]	% NaOH Lost Via Hydrolytic Side Reactions
1.97	73[f]	0.14	79°	-	6,120	-	16,224	2.6%
2.00	68[g]	0.15	83°	0.3	6,000	6,040	16,224	2.7%
2.00	73[h]	0.15	-	0.7	5,210	4,130	δ	4.9%
2.04	75[g]	0.17	-	0.5	8,360	8,920	δ	4.9%
2.10	81[g]	0.27	-	0.6	11,500	14,800	δ	6.8%

a) After sample stored in a dynamic vacuum overnight at 150°C.
b) Based on polystyrene calibration.
c) Determined by comparison of the integrals for the carbons ortho to the -OH end-groups vs. the carbons ortho to the -OCH₂O- linkages. Experimental error should be highest for the high molecular weight oligomers where end-groups become difficult to measure.
d) Assuming complete utilization of NaOH for forming -OCH₂O- linkages (i.e., no hydrolytic side reactions).
e) These are minimum values. Since substantial amounts of the lowest molecular weight oligomers were lost on precipitative workup, the "actual" $\bar{M}_n$'s of each polymer in its reaction mixture before isolation would have been less than the values obtained on the isolated samples. The differences, therefore, between the calculated $\bar{M}_n$'s and the "actual" $\bar{M}_n$'s should be larger than observed, indicating a greater loss of NaOH by side reactions. For the calculations, $\bar{M}_n$'s obtained by G.P.C. were used.
f) After one precipitation.
g) After two precipitations.
h) After three precipitations.

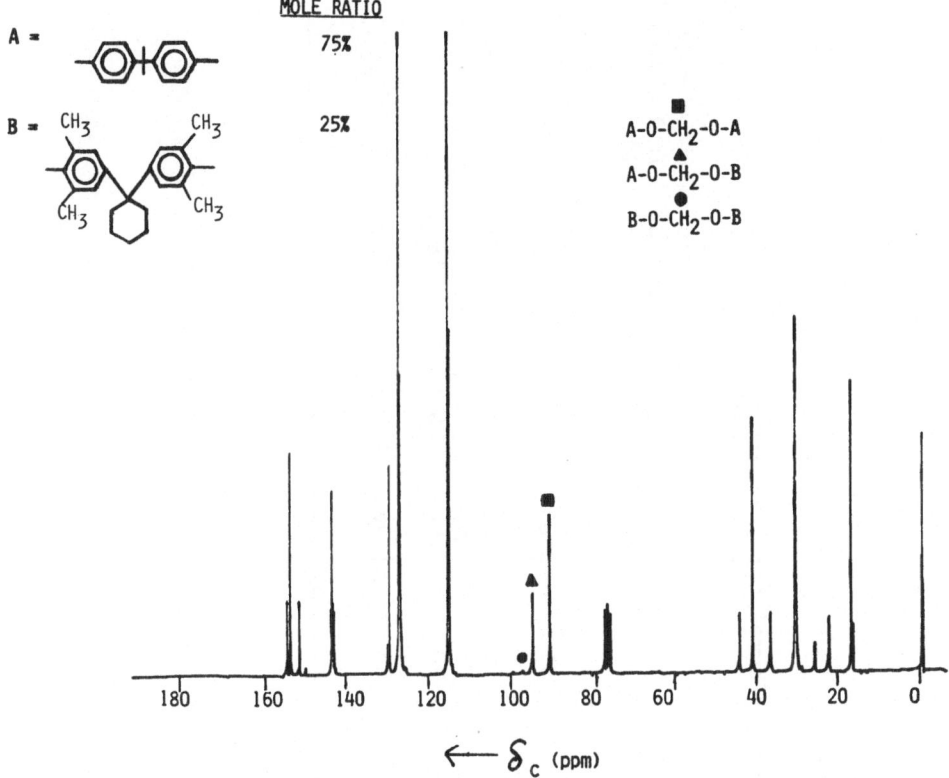

Fig. 4 ^{13}C NMR spectra for a 75/25 copolyformal based on
         BPA and bisphenol derived from B.

        Typical data for these materials is listed in Table 10.   The
agreement between the ratio of BPA in 6/BPA and the ratio of
polyformal/polycarbonate (as determined by ^{13}C NMR analysis) was
excellent.   The polymers 16 gave clear, tough, flexible films
on compression molding and a single T_g by DSC analysis.

Polymer Blends with BPA Polyformal

        A series of polymer blends was prepared with BPA polyformal
(6, [η] = 0.62 dl/g) by co-dissolving 6 and the other polymers in
methylene chloride and then co-precipitating them from methanol.
Blends of 6 with the following polymers were prepared (ratio
6/polymer):   BPA polycarbonate (80/20, 70/30, 60/40, 50/50, 40/60,
20/80, 15/85, 10/90, 5/95); polyphenylene oxide (17) (80/20, 65/35,
50/50, 35/65, 20/80); polystyrene (50/50); polysulfone (18) (80/20);
Ultem^R polyetherimide (19) (80/20), the polyformal 20 (80/20,
65/35, 50/50, 35/65, 20/80) and the polyformal 21 (75/25, 50/50,

Table 10. Block Copolyformal-carbonates (16)

Polyformal Oligomer ($\overline{M}_n$)[a]	Initial Weight Ratio of Oligomer:BPA	Initial Mole Ratio of BPA in Oligomer:BPA	% Yield[b]	[η] (HCCl3)	T_g (°C)	Mole Ratio of Polyformal:Polycarbonate in Isolated Material (13C-NMR)
6000	75:25	74:26	91	0.82	106 [c]	71:29
5200	50:50	49:51	94	0.52	112 [c]	49:51
6000	50:50	49:51	90	0.66	120 [c]	48:52
6000	25:75	24:76	85	1.06	130 [c]	22:78

a) $\overline{M}_n$ from GPC
b) After two precipitations.
c) Compression molding gave a tough, clear plaque.

25/75).   In all of these systems, we have observed two T_g's in the
DSC analysis and a lack of clarity in compression molded samples
which indicate inhomogeneous blends.   Most of these blends,
however, do give tough, flexible plaques on compression molding.

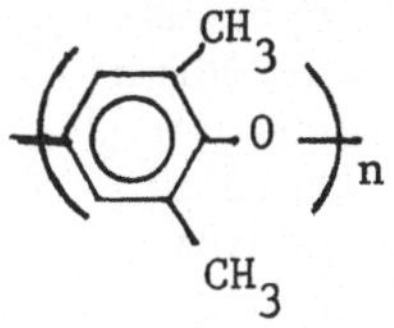

17

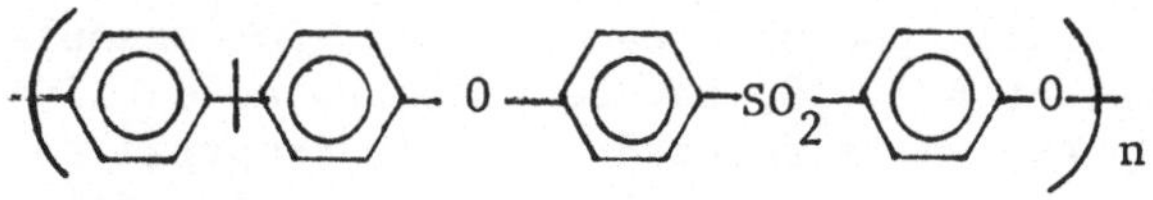

18

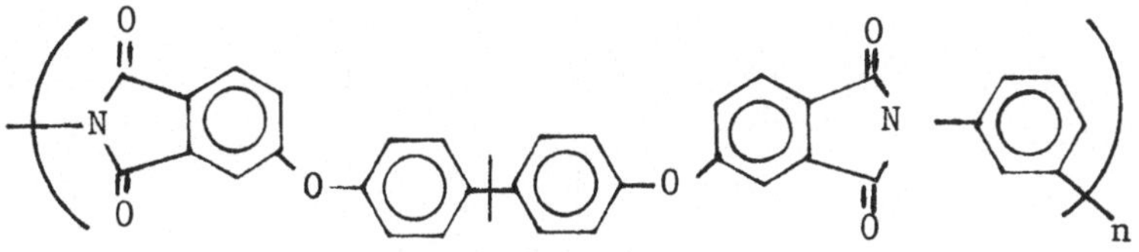

19

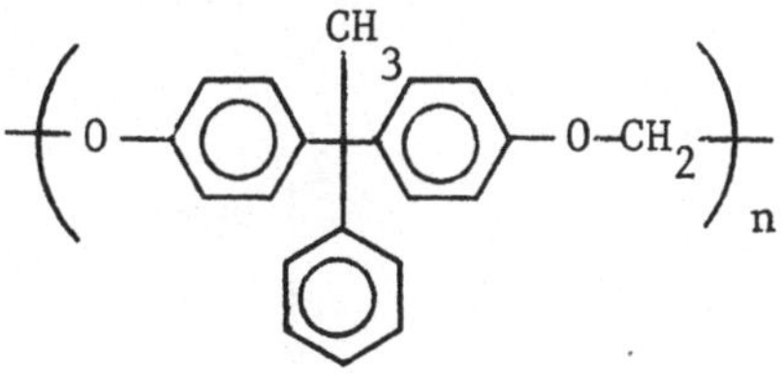

20

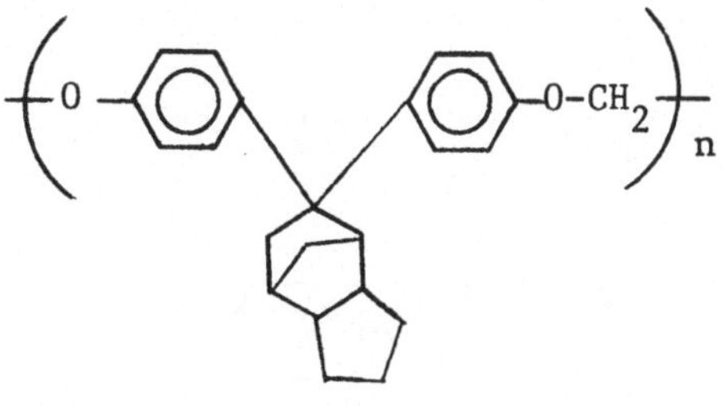

21

New Monomers and Polymers Containing the Formal Linkage

    In addition to using the reaction of bisphenoxides with
methylene halides to form polymer systems containing primarily
polyformal linkages, this displacement reaction can also be used
to synthesize new monomers or to couple other polymer systems
containing phenolic end groups. For example, the linear formal
dimer of BPA (10) can be prepared as outlined below (equation 9).
Bisphenol A is reacted with dihydropyran to give a mixture of
protected bisphenols from which 22 is isolated as the sodium
salt.   Reaction of 22 with methylene chloride in NMP gave 23 which
was hydrolyzed without isolation to give the monomer 10.   Polycarbonates were prepared by direct phosgenation of mixtures of 10
($[\eta]$ = 0.52 dl/g, T_g = 95°C) and by phosgenation of mixtures of
10 and BPA.   A polycarbonate prepared from a 1:1 molar mixture
of 10:BPA had an $[\eta]$ = 0.78 dl/g and a T_g = 125°C.

(9)

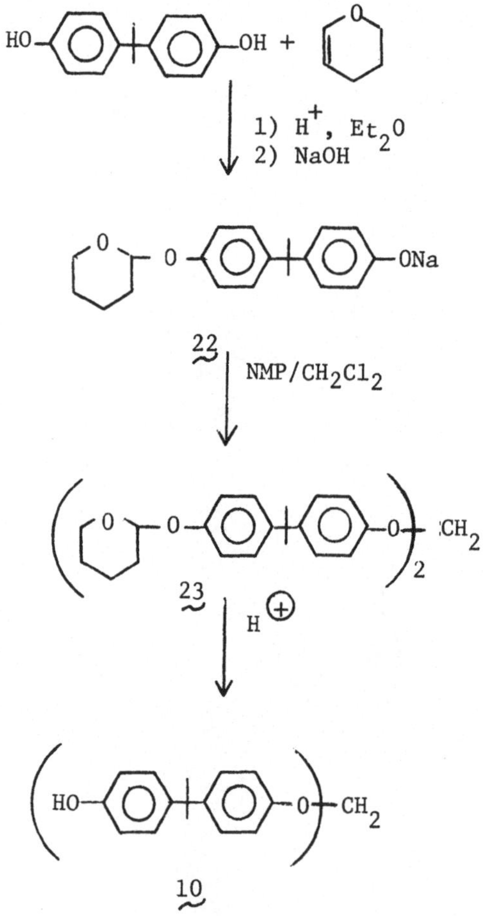

In a similar reaction, 4-hydroxythiophenol (24) was reacted with sodium hydroxide in a CH_2Cl_2/NMP mixture to give a material which was tentatively identified as the bisphenol 25 (equation 10). The structure of 25 was supported by ^{13}C NMR analysis.

(10)

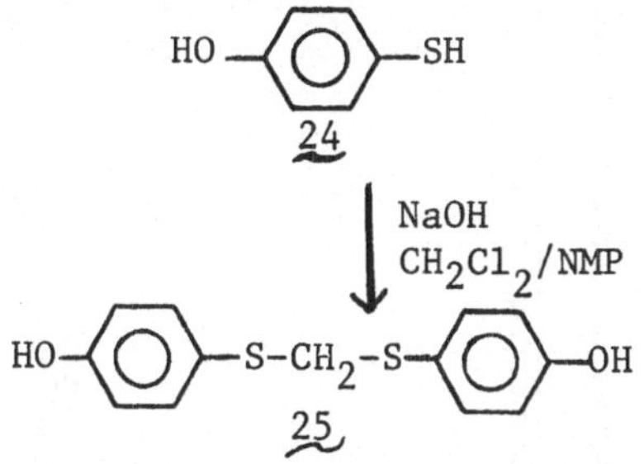

Another use of this type of displacement chemistry is illustrated by the synthesis of polyethersulfoneformals (26). Reaction of an excess of the disodium salt of bisphenol A with dichlorodiphenyl sulfone (DCDPS) (27) in DMSO results in a mixture of polysulfone oligomers which are terminated with phenolic end groups (as the sodium salt). This mixture can then be "zipped" together by the addition of methylene chloride, resulting in the formation of the polymers 26 (equation 11). The T_g of these polymers can be varied by controlling the ratio of BPA to the chlorosulfone 27. For example: using a 2/1 ratio of BPA/27, a polymer having a glass transition temperature of 148°C was obtained whereas a 1.34/1 ratio of BPA/27 gave a polymer having a glass transition temperature of 172°C.

(11)

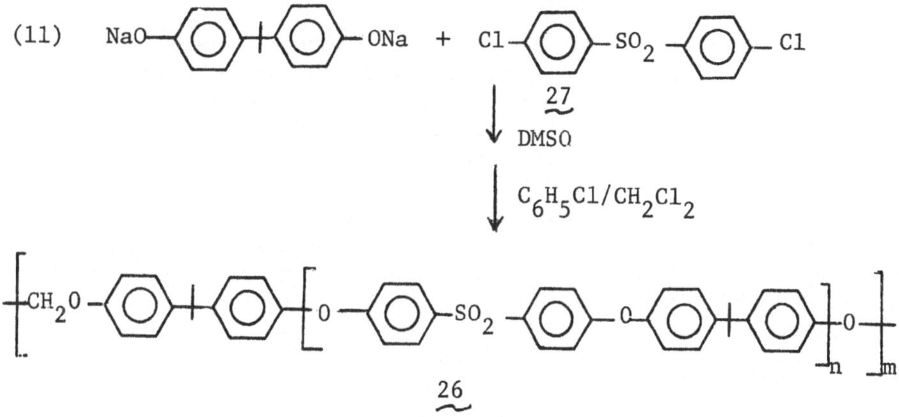

^{13}C NMR was once again a very powerful tool in characterizing these polymers as illustrated in Figures 5 & 6. Figure 5 shows the ^{13}C NMR spectrum of 26 using a 2/1 mixture of BPA/27 in the polymerization. The expanded region above the spectrum shows the three peaks at about 42 ppm corresponding to the quaternary carbon atom of the bisphenol A group. Three different environments are

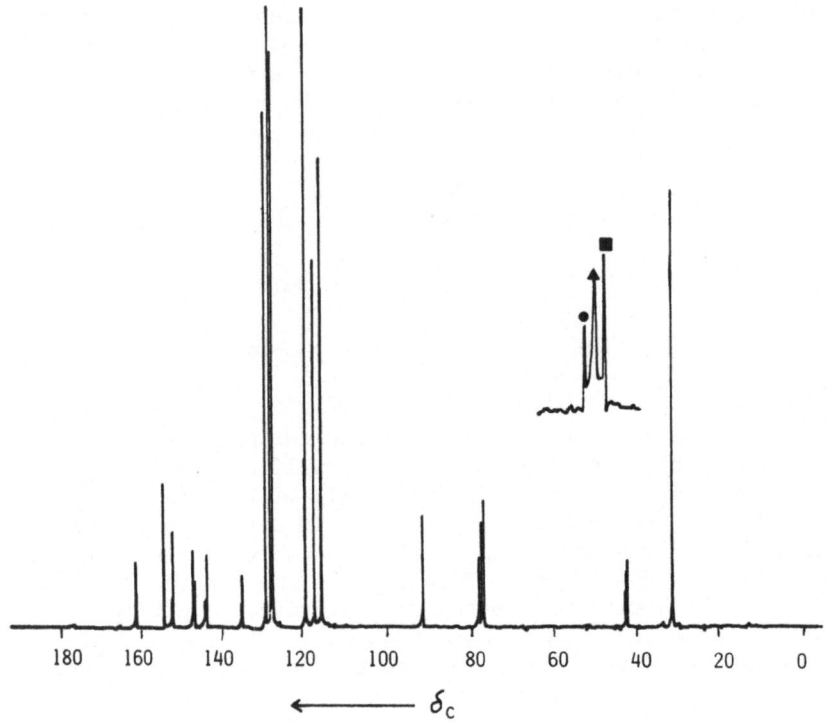

Fig. 5 ^{13}C NMR of polyethersulfone formals (26).

seen by this carbon atom depending upon whether a methylene group
or a phenylsulfone group has reacted with the phenolic oxygen.
Figure 6 better illustrates this point and shows the difference
in this part of the spectrum for two different polymers prepared
using different mole ratios of BPA/27.

     This reaction sequence offers the advantage of being able
to prepare linear oligomers under simplified reaction conditions
and then couple them with formal linkages to get polymers with
properties which are hopefully not too much different from the
homopolymers themselves.

EXPERIMENTAL

     The p-t-butylphenol and all the phase transfer catalysts and
crown ethers were purchased commercially. The NMP used was purchased from Burdick and Jackson; the methylene chloride, methylene
bromide, bromochloromethane and chlorobenzene were reagent grade.
The sodium hydroxide was Baker Reagent (97%) and the potassium
hydroxide was also Baker Reagent (86%). Sodium 4-methylphenoxide

Fig. 6 ^{13}C NMR spectra of polyethersulfone formal (26)
prepared from different ratios of BPA/27 (DCDPS).

was prepared by reacting 4-methylphenol with freshly prepared
sodium methoxide in methanol and removing the methanol under
vacuum.

All ^{13}C NMR spectra were recorded on either a Varian Associates XL 200 NMR spectrometer with ^{13}C observation at 50.3 MHz or
a Varian Associates FT-80A NMR spectrometer with ^{13}C observation
at 20 MHz. All spectra were obtained with complete proton de-
coupling. Chemical shifts were measured from internal tetramethylsilane or calibrated to this standard using known chemical
shifts of solvent peaks. Vapor phase chromatography (vpc) was
carried out on a Hewlett Packard 5750 instrument using a 6 ft
10% UC-W98 on 80/100 Chromosorb W column with temperature
programming between 200-300°C at 20°/min. Vpc yields were
obtained using an internal standard and correcting for response
differences. Integrations were done on a Spectra Physics SP4000.
Liquid chromatography was carried out on a Waters Associates Model
ALC-GPC 244 liquid chromatograph using a μC^{18} column. A 5 μl shot
of a 1% solution of the polymer in THF was normally used.
Standard conditions were: Prog. 7 reversed, 1 hr; Solv A THF
60 → 100%; Solv B H$_2$O 40 → 0%; U.V. detector at 280 nm and sensitivity = 0.05. Intrinsic viscosities were determined in chloroform.

Phase Transfer Reactions

Model Studies. General Procedure. The theoretical amount of
sodium or potassium hydroxide (corrected for the amount of water
present), an equivalent amount of 4-methylphenol, 0.05 equivalent
of the phase transfer catalyst, biphenyl (internal standard, 0.25
equivalents) and methylene chloride or a 3 to 1 mixture of chloro-
benzene/methylene chloride were heated at reflux. The amount of
solvent used was chosen to give a solution which would be 20% in
final product (g of formal/ml of solvent X 100 = 20%). Aliquots
were removed at timed intervals and were worked up with methylene
chloride and 1.2N HCl. These samples were analyzed by vpc and
the yields of the product were calculated, based on an internal
standard

Polymer Studies. General Procedure. The desired amount of
BPA (generally 45.35g), sodium or potassium hydroxide, phase
transfer catalyst (Aliquat 336 or tetrabutylammonium bromide),
p-t-butylphenol (0.75% as a chain stopper) and methylene chloride
or methylene chloride/chlorobenzene (1/3) were heated at reflux
for a period of 5 hr under nitrogen. At this point the reaction
mixture was diluted with methylene chloride and filtered through
Celite. The mixture was then either a) poured directly into
methanol to precipitate the polymer or b) split into two portions;
one portion was precipitated by methanol, the second was reverse
precipitated using methanol/acetone. (see below)

In the experiments using larger amounts of tetrabutylammonium
bromide (0.5 to 1.0 equivalents), the reactions were done on 0.1
to 0.2 times the scale of the previous work (i.e., 4.535g of BPA).
At the end of the reaction, the mixture was diluted with methylene
chloride and the solution was extracted with water and the organic
phase was dried over magnesium sulfate. The dried organic solution
was added to methanol to give the polymer which was collected,
dried, and analyzed. No further reprecipitation was required.

BPA-Polyformal (6) - NMP System

Into a 3ℓ Morton flask equipped with a mechanical stirrer and
reflux condenser were charged 453.5 g (1.99 mol) of BPA, 4.59 g
(.030 mol) of p-t-butylphenol, 618 ml of reagent grade methylene
chloride and 928 ml of Burdick and Jackson NMP. The mixture was
stirred under nitrogen until a homogeneous solution was obtained
(pale yellow). At this point 169.8 g (4.12 mol) of 97% NaOH
pellets were added under a thorough nitrogen blanket and the
mixture exothermed to 37°C. The mixture was then placed in a
73-76°C oil bath and heated at reflux for 5 hrs. After 5 hr. the
solution was cooled slightly and diluted with 8.0ℓ of chlorobenzene (9.5ℓ of chlorobenzene are required to give a 5% solution
needed for the reverse precipitation; the remaining 1.5ℓ was used

for rinsing during filtration).

The chlorobenzene mixture was allowed to cool to room temperature (overnight) and the white solid (NaCl) was removed by
filtration. The resulting filtrate was acidified with acetic acid
until a haze-free solution was obtained, and the polymer was
isolated by a reverse precipitation. This procedure involved the
addition of 1ℓ of a 50:50 (by vol) MeOH:acetone mixture to 1ℓ of
the polymer solution in a blender. The nonsolvent was added at a
uniform rate over a period of ten minutes. The resulting solid
was filtered and dried in a 65°C vacuum oven. In this fashion a
total of 440.80 (86% yield) of BPA polyformal was collected. This
material had an [η] (CHCl$_3$) = 0.54 dl/g and contained 1.0% cyclics
(C$_2$-C$_{17}$). ^{13}C NMR (CDCl$_3$) for $\underline{6}$; carbon atom, shift (ppm);
1, 154.7; 2, 144.5; 3, 127.7; 4, 115.6; 5, 91.0; 6, 41.7; 7, 30.9.

DMSO - Methylene Chloride System. Use of Preformed BPA Dianion

High Temperature. A mixture of 31.79 g of 50% aqueous sodium
hydroxide, 45.35 g of BPA, 180.2 ml of DMSO and 180.2 ml of toluene
was heated at reflux under nitrogen until all visible traces of
water had been removed (Dean Stark Trap). The condensed distillate
was then passed through a recirculating trap filled with molecular
sieves to complete the drying. The toluene was then distilled
from the system. The temperature of the mixture was maintained at
110°C while 25.26 g of methylene chloride was added dropwise over
a period of 20 minutes. A slight precipitate formed almost as
soon as the first drops of methylene chloride were added. The
mixture was stirred for an additional 15 minutes and it was then
added to methanol. The resulting solid ($\underline{6}$) was collected by
filtration and dried. [η] = 0.88 dl/g, 7.1% cyclics by L.C.
analysis.

Low Temperature. The BPA dianion was prepared exactly as
described above. After the toluene was removed, the system was
cooled to 25°C. To this heterogeneous mixture was added 72.1 ml
of methylene chloride and the mixture was then placed in an 85°C
oil bath and heated for 5 hr. (internal temperature 77-78°C).
The reaction mixture was added to methanol and the resulting
precipitate was collected and dried to give 42.32g of $\underline{6}$.
[η] = 0.38 dl/g, 66% cyclics by L.C. analysis.

Preparation of Di(4-methylphenoxy)methane (11)

A mixture of 2.19g of sodium 4-methylphenoxide, 11 ml of
methylene chloride and 11 ml of DMSO was heated in a 80°C oil bath
under nitrogen for 5 hr. The reaction mixture was cooled to room
temperature and diluted with 100 ml of methylene chloride. This
mixture was thoroughly washed 3X with 100 ml of 1.2N HCl and the
organic solution was dried over anhydrous magnesium sulfate.
The drying agent was removed by filtration and the solvent was
removed to give 1.85g (96%) of 11, mp 37-39.5°C. ^{1}H NMR (CHCl$_3$);
δ 2.25 (methyl, s,6), δ 5.60 (methylene, s,2), δ 6.95 (aryl,
s,8). ^{13}C NMR (CDCl$_3$) for 11: carbon atom, shift (ppm); 1, 20.5;
2, 131.6; 3, 130.0; 4, 116.5; 5, 115.0; 6, 91.7.

11

Block Copolyformalcarbonates (16)

Preparation of Oligomeric Mixtures of BPA-Polyformals.
Exactly 22.8 g (0.100 mole) of bisphenol-A (BPA), 31.1 ml of
CH$_2$Cl$_2$, and 46.7 ml of NMP were stirred under a stream of dry
nitrogen for 2 minutes and then 8.04 g of NaOH pellets (98% NaOH)
was added. The system was stirred, heated at reflux for 5 hours,
and cooled to ca. 25°.

After diluting with CH$_2$Cl$_2$ and filtering to remove inorganics,
the solution of oligomers was gradually diluted, with stirring,
with an equal volume of 50:50-acetone:methanol. The precipitated
oligomeric mixture was filtered, washed with some methanol, and
dried in vacuo at 60°C. Yield: 17.4 g; [η] = 0.13 dl/g in
chloroform; GPC: $\bar{M}_n$ = 6122, $\bar{M}_w$ = 11054, $\bar{M}_w/\bar{M}_n$ = 1.81; DSC: T_g =
79°C.

Exactly 17.1 g of the isolated material was redissolved in
CH$_2$Cl$_2$ and reprecipitated by addition to methanol. Yield: 15.9 g;
[η] = 0.15 dl/g in chloroform; GPC: $\bar{M}_n$ = 5999, $\bar{M}_w$ = 9569,
$\bar{M}_w/\bar{M}_n$ = 1.60; DSC: T_g=83°C. The ^{13}C-NMR spectrum of this
product was in complete accord with the proposed structure and,
from end group analysis, the average molecular weight was
calculated to be 6036, in good agreement with the GPC value.
The infrared spectrum showed weak phenolic -OH bands of 3540 and
3440 cm^{-1}. Liquid chromatography indicated the material was a
mixture of the linear BPA-polyformal oligomers with very minor

amounts of cyclic oligomers (total of C-2 through C-6 was only
ca. 0.3%).

Anal: Calc'd for $(C_{16}H_{16}O_2)_n$:   C, 80.0; H, 6.7.
                        Found: C, 79.5; H, 7.0.

Phosgenation of a Combination of 6 and BPA.   A mixture of
4.0 g of 6 (M_n = 6000), 4.0 g of BPA, 0.0649g of phenol, 66.85 ml
of CH_2Cl_2, 26.74 ml of water, and 0.083 ml of triethylamine was
placed in a 250 ml round-bottom flask fitted with a phosgene
inlet tube, condenser, overhead stirrer, pH electrode, and an
addition funnel. A 25% NaOH solution was added with stirring
until the pH was 11. Phosgene was bubbled into the vigorously
stirred reaction mixture for a total of 20 minutes while 25%
NaOH solution was added intermittently to maintain the pH between
10 and 11.5. The entire reaction mixture was then added to 500 ml
of methanol. The precipitated polymer was filtered and dried;
obtained 8.35 g. This material was redissolved in CH_2Cl_2,
filtered to remove some NaCl, and reprecipitated by adding to
methanol. The polymer was filtered and dried giving 7.6g of the
block copolyformal-carbonate; $[\eta]$ = 0.66 dl/g; DSC showed only
one T_g, 120°C; a compression molded plaque was flexible and
essentially clear. The ^{13}C-NMR spectrum was in complete accord
with the assigned structure 16 showing a polyformal/polycarbonate
ratio of 48:52 (theory, 49:51).

Linear Formal Dimer of BPA (10)

Preparation of 22. In a typical reaction, 57g (0.25 mole) of
BPA was dissolved in 300 ml of anhydrous ether, 0.3g toluene-
sulfonic acid was added and then 40 ml (0.43 mole) of dihydropyran
in 50 ml ether was added slowly over 30 min. The reaction was
exothermic and, after addition, was heated at reflux for 1 hr.
GPC analysis showed that the reaction was complete at the end of
addition.

To the ether solution, with rapid stirring, was then added
25% NaOH solution until precipitation was complete (requires
more than 1 equivalent). The ether was decanted and the solid
slurried with hot ether 3 times, then filtered. The dry filter
cake was twice slurried with 5% NaOH and filtered. The filter
cake was washed with ether once and dried. Yield of the trihydrate of 22 was 40.7g (0.105 moles): the yield based on BPA
was 42%, based on a 1:2:1 mixture of BPA:monocapped:dicapped, the
yield was 84%. The trihydrate was heated in a vacuum oven at 60°C
for 3 hr to give 22. ^{13}C NMR (DMSO-D_6) for 22: carbon atom, shift
(ppm); 1, 168.9; 2, 118.0; 3, 126.9; 4, 129.5; 5, 40.5; 6, 31.0;
7, 145.5; 8, 126.3; 9, 115.5; 10, 153.8; 11, 96.2; 12, 29.9;
13, 18.5; 14, 24.5; 15, 61.4.

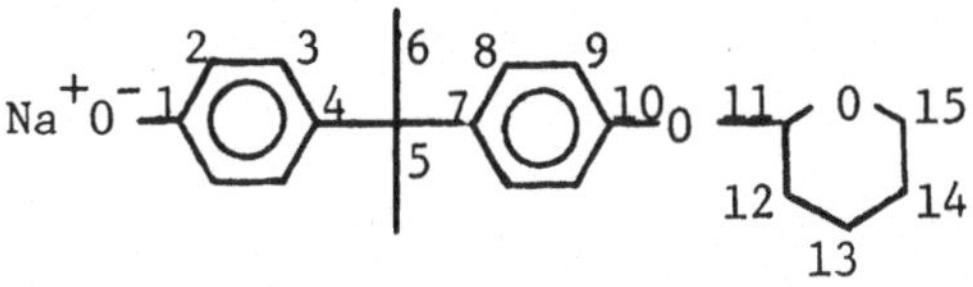

22

Preparation of 10

In a typical reaction, 66.8g (.172 mole) of 22, 1g ground
NaOH (to compensate for any base lost by hydrolysis of solvent)
and 240 ml of N-methylpyrrolidone (NMP) were mixed and warmed to
35°C until the bulk of the solids had dissolved. Then 160 ml of
CH_2Cl_2 was added and the mixture heated at 50°C for 1 hr. The
reaction was cooled, 50 ml H_2O added and the mixture acidified
with acetic acid. Total volume 500 ml. To this was added 100 ml
of 6N HCl and the mixture stirred at 37°C for 1 1/2 hrs. GPC
showed the hydrolysis to be 37% complete. After hydrolysis was
complete, the product was isolated by extracting into ether and
washing with water and bicarbonate until neutral. Removal of
the solvent gave an oil that was taken up in $CHCl_3$ and tri-
turated with hexane to give 36g (89%) of 10 as white crystals
that are 99% pure by HPCL analysis. Impurities are traces of
cyclic formal dimer and BPA. Recrystallized material has mp
130-131.5°C. ^{13}C NMR ($CDCl_3$) for 10: carbon atom, shift (ppm);
1, 153.19; 2, 115.93; 3, 127.83; 4, 144.93; 5, 41.72; 6, 30.92;
7, 143.12; 8, 127.73; 9, 114.76; 10, 154.81; 11, 91.55.

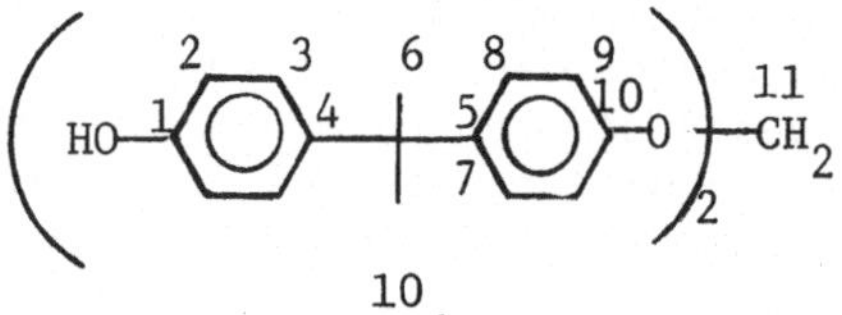

10

BPA-Formal Dimer/BPA Polycarbonates

A mixture of 6.2g of 10, 3.02g of BPA, 39 mg of phenol and
0.14 ml triethylamine was phosgenated under interfacial conditions
using aqueous sodium hydroxide and methylene chloride (see above).
The resulting polymer solution was washed to neutrality and precipitated into methanol to give a material which was further
washed with methanol and dried: yield: 8.61g; [η] = 0.78 dl/g,
T_g = 125°C.

Preparation of 25

A mixture of 5.00g of 4-hydroxythiophenol, 1.59g of sodium hydroxide pellets (1.0 equivalents), 6.0 ml of methylene chloride, and 9.0 ml of NMP (35% conc, 40/60) was stirred for 5 hr at 80°C (oil bath). Even before the NaOH was added, there was a vigorous exotherm which resulted in refluxing and foaming of the reaction mixture. After heating for 5 hr., the mixture was diluted with methylene chloride and washed with a 1% acetic acid solution and water. The organic phase was dried and concentrated to give a pale green oil. The oil was stirred with n-hexane for several hours to give 4.3g (82% yield) of 25 as a waxy white solid, mp 68-70°C. ^{13}C NMR (DMSO-d_6) for 25: carbon atom, shift (ppm); 1, 43.5; 2, 123.1; 3, 134.2; 4, 116.4; 5, 157.7.

25

Polyethersulfoneformals (26)

Into a 3-neck, 500 ml round bottom flask fitted with a mechanical stirrer, an addition funnel with nitrogen inlet and an adapter with thermometer and Dean-Stark trap was placed 10.272g (0.0451 mole) BPA, 25 ml DMSO and 100 ml of chlorobenzene. This mixture was heated to 70°C and 7.5g (0.0938 eq.) 50% NaOH was added. The white suspension was heated to reflux and when water ceased appearing in the distillate, the excess chlorobenzene was distilled out leaving a viscous oil at 155°C. Then, 6.46g (0.0225 mole) 4,4'-dichlorodiphenylsulfone in 35 ml DMSO was added over 10 min, causing a color change to deep green. The mixture was stirred at 160°C for 45 min, then cooled to 100°C and 20 ml of chlorobenzene containing 15g CH_2Cl_2 was slowly added. The green color discharged immediately and the viscosity rose rapidly. Phenol (50 mg) in 5 ml of chlorobenzene was added, the reaction cooled to room temperature, additional chlorobenzene added and the reaction filtered and precipitated into methanol. The stringy solid was washed, air dried, dissolved in CH_2Cl_2, filtered and reprecipitated, and dried in vacuum oven, to give 8.1g (53%) of 26; [η] = 0.43 dl/g; T_g = 147.5°C. See Figure 5 for ^{13}C NMR. TGA Analysis (5% weight loss): air = 480°C nitrogen = 475°C

SUMMARY

The reaction of bisphenols in the presence of hydroxide bases

with methylene halides is a straightforward and useful method for
the preparation of aromatic polyformals. The synthesis is best
carried out in the presence of a dipolar aprotic solvent (such as
NMP) or with large amounts (ca. 0.5 equivalents) of a phase
transfer catalyst. The reaction of bisphenol A with methylene
chloride gives a potentially interesting thermoplastic material (6).
Major disadvantages associated with the material are its low glass
transition temperature (94°C) and the accompanying formation of
7-10% cyclic oligomers even under optimized conditions. This
formation of an extended series of cyclic oligomers in the BPA
system (as well as with BPC) is extremely interesting and unusual
and certainly deserves further investigation.

A variety of new aromatic polyformals and copolyformals
with a wide range of T_g's have been prepared from other less
commercially attractive bisphenols. No additional information
regarding other properties or their propensity to form cyclics is
known for these systems. To date, we have been unsuccessful in
obtaining homogeneous blends of BPA polyformal with other polymer
systems.

A variety of other new polymer systems have also been synthesized based on this chemistry. These include block copolyformalcarbonates, polycarbonates using the linear formal dimer of
BPA, and polyethersulfoneformals.

The reaction of phenols, bisphenols, or phenolic end-groups
with methylene halides is clearly a very versatile method for
the preparation of a wide variety of new polymer systems. The
challenge remains for a system to be identified which has the
right combination of properties and economics to be of commercial
interest.

ACKNOWLEDGEMENTS

We would like to acknowledge the work of A. Factor and
W.L. Grosvenor on the oxidation of the formal linkage and
E.A. Williams for obtaining and interpreting the ^{13}C NMR spectra.

REFERENCES

1.   S.R. Sandler and W. Karo, "Polymer Synthesis", Vol. II,
     Chapter 6, Academic Press, Inc., New York (1977).
2.   J.W. Hill and W.H. Carothers, J. Amer. Chem. Soc, 57,
     927 (1935); D.P. Pattison, J. Org. Chem., 22,
     662 (1957).
3.   H.R. Musser and W.J. Jackson, U.S. Patent 3,875,257
     (1975) and U.S. Patent 3,809,681 (1974).
4.   See reference 1, Chapter 2, for a review of this subject.
5.   R. Barclay, Jr., U.S. Patent, 3,069,386 (1962).
6.   M. Matyner, A. Noshay and J.E. McGrath, Trans. Soc.
     Rheol., 21:2, 273 (1977).
7.   A. McKillop, J.C. Fiaud and R.P. Hug, Tetrahedron, 30,
     1379 (1974).
8.   E.V. Dehmlow and J. Schmidt, Tetrahedron Lett. 2,
     95 (1976).
9.   Aliquat 336 (95% active monomethyltricaprylyl ammonium
     chloride) obtained from General Mills Company.
10. For a preliminary report of this work see: A.S. Hay,
     F.J. Williams, et al, Polymer Preprints, 23, 117 (1982);
     A.S. Hay, F.J. Williams, et al, J. Polym.
     Sci, Polym. Lett. 21, 000 (1983).

     Also see:
     A.S. Hay, U.S. Patent 4,374,974 (1983); A.S. Hay,
     U.S. Patent, 4,254,254 (1981); G.R. Loucks and
     F.J. Williams, U.S. Patent 4,260,733 (1981);
     F.J. Williams and P.E. Donahue, U.S. Patent 4,136,087
     (1979); D.S. Johnson, U.S. Patent 4,163,833 (1979);
     J.C. Carnahan, U.S. Patent 4,216,305 (1980);
     J.C. Carnahan, U.S. Patent, 4,310,654 (1982).
     H.M. Relles and D.S. Johnson, U.S. Patent, 4,210,731
     (1980).

11. D.L. Head and C.G. McCarty, Tetrahedron Lett. 16,
     1405 (1973).
12. A. Maggiolo et al, U.S. Patent 3,139,440 (1964).
13. A. Factor and C.M. Orlando, J. Polym. Sci, Polym. Chem.
     Edit, 18, 579 (1980).

NEW POLYMERS PREPARED FROM N-CYANOUREA COMPOUNDS

Shiow C. Lin

Washington Research Center
W. R. Grace & Co.
7379 Route 32
Columbia, MD   21044

INTRODUCTION

This research was aimed at developing a novel class of monomers containing two N-cyanourea groups in a molecule. These monomers were found to polymerize to either thermoplastic or thermoset materials depending on the polymerization conditions.

The compound, N-cyano-N'-phenyl urea, was synthesized from phenyl isocyanate and aqueous alkaline solution of cyanamide with a high yield. This compound was reported to melt and to decompose in the range 127-128°C after recrystallization from acetone-petroleum ether mixture.[1]

Under basic conditions, cyanamide undergoes a dimerization and forms dicyandiamide at room temperature.[2,4] Based on this fact, a monomer terminated with two N-cyanourea groups would be expected to polymerize and form a thermoplastic at room temperature.

Upon heating, cyanamide was known to polymerize at temperatures approaching 100°C.[2,4] Therefore, the difunctional N-cyanourea monomer was also predicted to polymerize at elevated temperature. By controlling the polymerization conditions, either thermoplastic or thermoset materials should be obtainable from this novel class of monomers.

In this research, the linear polymer and the crosslinked polymer were synthesized from the difunctional N-cyanourea monomer. N-Cyano-N'-phenyl urea was used as a model compound to study the possible polymerization mechanisms.

EXPERIMENTAL

Preparation of Dicyanourea Monomer from Di(p-Isocyanatophenyl)methane (MDI)

To an aqueous solution containing 16.8 g of cyanamide, 16 g of sodium hydroxide and 200 ml of water was added dropwise 25.0 g of MDI dissolved in dioxane over an hour period at a temperature between 10 to 15°C. A 2N sodium hydroxide aqueous solution (200 ml) was also added slowly at a speed to finish addition at the same time as the addition of MDI solution. The aqueous solution was stirred at room temperature for an additional hour after the addition of reactants. In the presence of ice, the aqueous solution was acidified with concentrated hydrochloric acid to precipitate a fine crystal. After filtration, the dried fine crystal rapidly decomposes and forms yellow rigid foam upon heating at elevated temperatures.

Preparation of N-Cyanourea Terminated Resin

76 g of polycaprolactone diol (MW=530 g/mole) were added dropwise over a 6-hour period to a flask containing 50 g of tolyene diisocyanate in a nitrogen atmosphere. The reaction was continued with stirring overnight at room temperature. The resultant isocyanate terminated resin was heated to 50°C and mixed with 12 g of cyanamide. The reaction was cooled to room temperature as soon as cyanamide had dissolved. The IR spectrum of the resultant viscous liquid showed the disappearance of -NCO (2340 cm^{-1}) and a strong absorption at 2270 cm^{-1} (-C≡N). DSC thermogram showed an exotherm onset temperature at 137°C.

Polymerization of Difunctional N-Cyanourea Compound

At room temperature, 2.00 g of cyanamide were mixed with 5.96 g of P.P'-diphenylmethane diisocyanate and then dissolved in 80 ml of dry N-methyl pyrrolidone. A high viscosity solution was obtained in 72 hours. (This solution turned to gel after 96 hours of polymerization.) The polymer solution was added to 800 ml of acetone with vigorous stirring to precipitate the polymer in flake form. The polymer was collected by filtration and dried in vacuum. A new absorption at 2180 cm^{-1} was observed in IR spectrum.

Curing of N-Cyanourea Terminated Resin

The N-cyanourea terminated resin synthesized from polycapro-lactone-diol, tolyene diisocyanate and cyanamide was applied on an aluminum plate about 3 mil of thickness, heated at 135°C for 10 minutes and stood at room temperature overnight. The cured film had an MEK rub over 100, and an excellent adhesion on aluminum.

Thermal Stability Test on Polymer

The polymer powder prepared from cyanamide and MDI was mixed with KBr and compressed to form KBr pellet for IR spectrum. The IR spectra were taken after aging the sample at different time intervals and temperatures. The intensity of IR absorption at 2180 cm^{-1} decreased with time and finally completely disappeared. A new absorption at 2340 cm^{-1} was formed.

RESULTS AND DISCUSSION

Monomer and Model Compound Synthesis

The model compound, N-cyano-N'-phenyl urea, was prepared from phenyl isocyanate and an aqueous alkaline solution of cyanamide.[1] In general, the reactions can be summarized by the following equations.

$$\text{C}_6\text{H}_5\text{-NCO} + \text{H}_2\text{NCN} + \text{NaOH} \rightarrow \text{C}_6\text{H}_5\text{-NHCNNaCN} + \text{H}_2\text{O} \qquad (1)$$

$$\text{C}_6\text{H}_5\text{-NHCNNaCN} + \text{H}^+ \rightarrow \text{C}_6\text{H}_5\text{-NHCNHCN} \qquad (2)$$

The difunctional N-cyanourea monomer was also synthesized by the same procedures except a water soluble solvent such as dioxane or N-methyl pyrrolidone was used to prepare the diisocyanate solution. Usually, the monomer yield was above 90%.

Linear Polymer Synthesis

The linear polymer can be prepared either from a difunctional N-cyanourea monomer or directly from the mixture of a diisocyanate and cyanamide having a 1 to 2 mole feed ratio in a solvent such as N-methyl pyrrolidone at room temperature. The viscosity of the polymerization mixture increased with time and finally turned to gel upon stirring at room temperature. A high molecular weight polymer was obtained by adding the highly viscosity polymer solution to a large quantity of acetone after 72 hours of polymerization. For example, a fluffy white polymer was obtained from the polymerization of di(p-isocyanato-phenyl) methane and cyanamide at room temperature.

Preparation of Thermoset Materials

In general, the thermoset material was obtained by heating the N-cyanourea terminated monomer at a temperature above 100°C or at its melting temperature. Depending on the structure, the crosslinking reaction occurred within a few seconds to several minutes, and the final product varied from rigid foam to flexible film. Typically,

the N-cyanourea terminated resin was synthesized through the
following reactions. A diol was first reacted with two moles of a
diisocyanate to form an isocyanate terminated adduct which, in turn,
was reacted with two moles of cyanamide to produce the desired resin.
On heating, the resin changed to a crosslinked material. Various
kinds of thermosets having different properties were synthesized for
special applications.

$$
\overset{\text{O}\quad\text{O}}{\overset{\|\quad\;\|}{\text{HO-R-OH+2OCN-R'-NCO} \longrightarrow \text{OCN-R'NHCO-R-OCNH-R'-NCO}}}
\tag{3}
$$

$$
\overset{\text{O}\qquad\text{O}}{\overset{\|\qquad\|}{\text{OCN-R'-NHC-O-R-OCNH-R'-NCO+2H}_2\text{NCN} \longrightarrow}}
$$

$$
\overset{\text{O}\qquad\text{O}\;\text{O}\qquad\quad\text{O}}{\overset{\|\qquad\;\|\;\|\qquad\quad\|}{\text{NCNHCNHR'NHCOROCNHR'NHCNHCN}}}
\tag{4}
$$

Characterizations
_____

    All N-cyanourea compounds prepared in the study exhibit an
intense, resolved doublet at 2250 cm^{-1} and 2280 cm^{-1} on a single
absorption at 2270 cm^{-1}. For simplification, a typical IR spectrum
of the model compound, N-cyano-N'-phenyl urea is shown in Figure 1 to
illustrate the absorption of the cyano group. At room temperature
the difunctional N-cyanourea compound gradually shifts the absorption
of the cyano group to a doublet at 2160 cm^{-1} and 2190 cm^{-1} which
corresponds to the absorption of the cyano group of dicyandiamide.
Based on these facts, a polymer should be expected to form through
the dimerization of N-cyanourea group when the conversion of the
reaction is high enough.

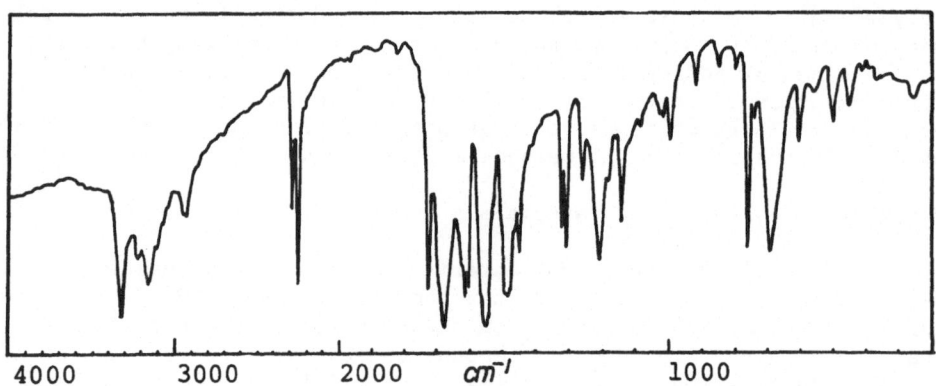

Figure 1. IR Spectrum of N-Cyano-N'-phenyl Urea

Using di(p-isocyanatophenyl)methane and cyanamide as monomers, in N-methyl pyrrolidone at room temperature, a high molecular weight polymer was obtained after three days. During the polymerization, the original absorption of the cyano group in the IR spectrum gradually reduced its absorption and a new doublet at 2160 cm^{-1} and 2190 cm^{-1} was formed. The high molecular weight polymer showed only an overlap absorption of the cyano group at 2180 cm^{-1} as indicated in Spectrum 1 of Figure 2. These facts imply that the polymerization mechanism, the same as the formation of dicyandiamide from cyanamide, should result in repeating segments containing a N,N'-biscarbamyl-N-cyanoguanidine unit in the polymer backbone. A tentative polymerization mechanism is shown in the following equations.

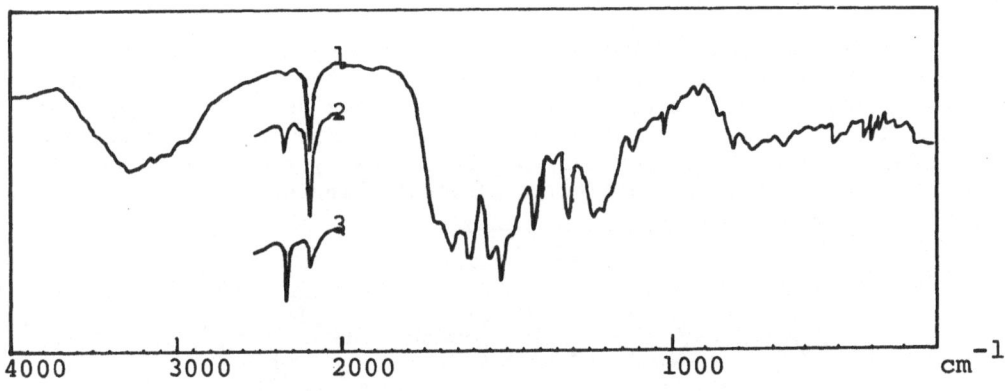

Figure 2. IR Spectra of Polymer Synthesized from Di(p-iso-cyanatophenyl) Methane (Spectrum 1), and the Polymer after Being Thermally Degraded for 20 Minutes (Spectrum 2) at 118°C and for Additional 25 Minutes at 148°C (Spectrum 3)

Upon heating at a temperature above 115°C, the intensity of IR absorption of this polymer at 2180 cm^{-1} decreased with time and finally completely disappeared. A new absorption at 2340 cm^{-1} was also formed and the tough, flexible polymer became brittle. Spectra 2 and 3 in Figure 2 shows the change of the cyano group in this polymer upon heating. Two possible reasons may be used to explain these results.

The first explanation is the depolymerization of N,N'-bis-carbamyl-N-cyanoguanidine units to its original N-cyanourea group. However, there is no observable increase in the absorption of the cyano group at 2270 cm^{-1} to support this assumption. Therefore, the other explanation, the rearrangement of N,N'-biscarbamyl-N-cyanoguanidine units, may account for the disappearance of the absorption at 2180 cm^{-1}. This is illustrated as the following reaction.

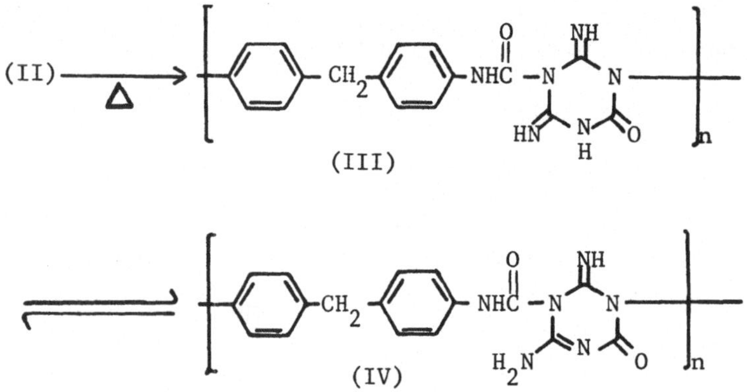

The generation of new absorption at 2340 cm^{-1} upon heating the polymer may be attributed to the formation of the isocyanate group through thermal degradation of the urea linkage as shown in Equation (8). This thermal degradation mechanism can also be used to explain the brittleness of the polymer and the viscosity reduction of the polymer solution after heating at elevated temperature.

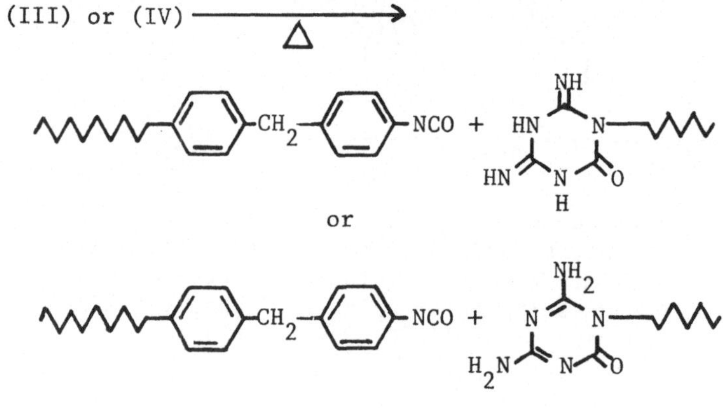

A yellow, apparently crystalline material was obtained upon heating N-cyano-N'-phenyl urea at 130°C. The material was a mixture of several components based on the spectroscopic studies. Some results and tentative conclusions regarding the nature of this mixture are given below:

- The yellow product is very insoluble in organic solvents.

- The IR and C-13 spectra indicate the loss of the cyano group.

- The proton and C-13 NMR spectra indicate a

$$\emptyset-NH\overset{\overset{\displaystyle O}{\displaystyle \|}}{C}-$$

  and one or more different phenyl moieties are present.

- The urea carbonyl signal in the spectrum of the original compound has disappeared and has been replaced with several new amide type carbonyl signals.

- The two -NH- signals 9.5 δ and 8.8 δ in the unheated sample are replaced with a series of weak signals over the range 9 to 12 δ, and a sharp signal at 8.6 δ which may be due to the

$$\emptyset-NH\overset{\overset{\displaystyle O}{\displaystyle \|}}{C}$$

  structure.

- The IR spectrum shows a weak, broad new signal at 2200 cm^{-1}. This may indicate that a small amount of a carbodiimide compound is present.

- The mass spectrum shows that CO_2 is the major gaseous product during the decomposition.

These observations indicate that this thermal decomposition involves several complicated reactions. The loss of the cyano group and the sharp peak at 8.6 in C-13 NMR may indicate that the major reactions during decomposition are the dimerization of N-cyano-N'-phenyl urea and the ring closure after dimer formation [Equation (9)]. This equation may also explain part of the observation, that one or more different phenyl moieties are present in the decomposition product.

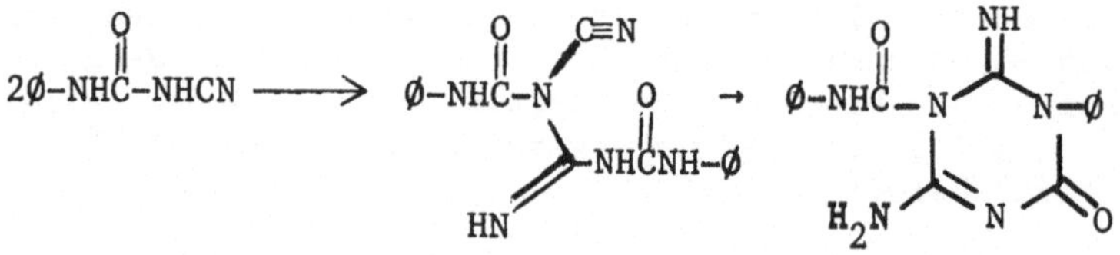

$$2\emptyset\text{-NHC(=O)-NHCN} \longrightarrow \emptyset\text{-NHC(=O)-N} \begin{array}{c}\text{C}\equiv\text{N}\\ \text{O}\\ \text{-NHCNH-}\emptyset\\ \text{HN}\end{array} \rightarrow \emptyset\text{-NHC(=O)-N} \begin{array}{c}\text{NH}\\ \text{N-}\emptyset\\ \text{H}_2\text{N} \quad \text{N} \quad \text{O}\end{array} \qquad (9)$$

The formation of carbodiimide group as indicated in IR spectrum at 2200 cm^{-1} may come from the isocyanate generated from the same mechanism as shown in Equation (8). Consequently, the isocyanate dimerizes and releases CO_2[3] under the catalyzation of chemical species in the decomposition product.

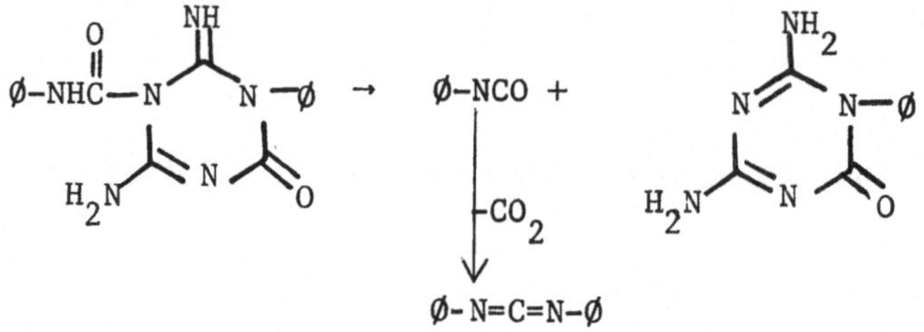

$$\emptyset\text{-NHC(=O)-N} \begin{array}{c}\text{NH}\\ \text{N-}\emptyset\\ \text{H}_2\text{N} \quad \text{N} \quad \text{O}\end{array} \rightarrow \emptyset\text{-NCO} + \begin{array}{c}\text{NH}_2\\ \text{N} \quad \text{N-}\emptyset\\ \text{H}_2\text{N} \quad \text{N} \quad \text{O}\end{array} \qquad (10)$$

$$\downarrow -CO_2$$

$$\emptyset\text{-N=C=N-}\emptyset$$

The trimerizations of N-cyano-N'-phenyl urea such as in Equations 11 and 12,

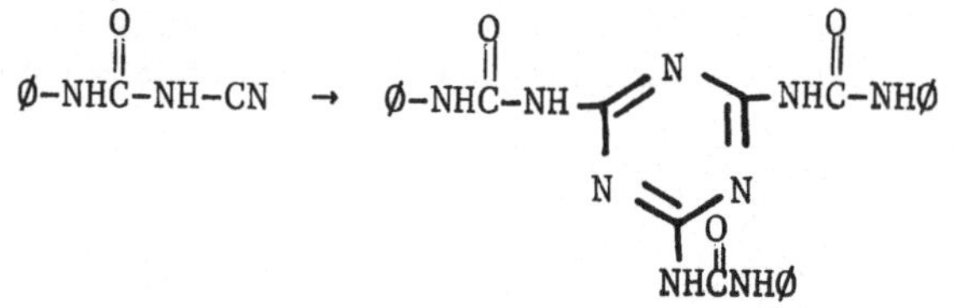

$$\emptyset\text{-NHC(=O)-NH-CN} \rightarrow \emptyset\text{-NHC(=O)-NH} \begin{array}{c}\text{N}\\ \text{NHC(=O)-NH}\emptyset\\ \text{N} \quad \text{N}\\ \text{O}\\ \text{NHCNH}\emptyset\end{array} \qquad (11)$$

$$3\ \emptyset\text{-NHC(=O)-NHCN} \rightarrow \emptyset\text{-NHC(=O)-N} \begin{array}{c}\text{NH}\\ \text{N}\\ \text{HN} \quad \text{N} \quad \text{NHC(=O)-NH-}\emptyset\\ \text{C=O}\\ \text{NH}\\ \emptyset\end{array} \qquad (12)$$

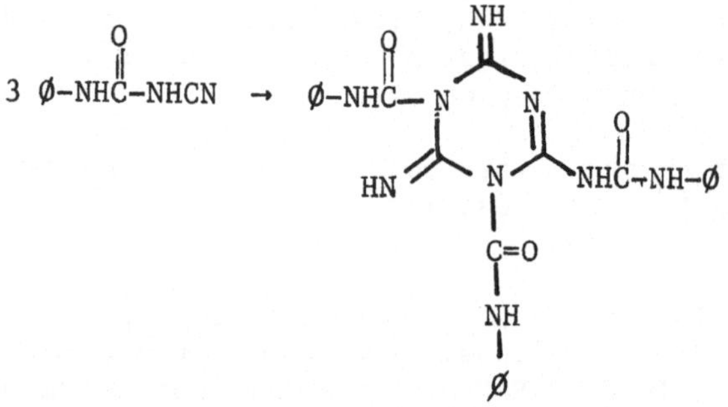

may also occur during the decomposition. Especially, cyanamide and substituted cyanamide were reported to rapidly form the melamine structure at elevated temperature[4] which is similar to the Reaction 11.

Based on these results, one can expect to obtain a fast curing system containing multi-functional N-cyanourea groups. The dicyanourea compound prepared from di(P-isocyanatophenyl) methane and cyanamide melts at 85°C and has an onset curing temperature at 122°C from DSC study. This compound after being heated at 125°C, turns to a yellow rigid form in a minute.

The difunctional N-cyanourea terminated resin synthesized from a polycaprolactonediol [MW=530 g/mole], tolylene diisocyanate and cyanamide was also used to study the feasibility of utilizing this chemistry to make thermoset plastics. The Spectrum 1 in Figure 3 shows the IR absorption of this fresh resin. Upon heating at 100°C, a yellow material was formed and showed the decrease of the cyano group absorption at 2270 cm^{-1} with the trace absorption at 2180 cm^{-1} generated in heating. This indicates that the ring closure of N,N'-biscarbamyl-N-cyanoguanidine unit is very fast at this temperature. The final cured product was not fusible upon heating and was insoluble in organic solvent. This implies that the resin was crosslinked to a thermoset plastic possibly through the trimerization of N-cyano groups and other side reactions.

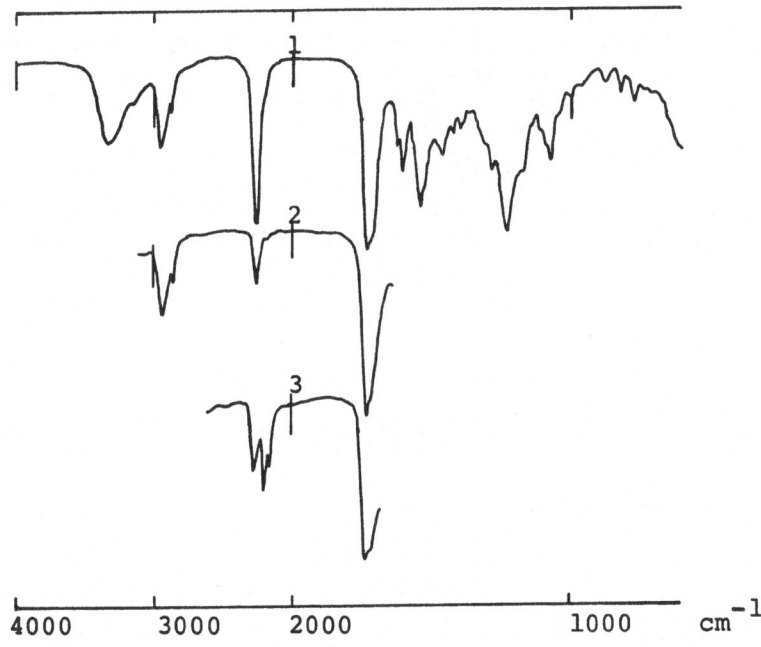

Figure 3.   (1)   IR Spectrum of N-Cyanourea Endcapped Resin Prepared from a Polycaprolactonediol (MW=530 g/mole) Terminated with 2,4-Tolylene Diisocyanate.   (2)   IR Spectrum of the Resin Being Cured at 100°C for 50 Minutes.   (3)   IR Spectrum of the Resin after Standing at RT for 30 Days

When this difunctional N-cyanourea terminated resin was left standing at room temperature, gradually the doublet absorption at 2160 cm^{-1} and 2190 cm^{-1} increased in intensity (Spectrum 3 in Figure 3) and formed an elastic thermoplastic having excellent physical properties.

CONCLUSIONS

An interesting class of monomers which polymerize into different polymeric materials at different temperatures was discovered. At room temperature, the linear polymer was obtained either from the polymerization of a dicyanourea monomer or directly from the polymerization of a mixture containing a diisocyanate and cyanamide. At elevated temperature (>100°C), the dicyanourea monomer or the mixture of diisocyanate and cyanamide crosslinked to a rigid foam or flexible film depending on the structure of the monomer. IR, DSC, and NMR were used for characterization of the polymers.

REFERENCES

1.   F. Kurzer and J. R. Powell, "Organic Syntheses," Coll. Vol. IV, p. 213 (1963).
2.   J. R. McAdam and F. C. Schaefer, "Kirk-Othmer Encyclopedia of Chemical Technology," 2nd Ed., Vol. 6, p. 563 (1965). D. R. May, "Kirk-Othmer Encyclopedia of Chemical Technology," 3rd Ed., Vol. 7, p. 291 (1979).
3.   F. Kurzer and K. Douraghi-Zadeh, "Chem. Rev." 67(2), 107 (1967).
4.   J. Migrdichian, "The Chemistry of Organic Cyanogen Compounds," Reinhold, N.Y., p. 117 (1947).

POLY (N-ACYL ETHYLENEIMINES) WITH POLARIZABLE AROMATIC SIDE CHAIN SUBSTITUENTS AND THEIR COMPLEXES: SYNTHESIS, STRUCTURE AND ELECTRONIC PROPERTIES

M.H. Litt, J. Rodriguez, H. Nava and J. Kim
Dept. of Macromolecular Science
T. McClelland, W. Gordon and M. Dyan
Dept. of Physics

Case Western Reserve University
Cleveland, Ohio 44106

INTRODUCTION

Our objectives in this work were to produce electrically interesting polymers with the following characteristics: 1. At some stage they should be formable, either through heat or solution; 2. The polymers should be regular and therefore crystallizable; 3. When crystallized, the π electrons of the electrically active portion of the molecule should overlap; 4. The active part should be a reasonably strong donor or acceptor, so it can be doped or complexed; and 5. The polymerization chemistry should be such that polymers can be made easily.

For this work, we chose to work with polymers whose active groups were in the side chain. There is one class known which has side chains attached to trigonal nitrogen and thus lacks the potential asymmetry of vinyl, epoxy, etc. polymers. The class is that of the poly (N-acyl ethyleneimines).[1] Since the polymers are regular, most crystallize readily and they all tend to crystallize in similar fashion.[2] Work up to now has shown that side chains alternate 180° around the backbone with a repeat of 3.15 to 3.3Å per monomer unit.[2,3] Thus the distance between side chains is 6.3 to 6.6Å. With this repeat, there are two types of overlap possible for aromatic groups attached to the side chains. We can represent the polymer formula as:

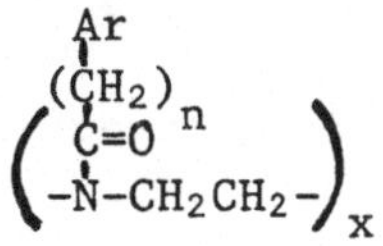

$$\left(\begin{array}{c} Ar \\ | \\ (CH_2)_n \\ | \\ C=O \\ | \\ -N-CH_2CH_2- \end{array} \right)_x$$

If n is odd, a three fused ring group such as anthracene or carbazole attached at the center atom, the 9 position in the two cited,
can twist to provide a π overlap of about 30%.[4] When n is even, the
aromatic system must be elongated to overlap; an example could be a
4 substituted biphenyl group. The polymerizing system is a basic
heterocycle which polymerizes cationically, the 2 substituted, Δ2,
1,3-oxazoline.[1,5,6] This is stable enough both as monomer and as
active center that monomers with very electron rich side chain substituents can be readily polymerized. [4b]

For this work we chose groups which could be excited to the
triplet state easily. One was the carbazole moiety. The second
was acridone. Both have acidic NH groups which can be used to
attach them to the appropriate side chain.

EXPERIMENTAL

Outlines of the synthetic routes used to obtain the desired
polymers are shown in figures 1 and 2. After each step each compound was purified to a sharp melting point. Each new compound
was identified by NMR and IR spectroscopy. NMR spectra were run
using a Varian A-60 spectrometer. IR spectra were run on KBr pellets
using a Digilab FTS-14 Fourier Transform Spectrophotometer.

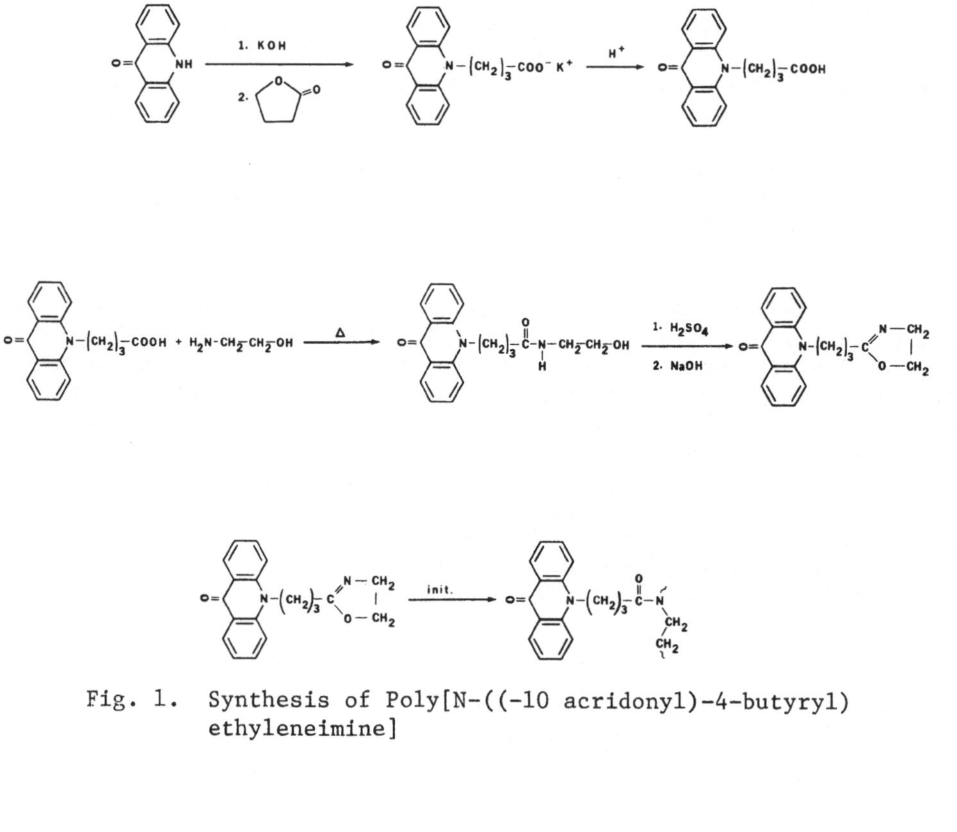

Fig. 1.   Synthesis of Poly[N-((-10 acridonyl)-4-butyryl)
          ethyleneimine]

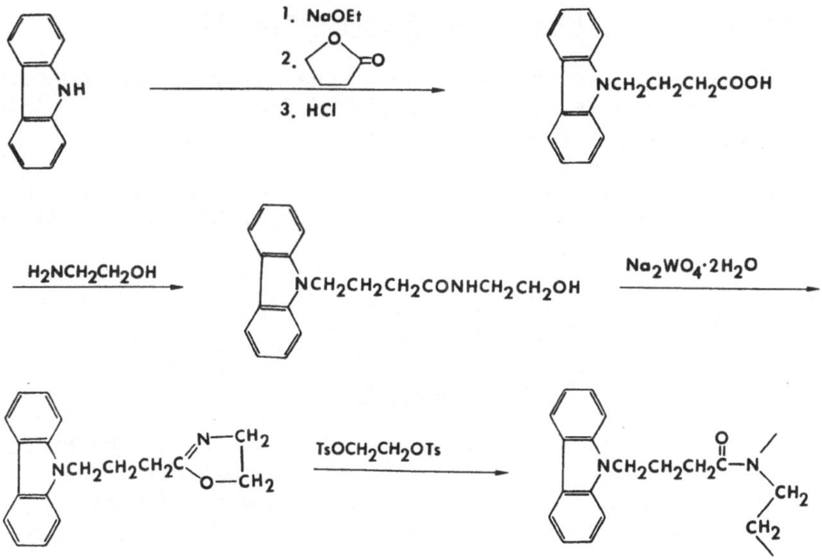

Fig. 2.   Synthesis of Poly[N((9-Carbazolyl)4-butyryl)
          ethyleneimine)

Ultraviolet-visible absorption spectra of polymers and model
compounds were measured on a Beckman ACTA MVI spectrophotometer.
Fluorescence spectra of compounds in dilute solution were recorded
using an Aminco-Bowman spectrophotofluorometer at 20°C. Differential scanning calorimetry (DSC) was performed using a Perkin-Elmer
DSC-2 scanning calorimeter. Melting points were determined on a
Perkin-Elmer DSC-II at a 5°C/min. heating rate. Elemental analyses
were by Galbraith Laboratories, Knoxville, Tennessee.

Synthesis of poly[N-((10-acridonyl)4-butyryl)ethyleneimine]

4-(10-acridonyl) butyric acid (I):

The potassium salt of acridone was made by reacting it with
one equivalent of KOH in ethanol The ethanol was evaporated and
the dried salt was refluxed in a large excess of γ-butyrolactone
for four hours. The excess γ-butyrolactone was removed under
vacuum. The remaining solid was dissolved in water and precipitated
by acidifying the solution. The crude yield of the reaction was
93%. The product was purified by recrystallization from chlorobenzene m.p. 218-220°C.

NMR(DMSO d_6) δ in ppm: 2.12 (m, $CH_2\underline{CH_2}CH_2$),2.64 (t, $-\underline{CH_2}COOH$)
4.53 (t, $-CH_2N$), 7.38-8.42 (m, 8 aromatic protons), 12.10 (s, COOH).

IR(cm^{-1}):   3000-2500 (ν OH), 1734 (ν C=O), 750 (ν aromatic CH).

N-(2-hydroxyethyl)-4-(10-acridonyl) butyramide (II):

II was obtained by the reaction of the acid, I, with monoetha-
nolamine.[7] The produce was recrystallized from water. The yield
was 75% pure II, m.p.: 205-207°C.

NMR(DMSO d$_6$) δ in ppm: 2.15 (m, -CH$_2$CH$_2$Ch$_2$⇌), 2.47 (t, CH$_2$CO),
3.40 (m, -NCH$_2$CH$_2$O-), 4.48 (t, -CH$_2$N), 4.70 (s, OH), 7.40-8.38 (m,
8 aromatic protons, NH).

IR(cm^{-1}): 3407 (ν OH), 3300 (ν NH), 1638 (Amide I), 1553
(Amide II), 750 (ν aromatic CH).

2-(10-acridonylpropyl)-2-oxazoline (III):

The oxazoline was obtained by cyclodehydration of the hydroxy-
amide with concentrated sulfuric acid.[1,8] The oxazoline was purified
by recrystallization from toluene. The yield was 63% pure III,
m.p.: 185-187°C.

NMR(CDCl$_3$) δ in ppm: 2.00-2.65 (m, -CH$_2$CH$_2$CH$_2$), 3.65-4.50
(m, NCH$_2$CH$_2$O and -CH$_2$N), 7.20-8.50 (m, 8 aromatic protons).

IR(cm^{-1}): 1667 (ν C=N), 1202 (ν C-O), 914, 951, 976 (ν CH
oxazoline ring).

Poly[N-((10-acridonyl)4-butyryl) ethyleneimine] (IV):

The polymerization of 2-(10-acridonylpropyl)-Δ2-oxazoline was
carried out using ehtylene glycol ditosylate initiator. The monomer
is high melting and attempts to polymerize in bulk yielded only
polymers of low molecular weight. Polymerizations in solution were
always accompanied by precipitation of the polymer. The best results
were obtained using dried, freshly distilled coumarin as solvent;
polymerization was carried out at 100°C for 12 hours and finished
by heating at 150°C for 2 hours. The yield was 100%. [η]=0.30 (in
N-methyl pyrrolidone at 25°C).

NMR(TFAA) δ in ppm: 2.6 (m, -CH$_2$CH$_2$CH$_2$-CO), 3.4-4.6 (broad,
-NCH$_2$CH$_2$N-), 4.6 (t, -CH$_2$-N), 7.7-8.7 (broad, 8 aromatic protons).

IR(cm^{-1}): 1637 (ν C=O), 754 (ν aromatic CH).

Synthesis of poly[N-((N-carbazolyl)4-butyryl) ethyleneimine]

4-(N-carbazolyl) butyric acid (V):

Carbazole sodium salt was formed by adding one equivalent of
sodium ethoxide in ethanol to a solution of carbazole in N-methyl
pyrrolidone. The ethanol was completely distilled out under vacuum,

an excess of γ-butyrolactone was added and the solution heated at 160°C for four hours. The reaction mixture was quenched in water and filtered to remove any carbazole. After acidification, the 4-(N-carbazolyl) butyric acid, V, was obtained in 93% yield; m.p.: 157°C. (Anal. calcd. for $C_{16}H_{15}O_2N$: C, 75.87%; H, 5.97%; N, 5.53%. Found: C, 76.01%; H, 6.05%; N, 5.63%.

NMR[$(CD_3)_2SO$, δ ppm]: 1.89-2.47 (m, CH_2CH_2CO), 4.43 (t, NCH_2), 7.05-7.69 (m, 6 aromatic protons), 8.18 (d, 2 aromatic protons), 12.17 (s, OH).

IR(KBr): 3128 (ν OH), 1729 (C=O), 1326 (ν NC aromatic), 750, 725 (δ CH aromatic).

N-(2-hydroxyethyl)-4-(N-carbazolyl butyramide) (VI):

V was reacted at reflux with an excess of ethanolamine (170°C, 5 hours). The excess ethanolamine was distilled off under vacuum. Water was added to reaction mixture and the precipitate was filtered and dried. The yield was 98%, m.p.: 122°C. (Anal. calcd. for $C_{18}H_{20}O_2N_2$: C, 72.95%; H, 6.80%; N, 9.45%. Found: C, 73.08%; H, 6.86%; N, 9.45%.

NMR($CDCl_3$, δ ppm): 2.04-2.27 (m, CH_2CH_2CO), 2.54 (s, OH), 3.25 (t, ali NCH_2), 3.54 (t, CH_2O), 4.33 (t, aro NCH_2), 5.73 (s, CONH), 7.02-7.45 (m, 6 aromatic protons), 8.06 (d, 2 aromatic protons).

IR(KBr): 3303 (ν NH), 1648 (Amide I), 1557 (Amide II), 1327 (ν NC aromatic), 749, 721 (δ CH aromatic).

2-[3-(9-carbazolyl propyl]- 2-1,3-oxazoline (VII):

5% by weight of $Na_2WO_4 \cdot 2H_2O$ was added to VI and the mixture distilled at 0.3 mmHg and 230°C. Originally a viscous oil was obtained in 72% yield. The crude product was recrystallized from hexane, m.p.: 53°C. (Anal. calcd. for $C_{18}H_{18}N_2O$: C, 77.67%; H, 6.52%; N, 10.06%. Found: C, 77.49%; H, 6.69%; N, 10.04%.

NMR($CDCl_3$, δ ppm): 1.95-2.30 (m, CH_2CH_2C), 3.55-4.13 (m, NCH_2CH_2O), 4.38 (t, aro NCH_2), 7.03-7.48 (m, 6 aromatic protons), 8.08 (d, 2 aromatic protons).

IR(KBr): 1664 (ν C=N), 1328 (ν NC aromatic), 1248 (ν C-O), 980, 952, 916 (ν CH oxazoline ring), 751, 724 (δ CH aromatic).

Poly[N-(9-carbazolyl 4-butyryl) ethyleneimine] (VIII):

Ethylene glycol ditosylate initiator was put into polymerization tube and VII (monomer/initiator mole ratio = 2093/1) was distilled into the polymerization tube at 170°C under vacuum (0.02 mm Hg) and the tube sealed. It was polymerized 24 hours at 110°C, then 5 hours at 130°C and finally 3 hours at 160°C. The intrinsic viscosity in chloroform at 25°C was 0.67.

NMR($CDCl_3$, δ ppm): 2.06 (CH_2CH_2CO), 3.01 (ali NCH_2CH_2), 4.18 (aro NCH_2), 7.20 (6 aromatic protons), 7.88 (2 aromatic protons).

IR(KBr): 1642 (ν C=O), 1328 (ν N-C aromatic), 1152 (ν CH_2N), 749, 723 (δ CH aromatic).

RESULTS AND DISCUSSION

Optical Properties in Solution:

A. Acridone side chain polymers - The optical properties of the acridone polymer were compared with those of a low molecular weight model compound, N-(2-hydroxyethyl)-4-(10-acridonyl) butyramide.

The UV-visible absorption spectra of polymer and model compound in NMP are very similar. Both show absorption maxima at 401, 382, 364, 307 and 294 nm; the extinction coefficients and spectral shapes are almost identical. The only difference is that the polymer shows a long wavelength absorption tail which extends to 460 nm. This long wavelength absorption disappears when lithium chloride is added to the solution (fig. 3). It probably arises as a result of inter-actions between the acridone moieties on adjacent residues of the polymer backbone. This does not exist in solutions of the model compound because the acridone molecules are not restrained by being attached to a polymer chain. The fact that the absorption tail dis-appears when LiCl is present in the solution can be rationalized in terms of the solubility of the polymer in these two systems. In pure NMP the polymer tends to precipitate out of solution on standing at room temperature for several days, but when LiCl is present, the polymer remains soluble. LiCl can complex with the tertiary amide groups of the polymer backbone and solvate the polymer chains. Therefore, when LiCl is present in the solution the macromolecular coils are more expanded than in its absence, the distance between acridone residues becomes larger; interaction between them is reduced and the absorption tail disappears. Another observation which sup-ports this hypothesis is that of the viscosity of the polymer in the two solvent systems. A polymer which has an intrinsic viscosity of 0.3 dl/g in pure NMP exhibits an intrinsic viscosity of almost 0.5 dl/g when LiCl is in the solution. This indicates that in LiCl solution the coils are more expanded.

In trifluoroacetic acid both polymer and model compound show the same spectrum, and the polymer does not exhibit a long wave-length absorption tail. This is probably because in acid solution the acridone residues, as well as the carbonyl groups of the polymer

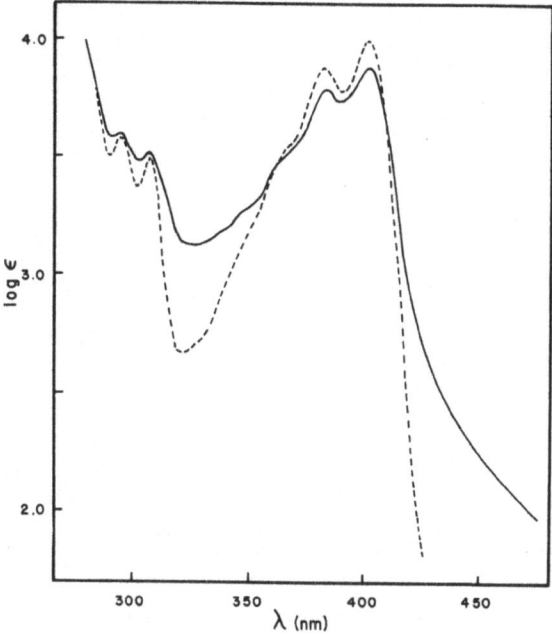

Fig. 3.   UV absorption spectra of the acridone substituted
          polymer in NMP (———) and in 4×10^{-4}M LiCl in
          NMP (-----).

backbone, are hydrogen bonded strongly to the acid. The polymer is
highly solvated and interactions between acridone molecules in adjacent residues of the polymer chain are decreased.

The fluorescence emission spectra of the model compound in NMP
and of the polymer in NMP with and without lithium chloride are
shown in figure 4. The spectral shapes are very similar; they show
emission maxima at 410 and 430 nm, which is usual for N-substituted
acridones in polar organic solvents.[9,10] The intensity of the fluorescent emission at these maxima is reduced for the polymer, specially
in pure NMP. Also the polymer shows a slightly higher emission at
longer wavelengths.

The reduction in intensity of the fluorescent emission of the
polymer is probably due mainly to self-quenching, since the amount
of reabsorption should be practically the same for both polymer and
model compound, and should be small at these concentrations.[11] This
self-quenching effect and the higher emission of the polymer at longer
wavelengths are probably due to interactions between adjacent residues
of acridone in the polymer backbone. The fact that the self-quenching

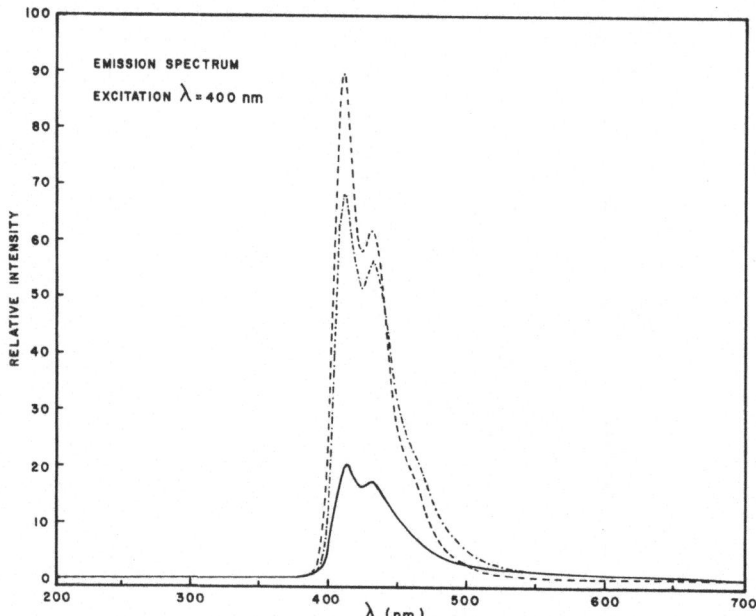

Fig. 4.   Fluorescence emission spectra in NMP: polymer (——,
          C=1.05x10^{-4} mole/1), polymer plus 4x10^{-4} M LiCl (–·–·–·,
          C=1.05x10^{-4} mole/1), model compound (-----, C=1.09x10^{-4}
          mole/1).

of fluorescence is reduced when LiCl is present in solution also
supports this explanation, since LiCl solvates the polymer and
increases the distance between the acridone residues, decreasing
their interactions.

     B. Carbazole side chain polymers - One expects to see changes
in solution spectroscopic properties in a polymer compared with
model compounds since the aromatic groups are constrained to be
near each other. There may be further effects since the mutual
orientation of near neighbors is also constrained. This can be
seen in the NMR spectra of the carbazole polymer and monomer in
CDCl$_3$. The aromatic protons of the monomer peak at δ=8.08 and
7.40 ppm from TMS. The corresponding peaks in the polymer are at
δ=7.88 and 7.20 ppm from TMS. Since both solutions have about the
same concentration, the extra shielding in the polymer is probably
due to the geometry of the side chains which tends to generate
overlap even in solution.

     The absorption spectrum of the polymer is at slightly longer
wavelengths than that of the model compound and has a long wavelength tail. This is also true for the fluorescence spectrum, as
described above for the acridone polymer.

Crystallization Behavior: Acridine Polymer

Semicrystalline films of very low molecular weight polymer ($[\eta]$=0.12) were obtained when pressed from the melt at 228°C. However, pressed films of higher molecular weight polymers were not crystalline. It was necessary to add small amounts of plasticizer to obtain some crystallinity. This is probably due to the fact that above a certain degree of polymerization, the increasing melt viscosity restricts the mobility of the chains and prevents rapid crystallization.

DSC traces for representative samples of this polymer with different thermal histories are shown in figure 5. Curve A is the heating curve of a sample obtained by pressing polymer ($[\eta]$=.12)

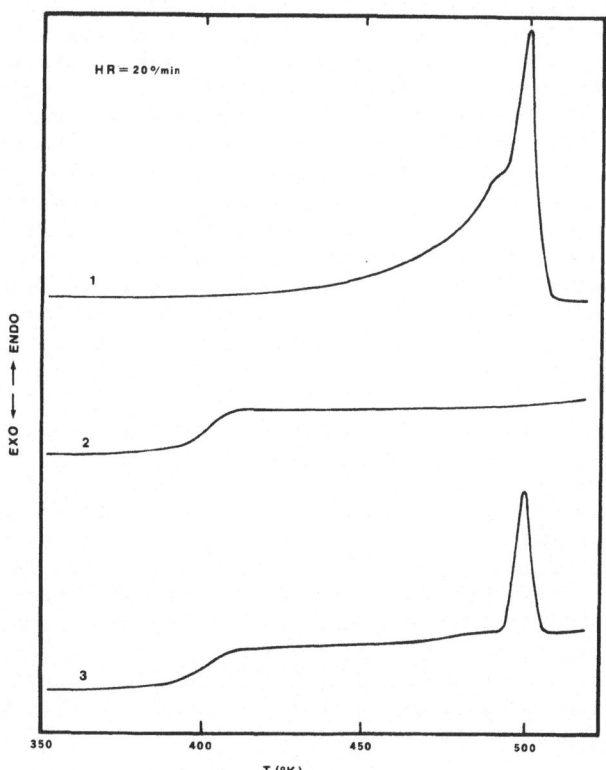

Fig. 5. DSC heating curves of the acridone substituted polymer with different thermal histories. A. – Compression molded at 228°C, slowly cooled to room temperature. B. – Heated to 550°K and cooled to 300°K. C. – Heated to 490°K and cooled at 20°/min. to 300°K.

at 228°C and cooling slowly to room temperature. It shows a sharp
melting peak at 498°K and a very broad melting range between 460
and 490°K. No glass transition temperature is observed in this
curve. Curve B is the DSC trace of the same sample after it was
heated to 550°K and cooled to 300°K. It shows a glass transition
at 400°K and no melting endotherm. This curve is always obtained
regardless of the cooling rate after the sample has been heated
above Tm. Cooling rates were varied from quenching to 0.32 °/min
and the same curve was obtained. C is the heating curve of a sample
of the same polymer after having been heated to 490°K, held for 15
minutes at this temperature, and then cooled at 20°/min to 300°K.
In this curve both Tg and Tm are present but the broad melting peak
between 460 and 490°K disappears.

        The fact that the polymer did not crystallize on cooling after
being heated above Tm, curve B, indicates that crystallization is
difficult for this polymer and the crystallization rate must be
slow. This is probably due to the large side groups which restrict
the mobility of the polymer chains and also make packing difficult.
However, if the polymer does not crystallize on cooling after being
heated above Tm, why did it crystallize when it was compression
molded at 228°C as shown by the heating curve A? A possible explanation is that the polymer powder used to prepare the sample had
already some crystalline nuclei which were not destroyed during the
processing since the molding temperature was just at Tm, and when
the sample was cooled slowly these nuclei started crystallization.
On the other hand, when the polymer was heated above Tm all the
nuclei were destroyed and upon cooling, crystallization did not
occur. This would mean that nucleation is the determining step in
the crystallization process.

        Curve A also implies that the crystallization was slow. There
is a broad melting range below the main melting peak. Since the
melting temperature range is indicative of the size and perfection
of the crystallites in the sample, this broad endotherm suggests
that there is a large amount of crystalline material with a high
level of imperfections and/or a broad distribution of lamellar sizes
in this sample. The broad melting range disappears when the sample
is heated to 490°K and cooled again, curve C. Thus, most of the
"imperfect" crystalline material melted and did not recrystallize
on cooling, nor was it annealed to higher melting materials. The
glass transition now appears because amorphous material was produced,
and a single sharp endotherm remains.

        Another interesting fact is that the polymer samples prepared
by compression molding did not show a glass transition (curve A),
but when the sample was heated and cooled again (curves B and C) a
glass transition was observed. This suggests that the intensity of
the glass transition in the semicrystalline polymer is substantially

lower than in the bulk amorphous polymer. It is possible that in
the semicrystalline polymer, even though there is a large amount of
"imperfect" crystalline material, there is almost none or very little
noncrystalline material left. The "amorphous" fraction probably
occupies the regions between lamellae and is restrained by ties to
the crystalline material; thus when the polymer is heated to Tg this
material cannot relax and no glass transition is observed.

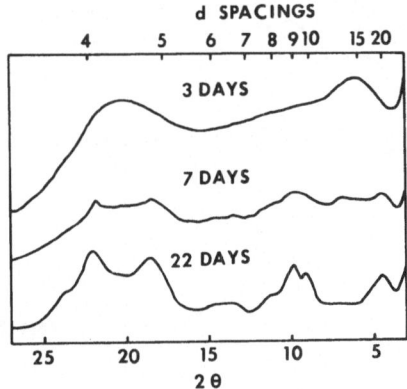

Fig. 6. Wide angle x-ray traces of polymer after different
         crystallization times.

Homopolymer Properties (Carbazole Polymer)

The carbazole polymer crystallized with difficulty and needed
plasticizer in order to crystallize. The polymer was crystallized
in saturated anisole vapor at 50°C. The crystallization was examined at a series of crystallization times by a polarizing microscope
and x-ray diffraction. The spherulites grew at a rate of about
3μ per day in diameter which is extremely slow compared with other
crystalline polymers. Fig. 6, a WAXD scan of the polymer as a
function on annealing time, shows that in the amorphous state

there are two broad peaks at d=4.4 and 15 Å; as the polymer crystal-
lizes, crystalline peaks appear at d=3.8, 4.0, 4.8, 6.5, 7.6, 8.8,
9.6 and 19.3 Å and become bigger and sharper with further annealing
while the amorphous broad peak at 15 Å disappears completely. The
DSC diagrams of amorphous and crystalline polymer are shown in
Figure 7. This shows Tg=112°C from the heating curve (Tg=101°C from

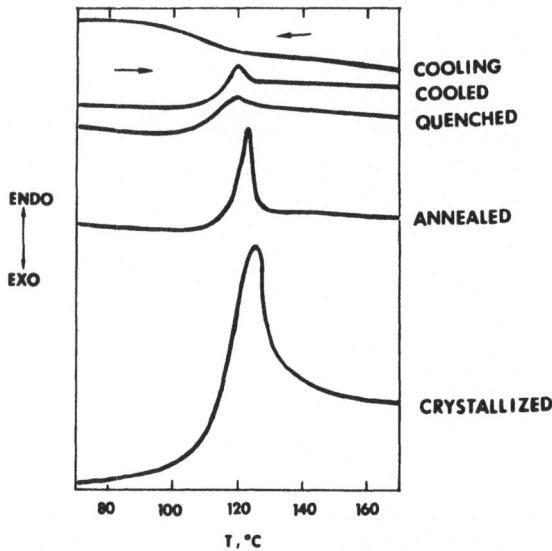

Fig. 7.   DSC spectra of carbazole polymer for different treatments.
          Annealed (90°C, 24 hrs): weight of sample, 3.02 mg;
          cooling and heating rates, 20°C/min. Crystallized (Anisole
          vapor, 50°C, 2 months): weight of sample, 5.30 mg; heating
          rate, 20°C/min.

the cooling curve) and Tm=125°C from the crystallized polymer curve
(ΔH=3.49 cal/gram of polymer). Tg and Tm are very close; this
explains why the polymer could not be crystallized without plasticizer. The curve for the sample annealed at 90°C (22°C below Tg)
for 24 hours showed an extraordinarily large endotherm peak (ΔH=1.22

cal/gram of polymer) even though the sample was amorphous by polarizing microscope and x-ray diffraction. Also, the sample which was
cooled at the rate of 20°C/min still showed an endotherm peak. This
might mean that at 10-30°C below Tg polymer can organize relatively
rapidly to a local structure which is too small in size to appear
in polarizing microscope and x-ray diffraction but is organized
enough to show an endothermic peak in the DSC.

COMPLEXES

Solution Properties:

Charge transfer spectra of the carbazole polymer with TCNQ
were measured in several solvents and compared to the model compound. The results are shown in Table 1. K and ε were calculated
using the Benesi-Hildebrand equation.[12]K is almost independent of
whether the material is polymer or model compound. ε is higher
for the polymer while λ max is unchanged. This is difficult to
explain at present. One hypothesis is that the polymer may form a
DDA complex, which will increase ε and may not increase λ max or K.

Solid State Properties:

On standing, a solution of TCNQ and carbazole polymer in dioxane
(polymer:TCNQ mole ratio was 1:2) gave a black crystalline precipitate. The complex had Tm=193°C and Tg=114°C (which is almost the
same as that of homopolymer). The elemental analysis of the complex
showed C, 74.22%; H, 5.60%, N, 14.35%. This indicates the mole %
of polymer, TCNQ and dioxane in the complex are 52.89%, 31.16%
and 15.95% respectively. The IR spectra of polymer-TCNQ mixture
(polymer: TCNQ mole ratio was 55:45) and the complex are shown in
Figure 8. The intensities as well as the positions of TCNQ bands
are changed when TCNQ was complexed with polymer. C≡N stretch bands
of the complex appear at 2218 and 2179 cm^{-1}which were shifted 6 and
45 wavenumbers from 2224 cm^{-1} which is the C≡N stretch band of pure
TCNQ. The 2218 cm^{-1} band is due to the charge transfer complex.
The 2179 cm^{-1} is due to the ion radical salt and is similar to the
values of the literature, e.g. 2200-2174 cm^{-1} for Et$_3$NH$^+$ TCNQ$^-$ [13]
and 2190 cm^{-1} for Li$^+$ TCNQ$^-$.[14] The 862 cm^{-1} , C-H out-of-plane
bending of TCNQ was shifted to 844 cm^{-1} in the complex. The polymer
band at 1209 cm^{-1} disappeared in the complex, which indicated that
there was almost no uncomplexed polymer. Four additional peaks in
the complex were found to be dioxane which was used as solvent when
the complex was formed. However, the relative intensities of
dioxane bands are also changed.

Dioxane plays an important role in forming the complex and is
involved in the crystalline phase. The polymer-TCNQ complex was
dried at 110°C for 24 hrs under vacuum to remove dioxane. IR of

Table 1. Charge Transfer complex constants for poly(N-(9-carbazole
          4-butyryl) ethyleneimine) and 9-carbazole butyronitrile.

SOLVENT	SAMPLE	K	$K\varepsilon$	ε	%SE EST
C_6H_5Cl	Polymer	7.48 (0.3)	23,200 (130)	3100 (120)	±0.5
	CBN	4.17 (0.71)	3,900 (30)	1025 (150)	±2.0
$C_2H_4Cl_2$	Polymer	3.89 (0.69)	14,750 (90)	3790 (670)	±1.0
	CBN	4.81 (1.42)	5,040 (74)	1050 (310)	±2-3
C_6H_5CN	Polymer	0.78 (0.24)	6,630 (87)	8525 (2590)	∿2.0
	CBN	1.11 (0.90)	3,350 (150)	3000 (2400)	∿3.0

CBN - Carbazolebutyronitrile

(SE) - Standard Error

a - λ is constant for all samples. It is 572 nm.

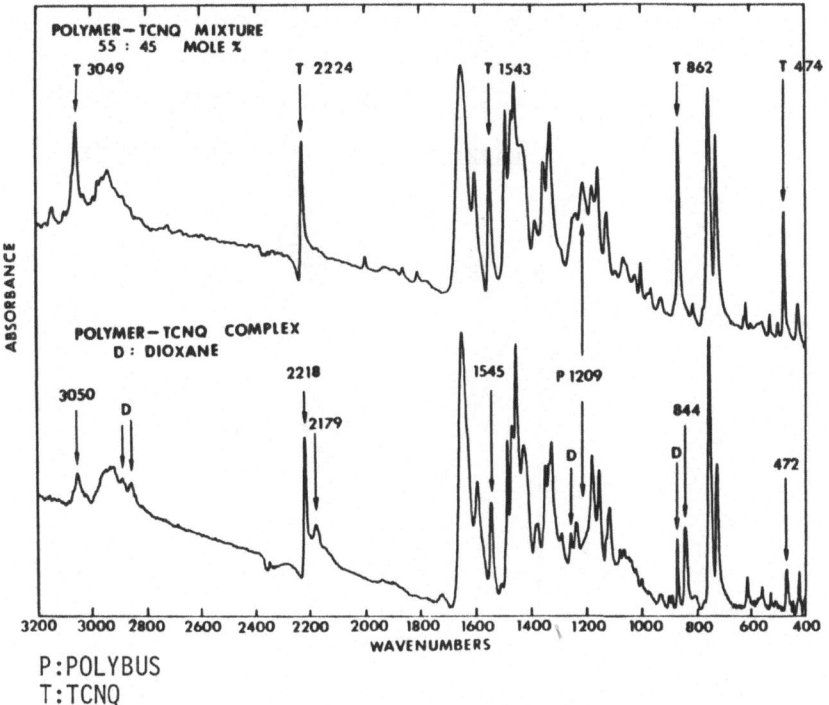

P:POLYBUS
T:TCNQ

Fig. 8.   IR spectra of polymer-TCNQ mixture and complex.

this sample showed that dioxane peaks were much reduced. Also the
ion-radical salt peak at 2179 cm^{-1} disappeared, while the polymer
peak at 1209 cm^{-1} reappeared. DSC of this sample showed two shoul-
ders at 120°C (related to the polymer Tg) and 149°C. The melting
peak was now at 177°C, much lower than the original Tm of 193°C.

Electrical Properties:

The carbazole polymer/TCNQ complex was pressed into a pellet
and conductivity measured as a function of temperature using
springloaded, unguarded electrodes. The results are presented in
Fig. 9. In spite of the lack of stochiometry in the carbazole/TCNQ
ratio, there is some conductivity but the pressed pellet is a very
poor semiconductor. The activation energy is lower than one would
expect from the pure polymer, which is an insulator, but still
high.

When heated above the melting point, the conductivity rose by
a factor of four. This factor was retained as the temperature was
lowered. The amorphous polymer, without dioxane, is a better conductor than the crystalline polymer. This is not surprising.
Pelletized materials are subject to large intergranular effects and

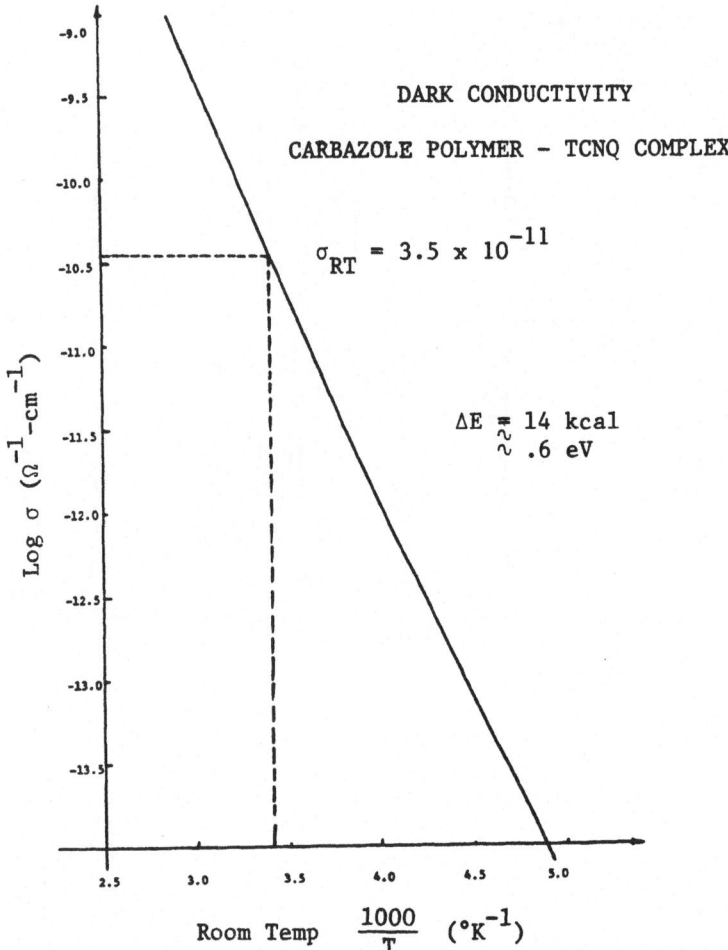

Fig. 9. Plot of the conductivity of a compressed pellet of
       carbazole polymer/TCNQ complex as a function of temperature.

melting the polymer fuses the particles. Thus, the effective conductivity can rise even if the conductivity within a particle drops
or remains unchanged.

    EPR data, Fig. 10, show that the interaction of TCNQ with the
polymer is qualitatively different from the interaction with model
compound. There is a large free electron signal from the polymer
complex (about 10^{18} spins/gm) and essentially none from the CBN
complex. This is in accord with the FTIR spectrum of the complex,
Fig. 8, which shows that some ion-radical TCNQ is present. However

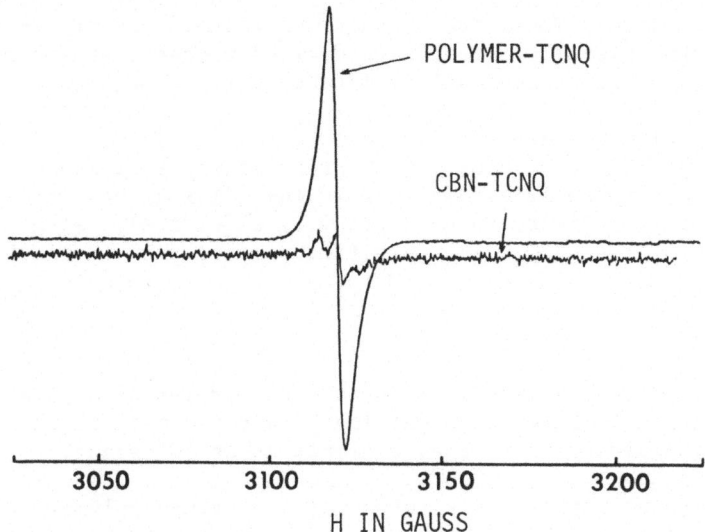

Fig. 10.   Electron paramagnetic resonance spectra for TCNQ complexes
of polymer and model compound.   Spectra taken at room
temperature.

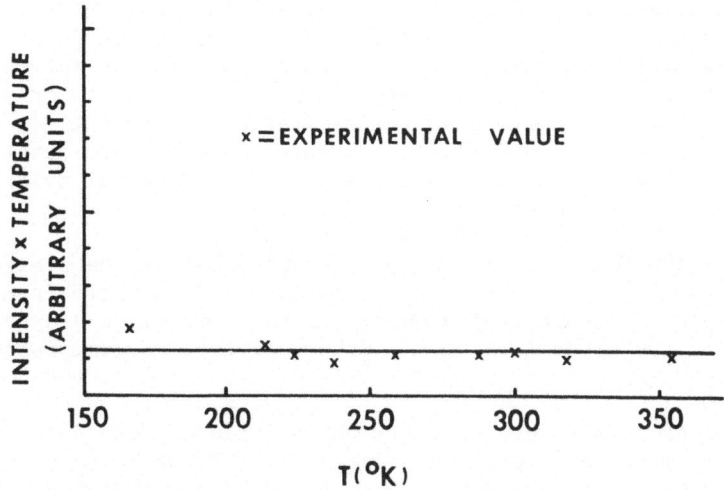

Fig. 11.   Temperature dependence of free spin concentration for
polymer/TCNQ complex.

the spins are trapped and thus cannot contribute effectively to the
complex conductivity. Fig. 11 shows the temperature dependence of
the number of spins. It is essentially constant over the whole
temperature range. Thus, even though ion radicals are present, the
nature of the crystallization, with dioxane present, prevents the
complex from being a reasonable semiconductor or conductor.

The packing geometry is critical to crystallization. Green-
black polymer-TCNQ complexes can be precipitated from many solvents.
We have found crystalline complex only when dioxane was the solvent.
Preformed polymer films can be doped very effectively using xylene
as the solvent for TCNQ. The resulting polymer film is black, but
non-crystalline.

CONCLUSIONS

The polymers are very interesting as examples of crystallizable
materials which have very low melting points relative to their Tg's.
For the carbazole polymer, they are only about 20° apart.

The nature of the initial organization of the polymers after
cooling is obscure; there might be liquid crystal character to the
organization. We are studying this at present.

The importance of geometry shows up most strongly in the complexing behavior. Normal model compounds should crystallize in
donor/acceptor stacks since carbazole is a relatively weak complexing
agent. However, the polymer gives some evidence for being arranged
in individual donor and acceptor stacks. The increase in ϵ for the
charge transfer with little or no increase in K can be interpreted
in terms of a D(DA) complex, where the second donor acts only to
solvate the first donor during the excitation, which will increase
the transition moment and thus ϵ. However, λ and K are determined
mainly by the ground state properties of the individual molecules
and thus did not change.

Also, the fact that the polymer complex showed the presence
of ion-radicals implies that it crystallized differently from the
model complex. The packing geometry is not perfect so dioxane was
included in the crystal lattice. Other solvents gave amorphous
complex.

In order to generate polymers which could conduct electricity,
two factors must be studied very carefully. First, a much stronger
electron donor must be used in place of carbazole. Second, the
complex geometry must be studied as a function of the nature of the
backbone and side-chain in order to tailor the polymer for complex
formation.

REFERENCES

1. T. G. Bassiri, A. Levy and M. H. Litt, Polymers Letters 5, 871 (1967).

2. M. H. Litt, Forrest Rahl and L. G. Roldan, J. Polymer Sci. A-2, 7, 463 (1969).

3. James W. Summers and M. H. Litt, J. Polymer Sci. Chemistry 11, 1353 (1973).

4. Unpublished estimations based on: a. M. H. Litt and J. W. Summers, J. Polymer Sci. A2, 11, 1339 (1973), and b. ibid p. 1359.

5. T. Kagia, S. Narisawa and K. Fukui, Polymer Letters 4, 441 (1966).

6. a. A. Tomalia and D. P. Sheetz, J. Polymer Sci. A1, 4, 441 (1966).

   b. W. Seeliger, Angew. Chem. 78, 613, 913 (1966).

7. H. Wenker, J. Amer. Chem. Soc., 57, 1079 (1935).

8. M. T. Leffler and R. Adams, J. Am. Chem. Soc., 59, 2252 (1937).

9. M. Siegmund and J. Bendig, Z. Naturforsch., 35a, 1076 (1980).

10. M. Siegmund and J. Bendig, Ber. Busenges, pysik. Chem, 82, 1061 (1978).

11. A. P. Kilimov and V. V. Voloshina, Optics and Spectroscopy (U.S.S.R.), 12, 362 (1962).

12. H. A. Benesi and J. H. Hildebrand, J. Am. Chem. Soc. 71, 2703 (1949).

13. L. R. Melby, R. J. Harder, W. R. Hertler, W. Mahler, R. E. Benson and W. E. Mochel, J. Am. Chem. Soc., 84, 3374 (1962).

14. A. Girlando, L. Morelli and C. Pecile, Chem. Phys. Letters, 22, 553 (1973).

AQUEOUS SOLUTION SYNTHESIS OF PLATINUM II POLYUREAS, POLYTHIO-UREAS AND POLYAMIDES

Charles E. Carraher, Jr., Tushar A. Manek,[a] David J. Giron,[b] Mary L. Trombley,[a,b] Kathy M. Casberg[b] and William J. Scott[a]

Departments of Chemistry[a] and Microbiology and Immunology[b], Wright State University, Dayton, Ohio 45435

INTRODUCTION

Malignant neoplasms are the second leading cause of death in the United States. Until the late 1960's little effort was made towards the synthesis and characterization of antineoplastic agents except for totally organic compounds. In 1964 Barnett Rosenberg and coworkers discovered by chance that bacteria failed to divide, but continued to grow giving filamentous cells in the presence of platinum electrodes (1). The cause of this inhibition to cell division was eventually traced to small concentrations (about 10 ppm) of cis-dichlorodiamineplatinum II (cis-DDP) and cis-tetrachlorodiamineplatinum IV (cis-TCP). This ability to inhibit bacterial cell division prompted their trial as antitumor agents. Both compounds were found to be active against Sarcoma 180 and Leukemia L1210 test tumors in mice with cis-DDP having greater activity than cis-TCP (2,3). In fact cis-DDP is active against a wide range of tumors in animals (4) and man (5) including several which are particularly resistant to treatment by other techniques such as ovarian and testicular carcinomas.

The success of platinum compounds has catalyzed the synthesis and characterization of a number of organometallic compounds, focusing on derivatives of cis-DDP, for antitumor activity.

Continued study of the effects of platinum compounds, both on animals and on man, has indicated that their positive attributes are coupled with a high level of negative side effects

133

which may be quite serious or even lethal (for instance 5-16).
Major complications include gastrointestinal, hematopoietic,
immunosuppressive, auditory (5,13) and renal dysfunction (5,9,14),
with the latter two being by far the most serious. One group
has also shown an allergic reaction in a man after a number
of doses of cis-DDP (10). Additional problems are continuing
to be discovered and dealt with (for instance 16,17).

Toxicity Minimized

     The first effort to overcome the toxicity of cis-DDP was
the hydration technique of Cvitkovic et al. (18). In this
method, the patient was given 1 to 2 liters of fluid overnight,
then cis-DDP was administered along with mannitol, a diuretic.
By flushing the drug along with large amounts of water past
the kidneys, renal toxic effects were substantially lowered.
This method allowed the administration of 3 to 4 mg of cis-
DDP per Kg body weight, about ten times the normal dosage,
while lowering damage to the kidneys. Though the antitumor
effects of cis-DDP at this high dosage seemed to be improved,
nausea and auditory losses were still observed and were probably
enhanced (19-21).

     A common method of minimizing the toxicity of antineoplastic
agents is to administer them along with a number of other antineoplastic agents at reduced dose levels. Thus, cis-DDP has been
administered in combination with adriamycin (ADM), cyclophospha-
mide (CTX), bleomycin (BLM), vinblastine (VBL) and actinomycin
D (ACD), among other agents.

     Merrin reported a 50% complete remission rate of advanced
testicular tumors when cis-DDP was administered in a three-
part program along with BLM, VLB, ACD and vincristine (21).
Einhorn and Furnas reported 100% partial or complete remission
of testicular tumors in 39 patients with a combination of cis-
DDP, VLB and BLM (22). Bruckner et al. (23), as well as Kwong
and Kennedy (24) have shown less spectacular, though quite
good, response rates on genitourinary tumors using cis-DDP
in combination with ADM. Other recent efforts are given in
references 25-27.

     A final method of lowering the toxicity of chemotherapy
using platinum is the synthesis of new compounds showing equal
or enhanced activity and toxicity in mice, as alluded to during
the discussion of structural requirements (28, 29, 33-36).
This effort has continued for some time and some derivatives
of cis-DDP are now beginning to enter phase I studies (30-32).
     More recently the research groups associated with Allcock
(37-41) and Carraher (42-50) have tackled the problem through

placement of the platinum-containing moiety in polymers. This
will be dealt with in the subsection "Rationale".

Structural Requirements

Most of the current efforts related to the synthesis of
platinum compounds for antitumor activity are aimed at new
compounds exhibiting antitumor activity comparable to cis-DDP
but with decreased toxicity or greater therapeutic properties
allowing decreased dosage level. Such efforts have produced
a wide variety of compounds, a number of which show antitumor
activity (51). From studying the structural features of such
compounds several features have emerged. Active compounds
are (a) neutral, (b) contain two inert and two labile ligands,
and (c) must have the ligands cis to each other.

One of the first points studied was the geometric arrangement
of ligands. Where trans compounds exist, they seem to be less
active in comparison to the cis isomer. Early antineoplastic
platinum complexes were characterized as having a pair of inert
amino ligands and a pair of labile chloride ligands, which
were believed to act as leaving groups. As suggested by the
well known trans effect, a chloride ligand is more reactive
in the trans isomer, trans-DDP hydrating about four times faster
(52) and undergoing ammination ten times faster (53) than cis-
DDP. This greater reactivity might imply a lowered reaction
specificity for the trans isomer. Thus, even though the two
isomers are of approximately equal toxicity, their therapeutic
levels differ vastly (33, 53, 54). This difference might be
due to the reaction of the trans isomer with various constituents
of the body prior to reaching the tumor site. Distribution
and excretion studies showed cis-DDP to be excreted much faster
initially (34). However, within five days the levels were
comparable at about 20% retention. Even at this point the
distribution of the two compounds differed radically, platinum
levels from trans-DDP remaining high in plasma at all times,
while levels from cis-DDP fall off markedly (15, 34). This
might be explained by suggesting that trans-DDP reacts with
some constituents in the blood, remaining there for some time,
while cis-DDP reacts somewhat later, thus being readily filtered
from the blood by the kidneys.

Activity of the platinum compound is highly dependent
on the leaving group. If the leaving groups are too reactive,
such as with the diaquo complex, the drug may chelate prior
to reaching the tumor site. If the leaving groups are not
reactive enough, such as with the diiodo complex, the drug
won't chelate with the proper cellular material to show antineoplastic activity.

More recently, antitumor activity has been observed with compounds containing bidentate leaving groups, such as malonate, sulfate or nitrate (for instance 35, 55-57).

The amine ligands have few basic requirements. The amine ligand should be inert toward replacement in the body, as is almost always the case. Secondly, it has been found that aliphatic amines work well (33, 34). Thirdly, the aliphatic substitution has a large effect on the water solubility of the drug (36). As the amine chain becomes longer, the resulting compound becomes less soluble (34, 36).

Suggested mechanistic pathways are reviewed in reference 58.

Rationale

The preference of including platinum-containing moieties within polymers compared with delivery via smaller molecules is clearly a debatable point, but there is sufficient evidence and potential to justify at least preliminary studies of such situations. The use of biological and synthetic macromolecules for delivery of tumor-suppressant drugs is well-known and has in some cases proved to be advantageous (for instance 58-61).

Specifically, advantages for synthesizing polymeric derivatives of cis-DDP include (a) restricted biological movement, (b) controlled release, and (c) increased probability of critical attachment.

Heavy metal toxicity related to the presence of large quantities of cis-DDP derivatives in the circulatory system (such as renal failure) is well established. Chain lengths of about 100 and greater are typically prevented from easy movement through biological membranes. Thus the location of the platinum drug can be somewhat restricted, for instance from the kidney, decreasing damage to the kidney-associated organs.

Studies by our group have established that most metal-containing polymers undergo hydrolysis when wetted. Polymeric derivatives of cis-DDP in DMSO solution also undergo hydrolysis when added to water. Thus, polymeric derivatives of cis-DDP can act as controlled release agents, releasing therapeutic quantities of the active drug.

Finally, if attachment - interstrand, intrastrand or otherwise - to DNA is essential for antitumor activity, and multiple attachments are required or advantageous (i.e. more than one attached platinum compound per DNA), then the fact that the platinum is itself an integral portion of a polymer is advantageous since the probability that successive attachments will be made on the same strand and adjacent strands is high after the first attachment.

Current Study

We recently reported the synthesis of polymeric platinum II derivatives of cis-dichlorodiamineplatinum II, cis-DDP (1; 1,2). These polyamines (2) exhibit good inhibition towards a wide range of tumor cells including L929, WISH and HeLa cell lines with good cell differentiation. Some extend the life of (induced) terminally ill cancerous mice by 50%. Further the polyamines exhibit activity towards virus at concentration levels where the cells are unaffected. Recent biological studies are aimed at evaluation of certain polyamines on inhibiting select tumors and viruses for specific disease control.

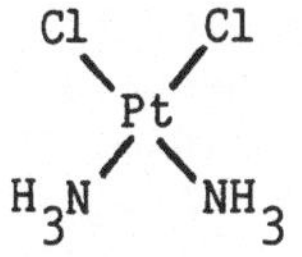

$$
\begin{array}{ccc}
Cl & & Cl \\
& \diagdown Pt \diagup & \\
H_3N & & NH_3
\end{array}
$$

$$\underline{1}$$

$$K_2PtX_4 + H_2N\text{-}R\text{-}NH_2 \longrightarrow \begin{array}{c} X \diagdown \diagup X \\ Pt \\ \diagup \diagdown NH_2\text{-}R\text{-}NH_2 \rightarrow \end{array}$$

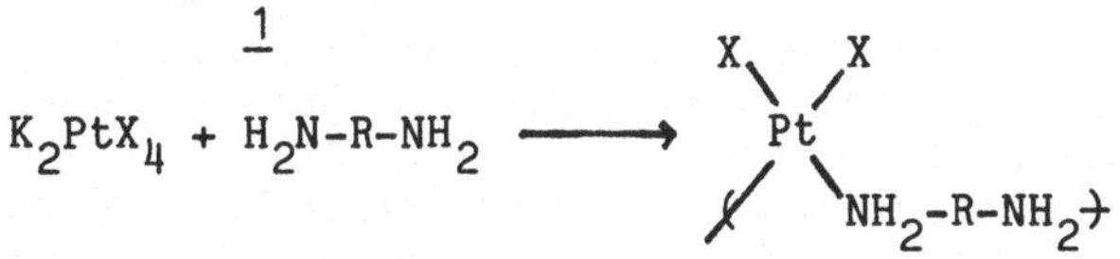

where X = Cl,Br,I $$\underline{2}$$

Recent chemical efforts are aimed at increasing the number of nitrogen-containing compounds synthesized, coupling diamines of known antitumoral activity, and the generation of kinetic data (synthesis and biological degradation). Here we report the synthesis of platinum II polyureas (3 where X=0) polythioureas (3 where X=S) and the analogous polyamides (4 where X=0,S).

$$K_2PtCl_4 + H_2N\overset{X}{\underset{\parallel}{C}}\text{-}NH_2 \longrightarrow \begin{array}{c} Cl \diagdown \diagup Cl \\ Pt \\ \diagup \diagdown NH_2\text{-}\underset{\underset{X}{\parallel}}{C}\text{-}NH_2 \rightarrow \end{array}$$

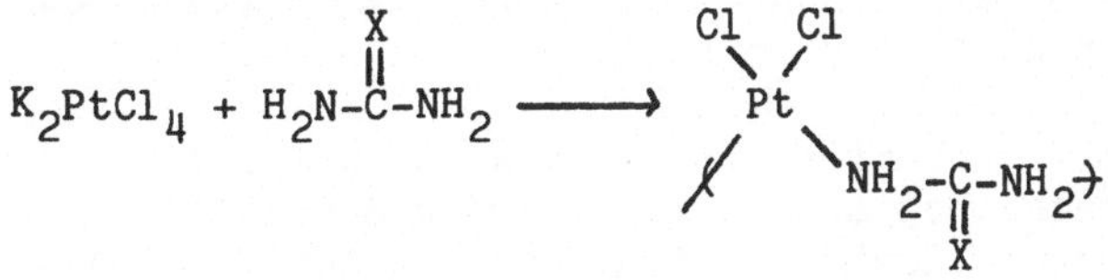

$$\underline{3}$$

$$K_2PtCl_4 + H_2N\overset{X}{\underset{\|}{C}}\overset{X}{\underset{\|}{C}}NH_2 \longrightarrow \underset{/}{\overset{Cl}{\diagdown}}Pt\underset{\diagdown}{\overset{\diagup}{Cl}}\overset{X}{\underset{NH_2-\underset{\|}{C}-\underset{\|}{C}-NH_2}{}}$$

$$\underline{4}$$

The first reason for desiring the synthesis of such compounds is that they include such biologically important moieties as purines, prymidines and select alpha-amino acids. Second, there are a number of antitumoral agents which possess no amine or one amine functional group but additional amide groups. Such compounds are typically expensive compared to the cost of urea, thiourea, oxamide and thiooxamide encouraging the use of the latter as model compounds. Third, hydroxyurea is a widely used antitumoral agent and its inclusion into platinum polyureas is consistent with our aim to couple known anticancer agents with platinum.

Finally thiourea is found to protect cells against toxic actions produced by platinum complexes including cis-DDP (for instance 62-64). Thus the survival of mouse leukemia L1210 cells treated with cis-DDP was enhanced by posttreatment incubation with thiourea. The mechanism(s) responsible for this protection is believed to involve reversing DNA-cis-DDP crosslinking. It is known that thiourea reacts with platinum-complexes as cis-DDP replacing various ligands from such complexes (65, 66). Both Zwelling and coworkers (63) and Filipski and coworkers (64) have demonstrated that the addition of thiourea subsequent to cis-DDP treatment leads to a reversal of DNA-cis-DDP crosslinks. Thus, synthesis of thiourea derivatives of cis-DDP may result in polymer with decreased toxicity, yet sufficient antitumoral activity.

EXPERIMENTAL

Syntheses are accomplished by mixing aqueous solutions containing the platinum compound and the diamide in an open Kimax glass beaker with constant stirring furnished by a magnetic stirrer. The product precipitates from the reaction mixture and is removed by suction filtration, washing repeatedly with ionized, doubly distilled water (utilized thoughout). The product is washed with water into a preweighed glass petri dish and permitted to dry in the air.

The following chemical reagents were employed as received: potassium tetrachloroplatinum II (J and J Materials), urea (Fisher), thiourea (Eastman), oxamide (Aldrich), malonamide (Aldrich), dithioxamide (Aldrich), and hydroxyurea (Aldrich).

Infrared spectra were obtained utilizing KBr pellets employing a Perkin-Elmer 457 Infrared Spectrophotometer. Elemental
analysis was done utilizing thermal degradation of the product
and employing a Perkin-Elmer 240 Elemental Analyzer assembly.
UV spectra were obtained using a Carey 14 Spectrophotometer.
NMR spectra were obtained in d_6-DMSO (using tetramethylsilane
as a reference) on a Varian EM360A spectrometer. Weight-average
molecular weights were obtained using a Brice-Phoenix 3000
Light-Scattering Photometer employing a Bausch and Lomb Refractometer (Model Abbe-31).

Bacterial tests were conducted in the usual fashion.
Tryptic soy agar plates were seeded with suspensions of the
test organism to produce an acceptable lawn of test organism
after 24 hours of incubation. Shortly after the plates were
seeded, the tested compounds (0.1 mg) were deposited as solids
on the plates as small round spots. The plates were incubated
and inhibition noted.

Tumoral cell lines were tested in the usual manner. The
cell lines were trypsinized, cells suspended and counted, then
diluted to 10^6 and plated onto Corning 1 ml well plates. The
plates were incubated overnight in a 100% humidity carbon dioxide
incubator and the next day the Dulbecco's Modified Eagle Medium
(DMEM) sucked off and substituted with DMEM containing various
microgram quantities of the platinum-containing polymer. Inhibition was tested for by both visual observation and employing
trypan blue (an exclusion stain).

RESULTS AND DISCUSSION

Structure

Structural evidence is derived from control reactions,
historical arguments, elemental analyses, spectral data (mainly
UV-VIS and IR) and molecular weight determinations. Histori-
cally, as noted in the introductory section, reaction with
nitrogen-containing compounds results in the formation, of
these compounds, through displacement of two chloride atoms,
through the nitrogen with the ligands cis to one another.
Control reactions, where one of the reactants is omitted, were
run and no precipitate formed consistent with the product containing moieties derived from both reactants.

Elemental analyses are consistent with the proposed structures depicted in 3 and 4. For instance, for the product from
thiourea - %C calc. 3.51, found 3.41, 3.46; %H - calc. 1.18,
found 1.13, 1.17; %N calc. 8.19, found 8.24, 8.31 and %Pt -
calc. 57, found 57.

Infrared spectra are similar to the spectra of the nitrogen-
containing reactant. In fact, it is known that the platinum
has little effect on the bond strengths of amines, other than
to weaken the N-H stretching vibrations (66, 67). Differences
are noted in the 420 to 550 cm^{-1} region attributed to the Pt-
Cl stretching mode. The number of Pt-N stretching vibration
bands is often taken as evidence of the geometry being cis
or trans with trans platinum amines showing one Pt-N band while
the presence of two bands indicates a cis geometry. While
the presence of the bands can be taken as evidence that the
structure is cis, the presence of only one band is not conclusive
proof that the product has trans geometry since the second
(cis-associated band) band is weak and at times missed and
not observed (66, 67). Fortunately, the products described
here exhibit two Pt-N bands and one Pt-Cl band. In each case
the lower (410-440 cm^{-1}) Pt-N band is weak (Table 1).

A further consideration involves the absence or presence
of bridging carbonyls which would result in the platinum existing
in an octahedral geometry. This possibility does not exist
for the synthesis of platinum polyamines   where the platinum
complex exists as a square planar structure.

Bridging-Octahedral Geometry

Normally two metal-associated carbonyl-related bands exist
for each set characteristic of bridging or nonbridging (68).

Table 1. Presence of infrared active Pt-N and Pt-Cl bands[a]

Nitrogen-containing Reactant	Pt-N	Pt-Cl
Urea	410, 455	338
Thiourea	430, 460	315
Oxamide	432, 472	335

[a]Values given in cm^{-1}.

Bridging carbonyl asymmetric stretching absorptions are found
about 1570 cm^{-1} whereas nonbridging groups exhibit a band about
1610-1720 cm^{-1}. The bridging carbonyl symmetrical stretching
band is found between 1410 to 1430 cm^{-1} whereas the nonbridging
bands occur around 1350 to 1390 cm^{-1}.

The infrared spectrum of the condensation product derived
from urea (Figure 1) exhibits bands at 1700 and 1390 cm^{-1} characteristic of a nonbridging structure as depicted in 3, a band
at 338 cm^{-1} attributed to Pt-Cl stretching and bands at 455
and 410, cm^{-1} attributed to Pt-N stretching. The product derived
from oxamide (Figure 1) has bands at 1650 and 1350 cm^{-1} attributed
to nonbridging carbonyls, 335 cm^{-1} attributed to Pt-Cl stretching
and 472 and 432 cm^{-1} attributed to Pt-N stretching. Thus the
products probably exist as nonbridged, square planar structures
as depicted in 3 and 4.

Results of the ultraviolet and visible spectra studies
appear in Table 2. Trans platinate compounds, including the
diaminodihaloplatinum II compounds, typically exhibit three
bands in the 150 to 450 nm region whereas the analogous cis
products show four bands in this region. The cis-platinum
II polyamines also exhibit four bands in this region (43, 44).
The present products also show the presence of four bands consistent with a cis-structure about the platinum atom.

Table 2. Ultraviolet and visible spectral bands of
various platinum compounds

Compound	Band peaks[a]			
$PtCl_4^{-2}$	177	210	255	302
cis-DDP	240	273	331	372
trans-DDP	-	268	317	367
Thiourea, $PtCl_4^{-2}$	280	320	350	395
Oxamide, $PtCl_4^{-2}$	280	305	360	b
Hexamethylenediamine,$PtCl_4^{-2}$	250	310	340	390

[a] All wavelengths in nano meters

[b] Several small peaks at 385, 395 and 408.

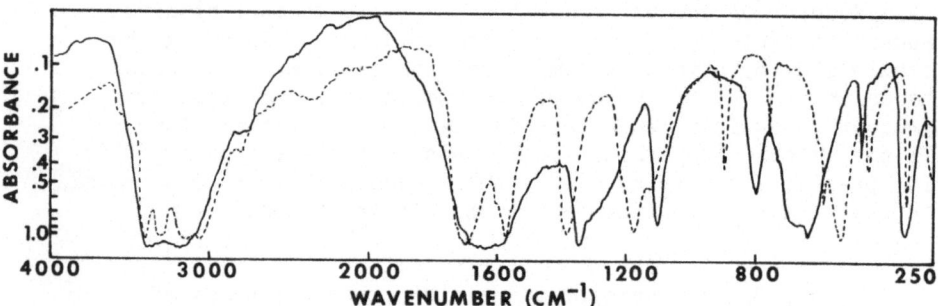

Figure 1.   Infrared spectra of the condensation products of $K_2 PtCl_4$ with urea (---) and oximide (——).

The products often yield highly colored solutions making molecular weight determinations by light scattering photometry not possible. Further their solubility in only dipolar aprotic solvents (as DMSO, DMF, HMPA) prevents their molecular weight determination employing other conventional approaches as membrane osmometry and HPLC (solvent incompatability with membranes and resins respectively). The products which give only mildly colored solutions are moderate ($\overline{DP}$=37) to high polymers ($\overline{DP}$ = 750).

Of interest is the great difference in product yield with regard to whether the nitrogen-containing reactant is sulfur or oxygen-containing with product yields being considerably greater for the sulfur compounds (Table 3). This is probably due to a decreased nucleophilicity for the oxygen compounds due to the greater electronegativity of oxygen compared to sulfur. This trend is consistent with other results where the product yield at constant time decreases as the electron density of the nitrogen decreases (Table 4).

Table 3. Polyurea and Polythiourea Synthesis and
Chain Length Results

Nitrogen-Containing Reactant	Yield (%)	Molecular Weight, $\overline{M}_w$	$\overline{DP}^b$
Urea	2	1.2×10^4	37
Thiourea	58	2.5×10^5	750
Oxamide	16	1.0×10^5	280
Malonamide	1	–	–
Dithioxamide	57	a	–

Reaction conditions: Potassium tetrachloroplatinate (2.00 mmoles) dissolved in 10 mls of doubly distilled-deionized water added to a solution containing the nitrogen-containing reactant (2.00 mmoles) contained in 10 mls of water at 25°C with stirring for 3 days.

a Too dark to permit light scattering photometry.

b Determined using light scattering photometry.

Table 4.   Approximate (to 50% yields) Reaction Rates with
           Tetrachloroplatinate

Nitrogen-Containing Species	Reaction Time
Diamines	2 Days
Dithioureas and Thioureas	3 Days
Diureas and Ureas	> 5 Days

(Initial) Biological Characterization

A brief survey of the inhibitory nature was done with
the products derived from thiourea and oxamide. The products
were not inhibitory to any of the tested bacteria (Escherichia
coli, Alcaligenes faecalis, Staphylococcus epidermis, Staphylococcus aureus, Enterobacter aerogenes, Neisseria mucosa, Klebsiella pneumoniae, Pseudomonas aeroginosa, Actinobacter calcoaceti-
cus and Branhamella catarohilis). This is in direct contrast
with the analogous platinum II polyhydrazines and polyamines
where there is found widespread antibacterial activity for
similar tests.

Next, two tumoral cell lines were tested. While the products
from urea and thiourea exhibit some tumoral inhibition, they
are less effective than analogous polyamines (Table 5; 43,
44, 48). Thus while the platinum II polyureas and polythioureas
exhibit some tumoral inhibition they are less active than analogous platinum polyamines.

Table 5. Cell Tumor Line Inhibition Test Results

Nitrogen Containing Compound	Concentration (ug/ml)						
	18	14	12	10	8	6	4
HeLa Cell Line							
Urea - 24 hrs.	3	2		2	1	0	0
Urea - 48 hrs.	3	2			0	0	
Thiourea - 24 hrs.		4	4	2	1	0	0
Thiourea - 48 hrs.		4	4	2	0	0	
4-Nitrophenylhydrazine - 24 hrs.			4	2	2	2	1
1,6-Hexanediamine - 24 hrs.[a]	0			0		0	
p-Phenylenediamine - 24 hrs.[a]	0			0		0	
L929 Cell Line							
Urea - 24 hrs.				2	1	0	0
Urea - 48 hrs.				0	0	0	
Thiourea - 24 hrs.	2	2		2	1		
Thiourea - 48 hrs.	1	1		0			
4-Nitrophenylhydrazine - 24 hrs.							
1,6-Hexanediamine - 24 hrs.[a]	0			0		0	
p-Phenylenediamine - 24 hrs.[a]	0			0		0	

4 = 100; 3 = 75; 2 = 50; 1 = 25; 0 = 0 inhibition (cell death)

[a]Activity begins at 30 ug/ml and is 100% by 50 ug/ml.

REFERENCES

1. B. Rosenberg, L. Van Camp and T. Krigas, Nature (London),
    205, 698 (1965).
2. B. Rosenberg, L. Van Camp, J. Trosko, and V. Mansour,
    Nature (London), 222, 385 (1969).
3. B. Rosenberg, Plat. Metals Rev., 15, 42 (1971).
4. B. Rosenberg, Cancer Chemother. Rep., Pt. 1, 59, 589 (1975).
5. J. Gottlieb and B. Drewinko, Cancer Chemother. Rep.,
    Pt. 1, 59, 621 (1975).
6. J. Howle, G. Gale and A. B. Smith, Biochem. Pharm., 21,
    1465 (1972).
7. J. Hill, E. Loeb, A. Machellan, N. Hill, A. Khan and J.
    King, Cancer Chemother. Rep., Pt. 1, 59, 647 (1975).
8. P. Kamalakar, A. Freeman, D. Higby, H. Wallace and L.
    Sinks, Cancer Treatments. Rep., 61, 835 (1977).
9. J. Ward, D. Young, K. Fauvie, M. Wolpert, R. Davis and
    A. Guarino, Cancer Treatment. Rep., 60, 1675 (1976).
10. A. Kahan, J. Hill, W. Grater, E. Loeb, A. MacLellan and
    N. Hill, Cancer Res., 35, 2766 (1975).
11. J. Holland, J. Rowland and M. Plumb, Cancer Res., 37,
    2425 (1977).
12. S. Sallan, N. Zinberg and E. Frei, New England J. of Med.,
    293, 795 (1975).
13. S. Stadnicki, R. Fleischman, U. Schaeppi and P. Merrimam,
    Cancer Chemother. Rep., Pt. 1, 59, 467 (1975).
14. J. Ward and K. Fauvie, Tox. and Appl. Pharm., 38, 535
    (1976).
15. C. Litterset, T. Gram, R. Dedrick, A. Leroy and A. Guarino,
    Cancer Res., 36, 2340 (1966).
16. K.P. Lewis, W.D. Medina, Cancer Treat. Rep., 64 (10-11),
    1162 (1980).
17. M.S. Aapro and D.S. Alberts, Cancer Chemother. Pharmacol.,
    7 (1), 11 (1981).
18. D. Hayes, E. Cvitokovic, R. Golby, E. Scheiner and I.
    Krakoff, Proceedings of American Assoc. Cancer Res., 17,
    169 (1976).
19. K. Chary, D. Higby, E. Henderson and K. Swinerton, Cancer
    Treat. Rep., 61, 367 (1977).
20. K. Chary, D. Higby, E. Henderson and K. Swingerton, J.
    Clinical Hematology and Oncology, 7, 633 (1977).
21. C. Merrin, Proceedings American Assoc. Cancer Res., 17,
    243 1976 and 18, 298 (1977).
22. L. Einhorn and B. Furnas, J. Clinical Hematology and Onco-
    logy, 7, 662 (1976).
23. H. Bruckner, C. Cohen, G. Deppe, B. Kabakow, R. Wallach,
    E. Greenspal, S. Gusberg and J. Holland, J. Clinical Hemato-
    logy and Oncology, 7, 619 (1977).
24. R. Kwong and B. Kennedy, Proceedings of American Assoc.
    Cancer Res., 18, 317 (1977).

25. J.A. Mabel, P.C. Merker, M.L. Sturgeon, I. Wodinsky and R.K. Geran, Cancer, 42 (2), 217 (1980).
26. M.H. Amer, R.M. Izbick, V.K. Vaitkerleins and M. AlSarraf, Cancer, 45 (2), 217 (1980).
27. H. Takita, F. Edgerton, P. Marabella, D. Conway and S. Harguindey, Cancer, 48 (7), 1528 (1981).
28. M. Cleare, Coord. Chem. Revs., 12, 349 (1974).
29. R. Speer, H. Ridgway, L. Hall, D. Stewart, K. Howe, D. Lieberman and J. Hill, Wadley Medical Bull., 5, 19 (1975).
30. R. Speer, H. Ridgway, L. Hall, A. Newman, K. Howe, D. Stewart, G. Edwards, and J. Hill, Wadley Medical Bull., 5, 335 (1975).
31. Y. Kidani, K. Indgaki, R. Saito and S. Tsukagoshi, J. Clinical Hematology and Oncology, 7, 197 (1977).
32. E. Loeb, J. Hill, A. Pardue, N. Hill, A. Khan and J. King, J. Clinical Hematology and Oncology, 7, 701 (1977).
33. M. Cleare and J. Hoeschele, Platinum Metals Rev., 17, 2 (1973).
34. M. Cleare and J. Hoeschele, Bioinorg. Chem., 2, 187 (1973).
35. P. Schwartz, S. Meischen, G. Gale, L. Atkins, A. Smith and E. Walker, Cancer Treat. Rep., 61, 1519 (1977).
36. M. Tobe and A. Khokhar, J. Clinical Hematology and Oncology, 7, 114 (1977).
37. H. Allcock, R. Allen and J. O'Brien, Chem. Comm., 717 (1976).
38. H. Allcock, Science, 193, 1214 (1976).
39. H. Allcock, Polymer Preprints, 18, 857 (1977).
40. H. Allcock, Organometallic Polymers (C. Carraher, J. Sheats and C. Pittman, Editors), Academic Press, N.Y. (1978), pgs. 283-288.
41. H. Allcock, R. Allen and J. O'Brien, J. Amer. Chem. Soc., 99, 3984 (1977).
42. C. Carraher, C. Admu-John and J. Fortman, unpublished results.
43. C. Carraher, D.J. Giron, I. Lopez, D.R. Cerutis and W.J. Scott, Organic Coatings and Plastics Chemistry, 44, 120 (1981).
44. C. Carraher, W.J. Scott, J.A. Schroeder and D.J. Giron, J. Macromol. Sci.-Chem. A15(4), 625 (1981).
45. C. Carraher, Organic Coatings and Plastics Chemistry, 42, 428 (1980).
46. C. Carraher, T. Manek, D. Giron, D.R. Cerutis and M. Trombley, Polymer Preprints, 23 (2), 77 (1982).
47. C. Carraher and A. Gasper, Polymer Preprints, 23 (2), 75 (1982).
48. C. Carraher, Biomedical and Dental Applications of Polymers (G. Gebelein and F. Koblitz, editors), Plenum Press, N.Y., 1981, Chpt. 16.
49. C. Carraher, D.J. Giron, T. Manek and D. Blair, unpublished results.

50. C. Carraher, H.M. Molloy, M.L. Taylor, T.O. Tiernan and W.J. Scott, unpublished results.

51. R. Speer, H. Ridgway, L. Hall, D. Stewart, K. Howe, D. Lieberman, D.A. Newman and J. Hill, Cancer Chemother. Rep., Pt. 1, 59, 629 (1975).

52. M. Tucker, C. Colvin and D. Martin, Inorganic Chem., 3, 1373 (1964).

53. C. Colvin, R. Gunther, L. Hunter, J. McLean, M. Tucker and D. Martin, Inorganic Chimica Acta, 3, 487 (1968).

54. T. Conners, M. Jones, W. Ross, P. Braddock, A. Khokharard, M. Tobe, Chemico-Biological Interactions, 5, 415 (1972).

55. K. Inagaki, Y. Kidani, K. Suzuki and T. Tashiro, Chem. Pharm. Bull. (Tokyo), 28 (8), 2286 (1980).

56. K. Okamoto, M. Noji, T. Tashiro and Y. Kidani, Chem. Pharm. Bull. (Tokyo), 29 (4), 929 (1981).

57. P. Ribaud, D.P. Kelsen, N. Alcock, G.E. Garcia, P. Dubouch, C.C. Young, F. Muggin, J. Burchenal and G. Mathe, Recent Results Cancer Res., 74, 156 (1980).

58. C. Carraher, W.J. Scott and D.J. Giron, "Bioactive Polymers," (Edited by C. Gebelein and C. Carraher) Plenum, 1983.

59. G. Rowland, G. O'Neil and D. Davis, Nature, 255, 487 (1975).

60. H. Ringsorf, Middland Macromolecules Meeting (Edited by H. Elias), Marcel Dekker, N.Y., 1978.

61. H.J. Ryser, Nature, 215, 934 (1967).

62. J.H. Burchenal, K. Kalaher, K. Dew, L. Lokys and G. Gale, Biochimie, 60, 961 (1978).

63. L.A. Zwelling, J. Filipski and K.W. Kohn, Cancer Research, 39, 4989 (1979).

64. J. Filipski, K.W. Kohn, R. Prather and W.M. Bonner, Science, 204 (13), 182 (1979).

65. F. Basolo and R.G. Pearson, Prog. Inorg. Chem., 4, 381 (1962).

66. A.A. Grinberg, M. Serator and M.I. Gel'fman, Russian J. Inorg. Chem., 13 1695 (1968).

67. G. Barrow, R. Krueger and F. Basolo, J. Inorganic and Nuclear Chemistry, 2, 340 (1956).

68. C. Carraher and J. Schroeder, Polymer Letters, 13, 215 (1975).

SYNTHESIS OF DICHLOROPALLADIUM II POLYAMINES

Charles E. Carraher, Jr.,[a] Andrew L. Gasper,[a] Mary L. Trombley,[a] Fred L. DeRoos,[b] David J. Giron,[c], George G. Hess[a] and Kathy M. Casberg[c]

Departments of Chemistry[a] and Microbiology and Immunology[c], Wright State University, Dayton, Ohio 45435 and Battelle Columbus Laboratories[b], Columbus, Ohio 53201

INTRODUCTION

The previous paper describes the synthesis of platinum II polyureas and polythioureas. Here we report the synthesis of palladium (II) polyamines (1) which may be considered analogs of cis-dichlorodiamineplatinum(II) cis-DDP (2), and the related platinum II polyamines (3).

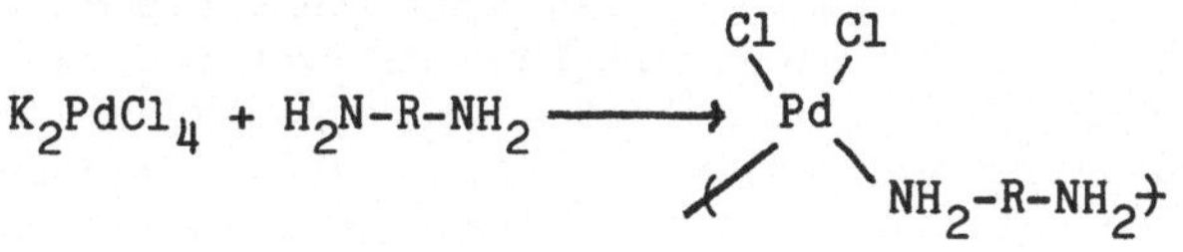
K_2PdCl_4 + H_2N-R-NH_2
Cl
Cl
Pd
NH_2-R-NH_2

1

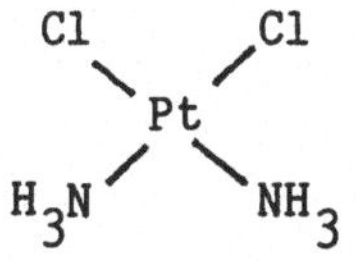
Cl
Cl
Pt
H_3N
NH_3

2, cis-DDP

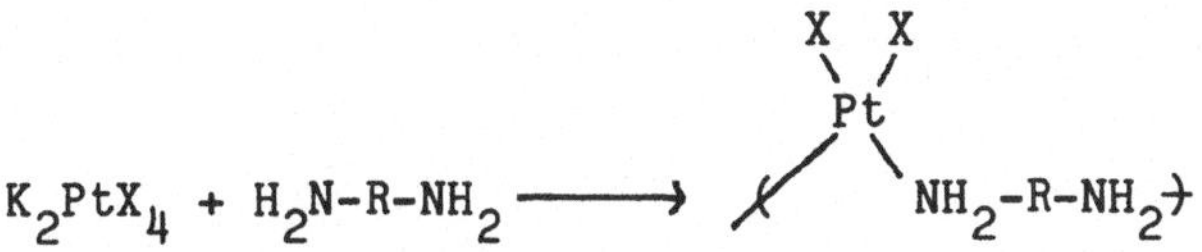
K_2PtX_4 + H_2N-R-NH_2
X
X
Pt
NH_2-R-NH_2

where X = Cl,Br,I

3

A great deal of activity has been expended on developing
palladium catalysts including the use of dipotassium and disodium
tetrachloropalladate(II) and cis and trans-diamminedichloropalla-
dium(II) (for instance 1).

Other interests involve the use of palladium-containing
compounds themselves in nonbiological applications. For instance,
the products derived from the condensation of K_2PdCl_4 with
polyvinyl pyridine form materials with low threshold voltages
for imagining and high-speed recording (2). Flexible, electrically conductive films have been prepared from the reaction
of K_2PdCl_4 with polyimides (3).

Another area of activity involves the use of palladium
compounds for biological applications. In particular both
the cis and trans palladium analogs of cis-dichlorodiamineplatinum
II are currently undergoing tests involving their ability to
inhibit tumors, with mixed results. Most show inhibitions
which are inferior to the analogous platinum derivatives (4).
Rosenberg and coworkers find that select analogs, such as 1,2-
diaminocyclohexanedinitratopalladium II (4) are about as effective as cis-DDP itself in preliminary studies involving animal
tumor systems (5).

$\underline{4}$

Thus investigations involving palladium amines are ongoing
and varied.

EXPERIMENTAL

Syntheses are accomplished by mixing aqueous solutions
containing the palladium compound and the diamine in an open
Kimax glass beaker with constant stirring furnished by a magnetic
stirrer. The product precipitates from the reaction mixture
and is removed by suction filtration, washed repeatedly with
ionized, doubly distilled water, washed onto a glass petri
dish and dried. The product is typically a light to dark brown
powder.

Infrared spectra were obtained using KBr pellets employing
a Perkin-Elmer 457 Infrared Spectrophotometer. Elemental analysis
was done utilizing thermal degradation of the product and employ-

ing a Perkin-Elmer 240 Elemental Analyzer for C,H,N analysis.

The mass spectra of the products derived from 1,6-hexane-diamine and p-phenylenediamine were obtained using a Finnigan Model 4000 mass spectrometer with a Model 4500 electronimpact ion source. Ballistic heating of the direct insertion probe provided a temperature increase of at least 50°C/min., up to about 475°C maximum. (Slow heating of the samples yielded few decompositions.) Mass spectral analysis of the product derived from 4,6-diamino-5-nitroso-2-phenylpyrimidine was performed using a direct insertion probe in a Kratos MS-50 mass spectrometer, operating in the EI mode, 8 KV acceleration and 10 sec/decade scan rate. The sample was ballistically heated to 350°C.

Weight-average molecular weights were obtained using a Brice-Phoenix 3000 Light-Scattering Photometer employing a Bausch and Lomb Refractometer (Model Abbe-31).

Solubilities were attempted in a wide variety of solvents including the dipolar aprotic solvents DMSO, DMF, HMPA, acetone and TEP by placing about 1 mg of product in 3 ml of liquid and observing for about a week.

The following chemical reagents were employed as received: p-phenylenediamine (Aldrich), 1,6-hexanediamine (Aldrich), 1,12-diaminododecane (Aldrich), 1,10-diaminodecane (Aldrich), 1,7-diaminoheptane (Aldrich), 4,6-diamino-5-nitroso-2-phenyl-pyrimidine (Aldrich), adenine (Aldrich), histamine (Aldrich), 1,8-diamino-p-menthane (Aldrich), potassium tetrachloropalladate (II) (Strem Chemicals), diamino-6-mercaptopyrimidine (Aldrich), diaminobenzanilide (Aldrich), L-(-)tryptophan (Aldrich) and 4,6-diamino-2-mercaptopyrimidine (Aldrich).

Bacterial tests were conducted in the usual fashion. Tryptic soy agar plates were seeded with suspensions of the test organism to produce an acceptable lawn of test organism after 24 hours of incubation. Shortly (within several minutes) after the plates were seeded, the tested compounds (0.1 mg of solid or designated polymer concentration if in solution) were deposited on the plates. The plates were incubated and inhibition noted.

For tumor cell lines, the cells were trypsinized, suspended and counted, then diluted to about 10^6 and plated onto Corning 1 ml well plates. The plates were incubated overnight in a 100% humidity carbon dioxide incubator and the next day the Dulbecco's Modified Eagle Medium (DMEM) is sucked off and substituted with DMEM containing various microgram quantities of the polymer. Inhibition was tested for employing both visual observation and trypan blue (an exclusion stain).

RESULTS AND DISCUSSION

Physical Characterization

The synthesis of palladium polyamines (2) is a straightforward extension of our previous work involving the synthesis of the analogous platinum polyamines (3;6,7). The geometrical structure of the units about the palladium atom is believed to be as depicted in 1 where the like units are cis to one another. This is supported by the trans effect (8) discussed in the previous paper and by spectroscopic measurements cited in the following paragraphs.

The trans effect is strongest for M=Pt and weakest for M=Ni, but still dominates even for Ni. In fact, the trans form of many palladium and platinum compounds is known but these products are made through utilizing different reactants in a process which may be depicted as follows (7).

$$\underset{H_3N}{\overset{H_3N}{\diagdown}}\underset{}{\overset{}{Pd}}\underset{NH_3}{\overset{NH_3}{\diagup}}{}^{+2} \quad + \ 2\ Cl^- \longrightarrow \quad \underset{H_3N}{\overset{Cl}{\diagdown}}\underset{}{\overset{}{Pd}}\underset{Cl}{\overset{NH_3}{\diagup}} \quad + \ 2\ NH_3$$

$$\underline{5}$$

The spectral evidence is in agreement with a structure of form 1. For instance, infrared spectra of the polymeric products are quite similar to spectra of the diamines themselves except for the presence of three new bands, one at about 340 cm^{-1} associated with Pd-Cl and two at about 400 to 460 cm^{-1} associated with the Pd-N moiety (9,10). The presence of two bands associated with the Pd-N group is generally taken to be evidence that the amines are cis to one another; and in cases where the nitrogens occupy the trans positions, only one band should appear. While this is typically found to be true, it must be noted that a few cis-platinum diamines exhibit only one band in the 400-600 cm^{-1} region, presumably due to the often encountered weakness of the second band. Thus the presence of two bands is good evidence that the products are cis, but the presence of only one band is not as strong at indicating the product is trans in geometry.

The infrared spectra of the product with p-phenylenediamine is reproduced as Figure 1. It is similar to that of p-phenylenediamine itself except for appropriate band shifts and accompany-

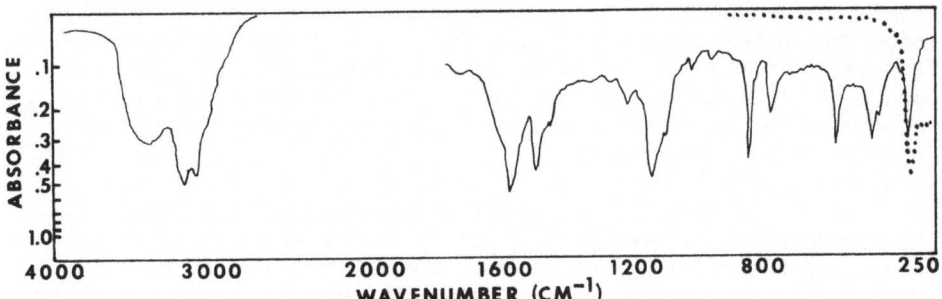

Figure 1.   Infrared spectra of K_2PdCl_4 ($\cdots$), and the condensation
product of K_2PdCl_4 with p-phenylenediamine (——).

ing new bands at 337 cm^{-1} (characteristic of the Pd-Cl stretching
vibration) and at 450 and 435 cm^{-1} (characteristic of the Pd-
N stretching vibration; 9,10).

Elemental analyses are in mild agreement with the repeat
unit described in 1. Sample results follow; for the product
from p-phenylenediamine - %C-calc. 25.2, found 24.2; %H-calc.
2.80, found 2.87; %N = 9.8, found 8.9; from 6-amino-purine-
%C-calc. 19.2, found 18.1; %H-calc. 1.6, found 1.8; %N-calc.
22.4, found 20.2.

A mass spectrum (ions with intensities $\geq$ 10%) of the polymer
derived from 1,6-hexanediamine is given in Table 1. The base
peak in both the polymer spectrum and that of pure 1,6-hexane-
diamine is m/e 30, $CH_2NH_2^+$. The other major ions in the amine
spectrum (m/e 28, 56, 87) are likewise present in the polymer
spectrum at substantial intensities. The masses 35-38 in the
polymer spectrum may be due in part to the Cl^+ and HCl^+, but
since the intensity ratios are incorrect, there must be interference from other ions. The spectrum of the polymer is more
complex than that of the pure amine, indicating substantial
decomposition during the heating process. In this polymer,
a peak at the amine molecular mass plus one is seen, suggesting
intramolecular protonation.

At masses above the molecular mass of the amine, there
are numerous small peaks (rel. int. $\leq$ 1%) extending to m/e
1000. Of these, the group between m/e 120 and 216 has higher
intensity, with m/e 216 being 0.94%. Its mass could be accounted
for by the amine dimer minus NH_2. Beyond m/e 300 the intensities
are less than 0.05%. Thus, while higher molecular weight fragments can be rationalized as derived from the amine, the abundance
of these ion fragments is low, and of minimal significance.

Table 1. Major Ions in the Mass Spectrum of Poly-(1,6-hexane-
          diaminepalladium(II) dichloride). The spectrum is
          an average of 5 scans taken after about 3 minutes
          of heating.

m/e	Rel. Int. (%)	m/e	Rel. Int. (%)	m/e	Rel. Int. (%)
117	46	55	22	37	41
100	31	45	16	36	67
99	10	44	27	35	44
98	25	43	26	31	24
87	11	42	33	30	100
70	22	41	38	28	54
69	10	39	31	27	36
56	41	38	47		

Table 2. Major Ions in the Mass Spectrum of Poly-(p-phenylene-
diaminepalladium(II) dichloride).

m/e	Rel. Int. (%)	Possible Structure
108	63	$H_2N-C_6H_4-NH_2^{+\cdot}$, $P^{+\cdot}$
107	18	
81	10	
80	23	
54	10	
53	12	
52	10	
38	34	$H^{37}Cl^{+\cdot}$
36	100	$H^{35}Cl^{+\cdot}$
35	14	$^{35}Cl^{+\cdot}$
28	24	

Table 3. Mass Spectrum of Poly-(4,6-diamino-5-nitroso-2-phenyl-
pyrimidinepalladium(II) dichloride).

Nominal m/e	Rel. Int. (%)	Atomic Composition (Mass dev., ppm)
216	12	$C_9{}^{13}CH_9N_5O$ (12.4) $(P+1)^{+\cdot}$
215	100	$C_{10}H_9N_5O$ (1.1) $P^{+\cdot}$
201	12	
197	23	$C_{10}H_7N_5$ (0.8) $(P-H_2O)^{+\cdot}$
143	18	$C_9H_7N_2$ (0.2)
128	10	$C_8H_4N_2$ (0.6)
104	100	C_7H_6B (-0.1)
103	95	C_7H_5N (2.1)
82	26	$C_3H_4N_3$ (1.5)
77	37	C_6H_5 (0.4)
76	35	
55	10	
51	14	

Further, while the spectrum is consistent with a product
of form 1, it bears only mild agreement with the spectrum of
1,6-hexanediamine itself and indicates that the polymer undergoes
thermodecomposition through a complex mode(s).

The mass spectrum of the polymer derived from p-phenylenediamine yielded three peaks in the reconstructed ion chromatogram
(RIC), at approximately 2.7, 3.8 and 4.2 mins. Mass chromatograms
of m/e 36, 80 and 108, the major masses, reflect quite accurately
the entire RIC. Mass spectra taken at various times show little
variation. A representative mass spectrum, which lists peaks
of relative intensity $\geq$ 20%, is given in Table 2. Above m/e
109, a number of small peaks with intensities $\leq$ 0.5% extend
to about m/e 600. The one exception is m/e 213 which has an
intensity of 1.5% in the spectrum given in Table 2. A possible
structure for this ion is $H_2N-C_6H_4-NHNH-C_6H_4-NH^+$, derivable
from two p-phenylenediamine units. It is interesting to note
that m/e 213 maximizes at 2.4 min., corresponding to a leading
shoulder on the 2.7 min. peak in the RIC, and is almost completely
gone by 3.0 min.

With the exception of m/e 36 and 38, obviously due to
HCl, the major peaks in the spectrum are in excellent agreement
with the mass spectrum of pure p-phenylenediamine.

A mass spectrum (masses scanned were m/e >50) of the substituted nitrosopyrimidine polymer is given in Table 3, again
showing only peaks with relative intensities $\geq$ 10%. Column
3 shows the atomic compositions of the peaks, as indicated
by the high resolution mass measurements, with the deviation
between measured and calculated mass given in ppm. Variations
in intensities of the major masses in three sequential scans
indicate that the decomposition of the polymer is yielding
different species with time. Thus, all the masses listed in
Table 3 need not be assumed to arise from the same species.
In particular, m/e 104 increases significantly in the three
scans, whereas m/e 215 remains the base peak throughout. The
m/e 215 ion fragment supports the presence of the nitrosopyrimidine moiety in the polymer.

None of the ion fragments are found to be due to Pd, PdCl
or $PdCl_2$ consistent with the use of thermodecomposition as
a method for the quantitative determination of palladium and
platinum (as the residue).

In summary, the ion fragments derived from the palladium
polyamines are consistent with the presence of chloride and
diamine.

Product yield and molecular weight results are given in
Table 4. Compared to the analogous platinum polyamines, reaction
is somewhat slower. Good (80% plus) yields of the platinum
polyamines typically occur (for chlorides) within 48 hours.
This is expected when considering the size of the metal atom to
be important since the smaller palladium atom "holds" its filled
and vacant orbitals closer to the nucleus making reactions
requiring polarization of the metal site more difficult.

A second difference is the poorer, more limited solubility
of the palladium compounds compared to the analogous platinum
polyamines. Most of the platinum polyamines thus far synthesized are soluble in a number of dipolar aprotic solvents with
some even soluble in chloroform. Only the palladium polyamines
derived from dissymmetrical diamines exhibit solubility in
any attempted solvent and here solubility is limited to only
dipolar aprotic solvents to an extent of 3% and less. Solubiliza-
tion in dipolar aprotic liquids is dependent on a number of
factors - the major one being the ability of the dipolar aprotic
liquid molecules to polarize the solute molecules. The smaller
size and corresponding poorer polarizability of the palladium
atom is probably responsible for this trend. A further, but
related factor, is the possible greater tendency for the palladium
polymers to form crystalline regions.

The condensation of tetrahalopalladium salts with diamines
is not new, yet it appears that the polymeric nature of the
products was not recognized (possibly the brown tar at the
bottom of the reaction flask discarded as unwanted) and/or
the products formed were indeed not polymeric. The latter
may result from the use of dissimilar molar quantities of reactants. The analogous reaction, except employing K_2PtX_4, requires
a near molar equivalence of the reactants to produce high chain
lengths (6,7).

Of additional interest is the product derived from histamine.
The reaction of K_2PdCl_4 with histamine forming soluble monomeric
histamine compounds (presumably of form 6) is known and utilized
in some studies as a method of determining the amount of histamine
in systems such as mast cells from rats (11,12). This complex
is six-membered and may account for the low yield of oligomeric
material obtained in our study. Further it is possible that
the six-membered ring product is initially formed and subsequently
forms longer chains through ring opening, with precipitation
being a driving force for chain formation.

6

Biological Characterization

Eleven of the palladium polyamines were tested as solids
against a group of ten bacteria. The ten bacteria were chosen
to give a broad range biological species, and they are responsible
for a wide range of biological responses. A brief description
of the biological responses associated with each bacteria follows
along with the abbreviations employed in Table 2: Actinobacter
calcoacetius (Actin) - associated with conjunctivities, keratitus
and chronic ear infections; Alcaligenes faecalis (Alcal) -can
cause urninary tract infections and, in debilitated individuals,
septicemia or meningitis; Branhamella catarrhalis (Bran) -associ-
ated with mucous membrane in glammations, venereal discharges,
meningitis and bacterial endocarditis; Enterobacter aerogenes
(Enter) - can cause urinary tract infections, endocarditis,
pneumonia and bacteremia; Escherichia coli (E. Coli) - may
cause cholecystitus, appendicitis, peritonitis, sinusitis and
summer diarrhea; Klebsiella pneumoniae (Kleb) - may cause lesions
in the body, pneumonia, chronic lung abscess, sinusitis and
upper respiratory infections; Neisseria mucosa (Neiss) - generally
nonpathogenic; Pseudomonas aeruginosa (Pseu) - opportunistic
pathogen, may infect wounds; contaminates burns, draining sinuses
and decubitus ulcers; can cause tract infections, eye infections
and meningitis; Staphylococcus epidermis (Staph E) - generally
causes mild infections; Staphylococcus aureus (Staph A) -causes
"pimples", abscesses, impetigo, wound infections, pyelitis,
cystitis, "food poisoning", pneumonia, meningitis and enteritis.

Table 6 contains tabulated results derived from Table
5. In general it appears that if a polyamine inhibits one
or two bacteria, they will inhibit a majority of others, i.e.,
the polyamines appear to be largely noninhibitory or widely
inhibitory. Second, there is a distinct trend with respect
to length of the aliphatic chain such that the polyamines derived
from 1,6-hexanediamine and 1,7-diaminoheptane exhibit wide
inhibition whereas those derived from 1,10-diaminodecane and
1,12-diaminododecane are noninhibitory.

Table 4. Results for Palladium(II) Polyamines

Diamine	Reaction Time(hrs)	Yield (%)	$\overline{M}_w{}^a$	$\overline{DP}$	dn/dc
p-Phenylenediamine	45	29			
	50	38			
1,6-Hexanediamine	50	19			
	76	26			
1,7-Heptanediamine	45	78			
1,10-Decanediamine	46	81			
1,12-Dodecanediamine	29	62			
Diamino-5-nitroso-2-phenyl Pyrimidine	168	31			
Adenine(6-Aminopurine)	96	100			
Histamine	120	24	2.6×10^3		1.7
Diamino-p-methane	116	70	3.3×10^6	9,000	1.05
Diamino-6-mercaptopyrimidine	96	82			
Diaminobenzanilide	96	24	2.7×10^7	88,000	0.18
L-(-)Tryptophan	144	88	3.0×10^6	7,800	0.055
4,6-Diamino-2-mercapto-pyrimidine	192				

Reaction Conditions: K_2PdCl_4 (1.00 mmole) dissolved in doubly distilled-deionized water (15 ml) added to aqueous solutions (15 ml) containing the diamine at 25°C with constant stirring for the cited reaction times.

^{a}Determined using light scattering photometry employing DMSO as a solvent.

Table 5.   Inhibition of bacteria by palladium polyamines

Diamine	Bran	Actin	Enter	Kleb	Staph E	Alcal	Staph A	Neiss	Pseu	E Coli
1,6-Hexanediamine	I	I	I	N	I	S	I	S	I	I
1,7-Diaminoheptane	N	S	I	N	I	S	I	I	I	I
1,9-Diaminononane	N	N	N	N	N	N	N	N	N	I
1,10-Diaminodecane	N	N	N	N	N	N	N	N	N	N
1,12-Diaminododecane	N	N	N	N	N	N	N	N	N	N
p-Phenylenediamine	I	S	S	S	S	S	S	S	S	S
2-phenyl pyrimidine										
1,8-Diamino-p-menthane	7	7								
1,8-Diamino-p-menthane	N	I	N	N	I	I	I	I	I	I
Histamine	I	S	I	N	N	S	N	I	I	I
6-Aminopurine	N	N	N	N	N	N	N	S	N	N
2,4-Diamino-6-mercapto-pyrimidine	I	N	I	I	N	N	N	N	N	N

I = Inhibition
S = Slight Inhibition
N = No. Inhibition

Table 6.   Tabular summary of bacteria results from Table 5.

Diamine	Number of Bacteria Showing Some Inhibition	Number of Bacteria Showing Total Inhibition
1,6-Hexanediamine	9	7
1,7-Diamineoheptane	8	6
1,9-Diaminononnane	1	1
1,10-Diaminodecane	0	0
1,12-Diaminododecane	0	0
p-Phenylenediamine	10	1
4,6-Diamino-5-nitroso-2-phenyl pyrimidine	7	7
1,8-Diamino-p-menthane	7	7
Histamine	7	5
6-Aminopurine	1	0
2,4-Diamino-6-mercapto-pyrimidine	3	3

REFERENCES

1. W. Keim and M. Roeper, J. Org. Chem., <u>46</u> (18), 3702 (1981).
2. K. Matsumoto, K. Takeda and M. Nagata, Jpn. Kokai Tokkyo
    Koho 79, 121, 748 (1978); CA:92:50087d.
3. L. Taylor, A. St. Clair, V. Carver and T. Furtsch, U.S.
    Pat. Appl. 135,058 (1980); NTIS-PAT-APPL-135 058; CA:94:
    48365r.
4. M. Cleare and J. Hoeschele, Bioinorg. Chem., <u>2</u>(3), 187
    (1973).
5. C. Carraher, W.J. Scott and J. Schroeder, J. Macromol.
    Sci.-Chem., <u>A15</u>(4), 625 (1981).
6. C. Carraher, D.J. Giron, I. Lopez, D.R. Cerutis and W.J.
    Scott, Organic Coatings and Plastics Chemistry, <u>44</u>, 120
    (1981).
7. J.E. Huheey, "Inorganic Chemistry: Principles of Structure
    and Reactivity," 2nd Ed., Harper & Row, N.Y., 1978.
8. J.S. Coe and A.A. Malik, Inorg. Nucl. Chem. Lett., <u>3</u>(3),
    99 (1967).
9. J.R. Durig and B.R. Mitchell, Appl. Spectrosc., <u>21</u>(4),
    221 (1967).
10. I. Zakharova, V. Tomilets and V. Dontsov, Inorg. Chim.
    Acta, <u>46</u>(1), L3 (1980).
11. S. Valladas-Dubois and M. Cain, C.R. Acad. Sci., Ser.
    C, 276(12), 1003 (1973).

Acknowledgement: The mass spectrum of the nitrosopyrimidine
polymer was obtained at the Midwest Center for Mass Spectrometry,
supported by NSF Regional Instrumentation Facility Grant #CHE
78-18572.

STAR AND COMB SHAPE POLY(OXYETHYLENE-g-ETHYLENEIMINES)

Samuel J. Huang and Peter T. Trzasko

Department of Chemistry and
Institute of Materials Science
University of Connecticut
Storrs, Connecticut 06268

INTRODUCTION

The grafting of polymers has been firmly established as a
means of modification of polymer properties (1-3). However, most
of the grafting of polymers utilizes free radical initiated reaction
of vinyl monomers which generally give difficult to separate
mixtures of homopolymers with the desired graft copolymers (3).
In addition, free radical initiated grafting of vinyl monomers
results in graft copolymers with graft segments of various sizes
and non-uniform frequency of attachment. For these reasons the
study of structure-property correlation of graft copolymers is
often very difficult. Recently developed grafting of vinyl monomers through ionic mechanisms has greatly improved the situation
(4-8). Grafting of preformed segments of uniform sizes onto
polymers through condensation mechanisms offers another route to
obtain graft copolymers of reasonably well defined structures.
The ability of linear (9-12) and cyclic (13,14) polyoxyethylenes
to complex metal ions has been of great interest in recent years.
The interaction between polyoxyethylenes with polyacids (16-18)
and polyoxyethylenes with proteins also have received increasing
attention recently (19). In all these cases the electron-donating
ability of the polyethers is mainly responsible for their interesting property. Currently available non-linear polyethers are
mostly graft copolymers of hydrophobic carbon back bone containing
graft segments of polyoxyethylene (18-22). We reasoned that
linear, star, and comb shape polyether segments together with
polyamine segments (stronger electron-doner than polyethers)
should have very interesting properties. We report here the

163

synthesis and properties of poly(oxyethylene-g-ethyleneimines).

EXPERIMENTAL

Starting materials 1,6-diisocyantohexane and poly(ethylene glycol mono methyl ether) PEGME, mol. wt. ave.-350,500,750,1900 and 5000- were obtained from Aldrich Co. Polyethyleneimines 6, 18, 200, and 1000 with Mn of 600, 1800, 10-20,000 and 50-100,000 respectively, were used as obtained. Simple amines were distilled before use.

General Procedures:

Preparation of Preformed Grafts. In a 200-ml 3-necked flask was placed 25gms of PEGME 750(0.33 mol/OH groups). The flask was heated to 70°C under vacuum for 4 hrs. to remove water. The vacuum was then replaced by a dry gas flow and condensor. 0.6 ml of triethylamine and 5.71 ml (0.34 mol) of 1,6-diisocyantohexane was added and the reaction run at 65°C for 24 hrs. The PEGME isocyanate was then washed with dry hexane and dried under vacuum.

Model Grafting. 1.76 ml (0.0165 mol) of N,N'-dimethylethylene diamine was dissolved in 50 ml of chloroform. This solution was then added dropwise over a 30 min. period to the PEGME 750-isocyanate prepared above. The reaction mixture was stirred for 24 hrs. at room temperature. The product was precipitated into hexane and dried in vacuum.

Grafting of PEI 6. 1.42 gms. (0.33 mol x 34 (based on repeat unit $-CH_2-CH_2-NH-$)) of PEI 6 was dissolved in 100 ml of chloroform. This solution was then added dropwise over a period of 1 hr. to the PEGME 750-isocyanate. The reaction mixture was stirred for 24 hrs. at room temperature. The product was precipitated into hexane and dried in vacuum. DMSO was used as solvent for grafting of PEGME-isocyanate onto higher MW PEI 200 and PEI 1000.

Phase Transfer Studies. Phase transfer studies were carried out by using 10 ml of aqueous picrate solution (7×10^{-5}m) with (0.10m) potassium hydroxide and 10 ml volume of CH_2Cl_2 with different concentrations of polymer. Picrate concentrations were determined on a Cary 17D Spectrophotometer.

RESULTS AND DISCUSSION

In this investigation uniformly grafted polymers which contain polyethyleneimines as the backbone component and polyoxyethylene as the graft component were prepared by condensation reactions shown in Scheme 1. Preformed grafts were first prepared by the reaction of poly(ethylene glycol monomethyl ether), PEGME, with 1,6-diisocyanatohexane to give polyoxyethylene with a reactive isocyanate end group. Five differenct molecular weight PEGME-iso-cyanates were prepared. These were in turn allowed to react with polyethyleneimines, PEI, of various molecular weight to give poly (oxythylene-g-ethyleneimines) with comb like structures. Reactions of PEGME with N,N'-dimethylethylenediamine gave linear (two grafts) polymers whereas reactions of PEGME with diethylenetriamine and triethylenetetraamine gave star shape polymers with five and six grafts respectively.

The progress of the reactions was followed spectroscopically with an infrared spectrophotometer. The reaction of PEGME with 1,6-diisocyanatohexane yielded the following absorption bands which are indicative of the urethane linkage and the isocyanate terminal: 3300 cm^{-1} -N-H stretching; 1730 cm^{-1} C=O stretching 1530 cm^{-1} -amide, and 228 cm^{-1} and the appearance of bands characteristic of urea. 3400 cm^{-1} -NH stretching, 1640 cm^{-1} -C=O stretching. Reactions were stopped when samples withdrawn from the systems showed no more change. In most of the cases there was no detectable isocyanate group left by the end of a 12-hr. reaction period. The graft co-polymers were purified by precipitation from chlorobenzene solution with hexane and evaporation under nitrogen to give viscous oils (at r.t.).

The viscosity of the graft copolymers were measured in benzene (for comparison with the original PEGME segments) and water. The viscosities of the linear and star shape graft copolymers, as ex-pected, increase as the number of branches increase and also as the size of the branches increases, Table 1.

Table 1. Intrinsic Viscosities of Linear and Star Shape Poly(oxyethylene-g-ethyleneimines) in Water at R.T.

PEI Backbone	Max. No. Branches	[η] 350	550	750	1900	5000
None	0	0.030	0.032	0.038	0.083	0.135
CH_3-$NHCH_2CH_2NHCH_3$	2	0.034	0.040	0.046	0.141	0.245
$H_2N(CH_2CH_2NH)_2H$	5	0.044	0.044	0.054	0.156	0.286
$H_2N(CH_2CH_2NH)_3H$	6	0.055	0.051	0.064	0.210	0.305

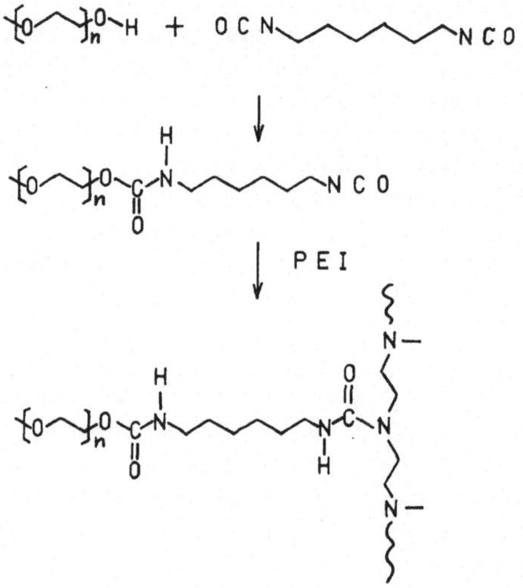

Synthesis of Poly(oxyethylene-g-ethyleneimines)

The comb shape graft copolymers can have a maximum of ∿13 branches for PEI 6 to ∿3000 branches for PEI-1000. The intrinsic viscosity of these graft copolymers are shown in Tables 2 and 3. Their values show a more complicated situation. There is little difference between the graft copolymers with 350, 550, and 750 molecular weight PEGME side chains. However, a radical difference is seen for the graft copolymers prepared from larger PEGME segments (1,900 and 5,000). All these graft copolymers were found to form gel in common solvents such as chloroform, chlorobenzene, DMSO, and water. No gel formation was observed when high molecular weight polyoxyethylene solutions were mixed with PEI solutions, suggesting that the gel formation is an unique property of the graft copolymer. Insoluble complexes are formed when aqueous solution of poly(acrylic acid) is mixed with poly(vinylbenzo-(8-crown 16) or polyvinylbenzoglymes (18). Complex formation has been attributed to be the results of hydrogen-bondings between the unionized carboxyl groups and crown ether- or glyme-oxygen atoms as well as that of the hydrophobic interactions. Apparently intermolecular hydrogen-bondings between the urethane and urea groups with the polyether-oxygen atoms in our graft copolymers are sufficiently strong in the cases of long polyether graft segments containing copolymers to cause gel formation but not in the cases of short graft segments containing copolymers. It is also interesting to note that the observed viscosities of the graft copolymers are in general lower in benzene than in water. The hydrophilic copolymers are solvated better in water due to hydrogen-bondings than are in benzene. The conformation of copolymers are thus looser in water than in benzene resulting in higher observed viscosities in water than in benzene.

Table 2. Intrinsic Viscosities of Comb Shape Poly(oxyethylene-g-ethyleneimines) in Water at R.T.

PEI Backbone $\overline{MN}$	Max No. branches	[η] 350	550	750	1900	5000
PEI 6 600	13	0.083	0.069	0.074	gel	gel
PEI L8 1,800	30	0.072	0.052	0.061	gel	gel
PEI 200 2×10^4	500	0.077	0.072	0.064	gel	gel
PEI 1,000 10^5	3,000	0.301	0.328	0.152	gel	gel

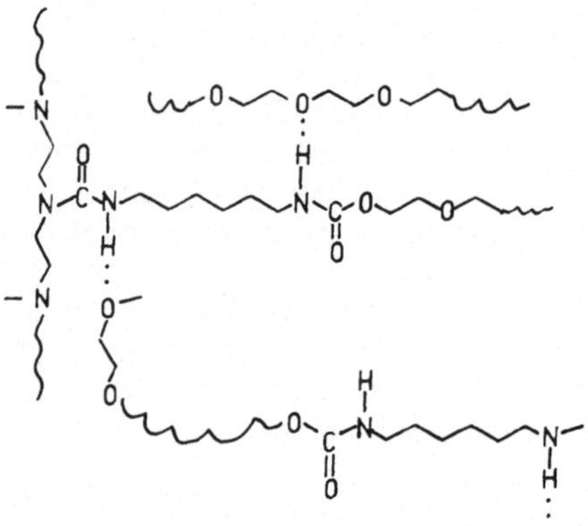

Poly(oxyethylene-g-ethyleneimine) gels

Table 3.   Intrinsic Viscosities of Comb Shape Poly-
(oxyethylene-g-ethyleneimines) in Benzene.   R.T.

PEI Backbone	$\bar{M}n$ of PEGME Graft Segment [η]		
	350	550	750
PEI 6	0.062	0.053	0.066
PEI 18	0.058	0.054	0.067
PEI 200	0.051	0.061	0.064
PEI 1000	0.100	0.114	0.082

The ability of polyethers to act as phase transfer catalysts
is behind many of the current studies on polyethers, especially
those containing cyclic (crown) and branched (graft) structures.
The abilities of the graft copolymers to act as phase transfer
catalysts were examined by studying the transferring of picrate
salt from aqueous solution to chloroform and methylene chloride
solutions of the graft copolymers.   Results of these experiments
are listed in Tables 4 and 5.   The picric acid concentration was
held constant while the amount of polymer in the organic layer was
varied.   Three relationships were investigated.   The first comparison is between graft copolymers which have the same length
graft, 750, but vary in the size of the backbone chain and hence
the amount of branching.   The complexing ability of these three
graft copolymers is found not to differ appreciably.   Secondly,
when these branched poly(oxyethylenes) are compared to the linear
PEGME 750 it appears that the linear system has similar complexing
ability.   The third relationship which was explored was the complexing ability of graft copolymers which had the same backbone
but differed in the size of the side graft.   The three graft copolymers investigated showed that as the size of the side graft
decreases so does the copolymers complexing ability.   There is no
observed selectivities for extraction between potassium and sodium
picrates.

Further investigations into the effect of the size of the side
chains on polymer properties as well as the ability of these
branched poly(oxyethylenes) to complex other metal ions is currently
in progress.

Table 4. Extraction of Potassium Picrate into CH_2Cl_2
layer by Poly(oxyethylene-g-ethyleneimines)[a]

% Picrate in CH_2Cl_2 layer
Grams of Polymer/ml.

Polymer	0.1	0.05	0.02	0.01	0.005	0.001
PEG750-g- PEI 6	97	96	76	67	48	16
PEG750-g- PEI 18	96	93	77	64	49	20
PEG750-g- PEI 200	96	96	76	62	43	14
PEGME 750	96	96	88	77	62	28
PEG550-g- PEI 18	96	92	73	63	44	14
PEG350-g- PEI 18	90	85	69	49	34	12

[a]10 ml of aq. $7x10^{-5}$M potassium picrate was shaken with 10ml of
CH_2Cl_2 solution of graft copolymer. The amounts of picrate
in the CH_2Cl_2 layers after equilibrium were determined by a
Carry 17D Spectrophotometer at 375 nm.

Table 5. Extraction of Sodium Picrate into CH_2Cl_2.
Layer by Poly(oxyethylene-g-ethyleneimines)[a]

Grams of Polymer/ml.

Polymer	0.1	0.05	0.02	0.01	0.005	0.001
PEG750-g- PEI 18	100	97	84	69	50	18
PEG550-g- PEI 18	100	100	82	63	48	28
PEG350-g- PEI 18	90	86	66	47.	32	12
PEG750	100	100	85	72	57	27

[a]10 ml of aq. $7x10^{-5}$M sodium picrate was shaken with 10 ml of
CH_2Cl_2 solution of graftcopolymer. The amounts of picrate
in the CH_2Cl_2 layers after equilibrium were determined by
a Carry 17D Spectrophotometer at 375 nm.

CONCLUSIONS

Poly(oxyethylene-g-ethyleneimines) have been prepared successfully from poly(ethylene glycol monomethyl ethers), 1,6-diisocy-
anatohexane, and polyethyleneimines. These graft copolymers
contains various structures varying from linear, star shape to
comb shape. Polymers with long graft segments form gels in various
solvents. All graft copolymers were found to be efficient complexing and phase transfer agents for metal ions. Potential
application of those polymers as viscosifier and blending agent
for polymers are being explored.

ACKNOWLEDGEMENT

The authors thank the National Science Foundation (Grant
DMR8013689) for support of this research.

REFERENCES

1.   V. Stanett and T. Memetea, J. Polym. Sci.: Polym. Symp.
     64:57 (1978).
2.   J. P. Kennedy, J. Polym. Sci.: Polym. Symp. 64:117 (1978).
3.   J. P. Kennedy, in "Recent Advances in Polymer Blends Grafts
     and Blocks", L. H. Sperling, Ed., Plenum, New York, p. 3.
4.   H. A. J. Battaerd and G. W. Tregear, "Graft Copolymers",
     Interscience, New York, 1967.
5.   W. J. Burland, "Block and Graft Copolymers, "Reinhold,
     New York, 1960.
6.   R. J. Ceresa, "Block and Graft Copolymers", Butterworths,
     London, 1962.
7.   R. J. Ceresa, Ed., "Block and Graft Copolymers" Wiley,
     New York, Vol. I, 1973; Vol. 2, 1976.
8.   J. P. Kennedy, Ed., "Cationic Graft Copolymerization",
     J. Appl. Polym. Sci., Appl. Polym. Symp., 30 (1977).
9.   A. A. Blumberg and S. S. Pollack, J. Polym. Sci. A,
     2:2499 (1964).
10. R. D. Lundberg, F. E. Bailey, and R. W. Callard, J. Polym.
     Sci. A-1, 4:1563 (1966).
11. D. E. Fenton, J. M. Parker, and P. V. Wright, Polymer,
     14:589 (1973).
12. S. Yanagida, K. Takahashi, M. Okahara. Bull. Chem. Soc.
     Japan, 50:1386 (1977).
13. C. C. Lee and P. V. Wright, Polymer 23:681 (1982).
14. C. J. Pedersen, J. Am. Chem. Soc., 89:7017 (1967).
15. C. J. Pedersen, J. Am. Chem. Soc., 92:389 (1970).
16. F. E. Bailey, Jr., R. D. Lundberg, and R. W. Callard,
     J. Polym. Sci. A, 2:84 (1964).

17. Y. Osada and M. Sato, J. Polym. Sci.: Polym. Lett.
     Ed., 14:129 (1976).
18. G. D. Jaycox, R. Sinta, and J. Smid, J. Polym. Sci.:
     Polym. Chem. Ed., 20:1629 (1982).
19. S. C. Lin, P. Beahan, D. Hull, Proc. 8th Annual Meeting
     of Soc. Biomaterials, 86 (1982).
20. H.-B.-Gia, R. Jerome, and Ph.Teyssie, J. Polym. Sci.:
     Polym. Chem. Ed., 18:3483 (1980).
21. L. J. Mathias, J. B. Canterberry and M. South, Proc.
     IUPAE 28th Macromol. Symp. 212 (1982).
22. L. J. Mathias and J. B. Canterberry, "Cyclopolymer-
     ization and Polymers with Chain-Ring Structures", G.
     Butler and J. E. Kresta, Eds. ACS Symp. Series 195,
     (1982) pp. 139-148.

POLY[AZOALKYLENE-N,N'-DIOXIDES], 2.α-4-NITROISOPROPYL-IDENEBICYCLOHEXYL-ω-NITROPOLY[AZO-1,4-CYCLOHEXYLENE-ISOPROPYLIDENE-1,4-CYCLOHEXYLENE-N,N'-DIOXIDE] AND α-4-NITROBICYCLOHEXYL-ω-NITROPOLY[AZO-4,4'-BICYCLOHEX-YLENE-N,N'-DIOXIDE].

D.K. Dandge and L.G. Donaruma*

Departments of Chemistry, New Mexico Institute of Mining and Technology, Socorro, NM 87801 and Polytechnic Institute of New York Brooklyn, NY 11201 and the New Mexico Petroleum Recovery Research Center, Socorro, NM 87801

INTRODUCTION

Simple nitroso groups when attached to carbon show a distinctive tendency towards dimerization.

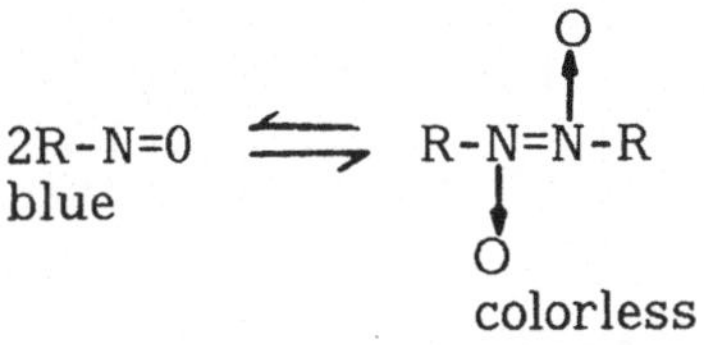

$$2R\text{-}N\text{=}0 \rightleftharpoons \underset{\underset{O}{\downarrow}}{\overset{\overset{O}{\uparrow}}{R\text{-}N\text{=}N\text{-}R}}$$

blue   colorless

However, stability of a dimer relative to its monomeric form depends on a variety of factors, which have been reviewed in an earlier paper.[1] The same paper also described a novel method of catenation exemplified by the preparation of an oligomeric nitrosoalkane from 1,4-dinitrosocylohexane. Thus, bifunctional nitrosoalkanes with suitable structural features may have the potential to form poly[nitrosoalkanes] by the dimerization reaction. It was the objective of the present work to attempt poly-[nitrosoalkane] formation by using different bifunctional nitro-soalkanes and further try to obtain polymers of higher molecular weight.

RESULTS AND DISCUSSION

Scheme 1 represents the synthetic sequence followed to prepare polymers 11A and 11B. Dinitroalkanes 6A and 6B were the isolable precursors since dinitrosoalkanes 10A and 10B would be generated in situ and the final products would be polymeric. The following criteria thought important and essential to preparing a stable polymeric dinitrosoalkane[1] were considered in the selection of precursors for 10A and 10B. Both 10A and 10B would have (i) no nitroso groups bonded to aromatic nuclei, (ii) no electronegative or inordinately bulky substituents bonded to the nitroso bearing carbon atoms, and (iii) the nitroso functions located in such a fashion that a five or six membered ring would not form via dimeriztion.

As is obvious from Scheme 1, to obtain dinitroalkanes 6A and 6B, a sequence of reactions was followed. The intermediate compounds obtained in this process were known compounds. These were characterized by means of physical constants and infrared (IR) and proton magentic resonance (PMR) spectroscopy. Compounds 6A, 6B, 7A, 8A and 8B were found to be new compounds.

Peroxytrifluoroacetic acid oxidation of 5A yielded mainly a pure isomer of 6A. An isomeric mixture having a broad melting range (110-145°C) also was obtained. The isomers were not separated at this point although the stereochemistry might influence the nature of the ultimate product, a poly[nitrosoalkane]. 7A also was obtained from oxidation of 5A. Its melting point had a wide range which could be due to the presence of stereoisomers in the product. 7A on oximation by the procedure of Campbell[2] gave 8A. 8A was characterized by IR and PMR spectroscopy and elemental analysis. The hydrolysis[3] of 8A back to 7A provided additional proof for the structures of both compounds.

It was noticed that when 5B was oxidized using much less than stoichiometric quantities of oxidizing agent, a high melting isomer of 6B resulted, although a comparatively large amount of unreacted 5B was recovered. In contrast, use of stoichiometric or a slight excess of oxidizing agent produced predominantly a low melting isomer of 6B, along with some 8B which were obtained by column chromatography. We however, did not succeed in obtaining 7B as one of the products of oxidation. The IR spectra of both isomers of 6B were identical except for a sharp additional peak at 840 cm[1] in the case of the low melting isomer.

Reduction of nitro compounds with zinc and ammonium chloride frequently is known to yield the corresponding hydroxylamines.[4] When this reduction procedure was applied to 6A and

SCHEME 1

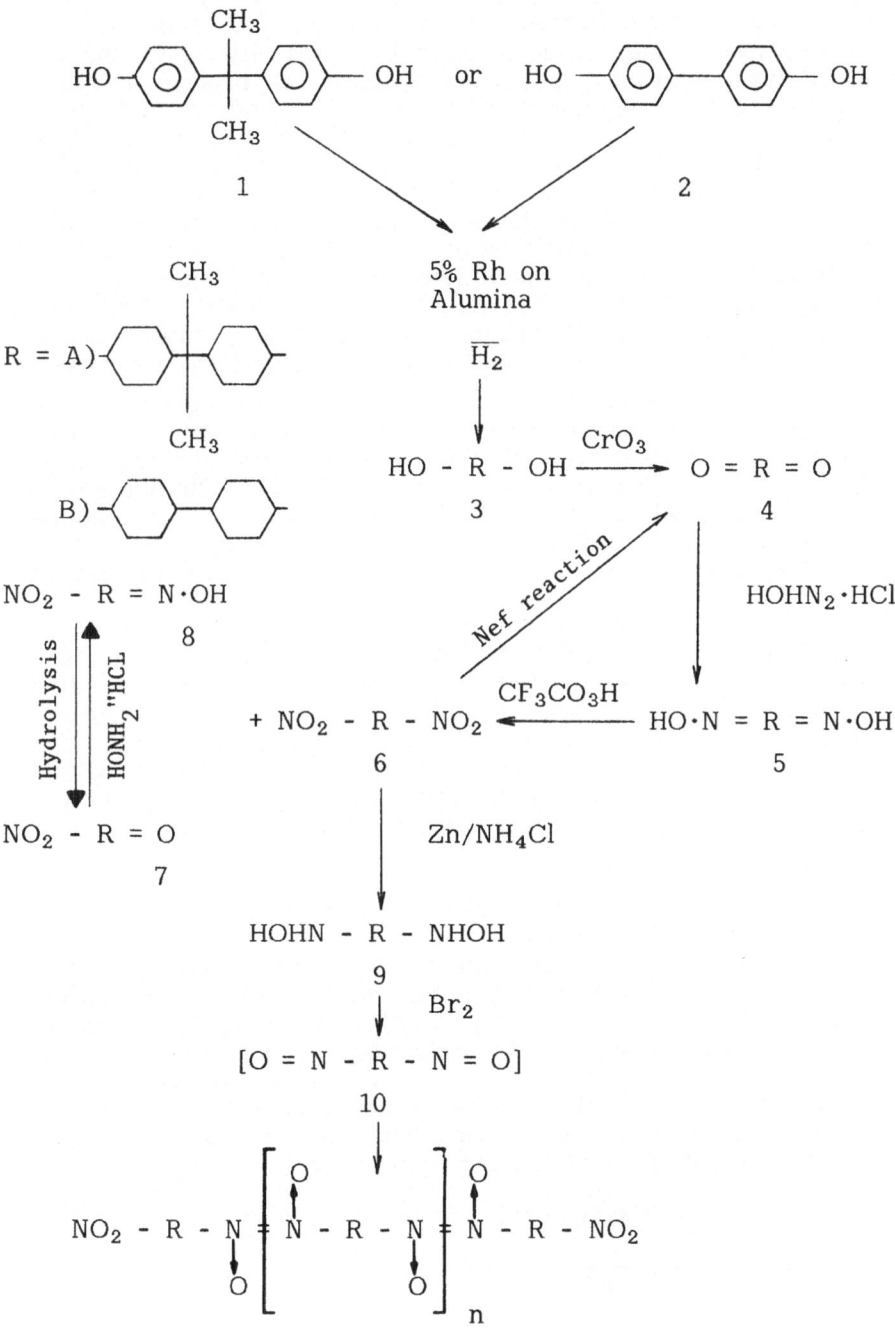

6B, products giving positive Tollen's Reagent tests were obtained. No attempts were made to isolate either 9A or 9B since previous work[1] had indicated that such compounds were subject to decomposition, and bromine oxidation was attempted on the crude reaction mixture. For both systems, as bromine addition was carried out, a transient blue color was noticed and white precipitates appeared in the reaction mixtures. In order to obtain 11A and 11B in purer form, the precipitated crude polymer was subjected to Soxhlet extraction with ethanol. This was followed by azeotropic distillation of benzene over the insoluble products. These two purification steps removed most of the non-polymeric by-products including unreacted or partially reacted nitro compounds as well as water and ethanol. Both 11A and 11B showed strong IR absorption at 1200 cm^{-1} characteristic of the <u>trans</u> nitroso dimer bond,[5] and a medium sharp absorption at 1570 cm^{-1} indicating the presence of the nitro group. The latter function also was indicated by positive blue tests[6] for both polymers. The quantitative alkaline hydrolysis pattern of a nitro terminated poly[nitrosoalkane] was established by Childress and Donaruma,[1] and when 11A and 11B were subjected to alkaline hydrolysis, we obtained oximes 5 and 8 as expected. (see Scheme 2), which were identified by infrared and ultraviolet spectroscopy.

The infrared data for 11A and 11B, the blue test and identification of hydrolysis products confirmed the presence of nitro end groups and the repeat units in the polymers. Scheme 2 infers how the nitro function containing end groups may have acted as capping agents thereby restricting the growth of the polymer chain. Apparently, the presence of nitro groups in the polymers was the result of partial reduction of 6 as shown in Scheme 2.

The determination of degree of polymerization was done following the procedure of Childress and Donaruma[1], which is an indirect method of end group analysis yelding number average molecular weights. Both 11A and 11B were insoluble in common organic solvents thereby making it difficult to apply the usual methods of molecular weight determination. Since it was already established that 5 and 8 were the products of isomerization, it was possible to obtain the degree of polymerization by knowing the relative amounts of each oxime produced (see Scheme 2) by solving the following equations simultaneously:

$$M_a X + M_b Y = C$$

$$E_a X + E_b Y = A$$

SCHEME 2

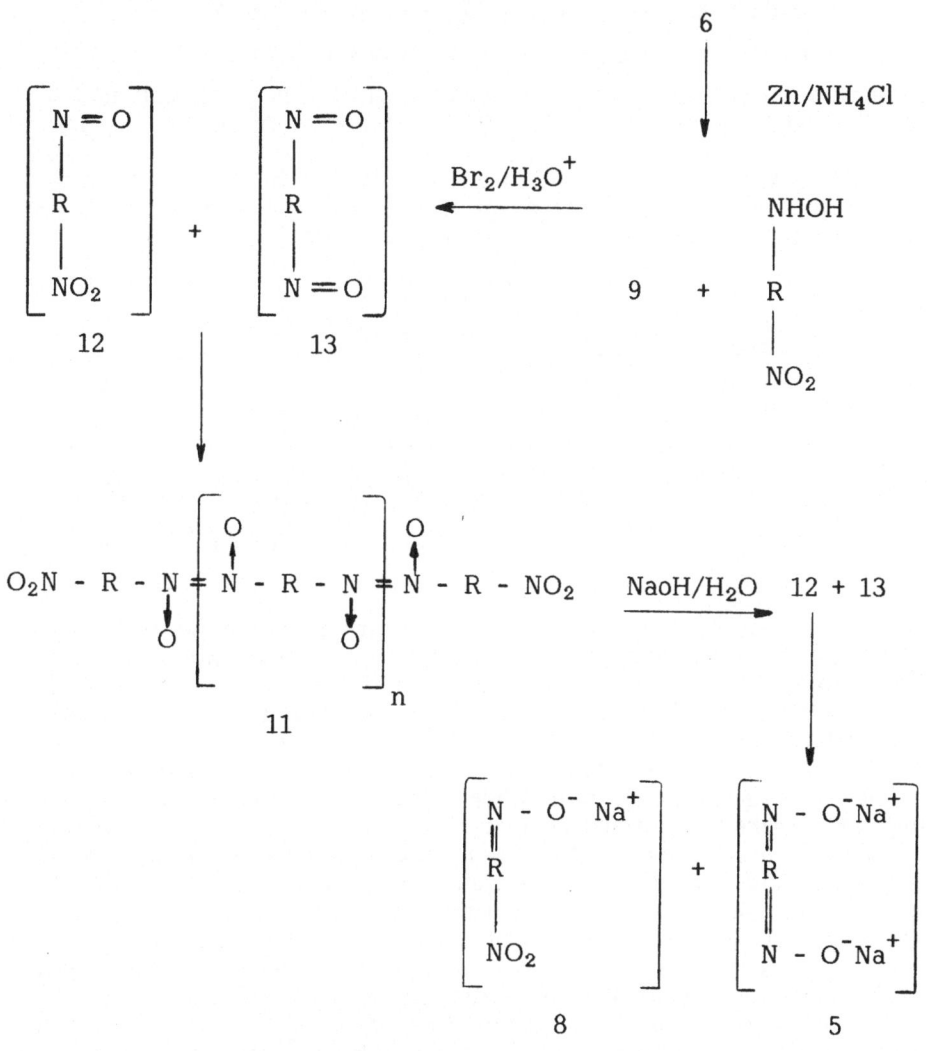

where X = molar concentration of 8, Y = molar concentration of 5, M_a = molecular weight of 8, M_b = molecular weight of 5, E_a = molar extinction coefficient of 8 at wave length λ, E_b = molar extinction coefficient of 5 at wavelength λ, A = absorbance of solution at wavelength λ, and C = concentration of 11, in g/L.

Once values of X and Y were obtained, the degree of polymerization (n) was expressed by n = 2Y/X. The data on molecular weight determination are shown in Table 1.

TABLE 1

Polymer	Conc. of Polymer g/L	Absorbance A at λ=232 nm	E_a	E_b	n	Mol.wt. M_n
11A	0.05	1.120	17,522	5699	87.38	23,243
11B	0.05	1.447	16,345	4682	10.48	2,240

The most probable reason for the higher molecular weight of 11A was the presence of large amounts of bifunctional nitrosocycloalkane(10A) on bromine oxidation. This in turn was the result of reduction of dinitrocompound 6A giving higher yields of 9A. The relative viscosity of 11A was determined in conc. H_2SO_4 at 25°C and found to be 1.03. A part of the sulfuric acid solution of 11A was immediately neutralized and a yellowish powder showing an infrared spectrum comparable to 11A prior to solution was obtained. Also, its melting point remained unchanged. When the acid solution of 11A was neutralized after standing for several days, a dark brown powder was obtained, the infrared spectrum of which showed some changes compared to that of 11A. Two prominent peaks had appeared at 3400-3250 cm^{-1} and 1720 cm^{-1} indicating the presence of oxime and ketone in the recovered product and the intensities of peaks due to nitro and nitroso-dimer functions at 1560 and 1200 cm^{-1} respectively had diminished some. These observations suggested that protonation only was occurring at first, followed by degradation upon prolonged exposure of polymer to concentrated sulfuric acid. This degradation probably was in the form of dissociation of dimeric linkages into monomeric nitroso species which under the strongly acidic conditions employed, caused isomerization to oximino groups some of which then were hydrolyzed to carbonyl groups in the acid medium.

In conclusion, we have confirmed earlier work[1] which indicated that bifunctional C-nitroso compounds might form stable polymers. Further we succeeded in preparing a higher molecular weight polymer (11A). 11B was obtained only as an oligomer, yet its degree of polymerization was greater than that of the first oligomer prepared[1]. Clearly, a new class of polymers has been shown to exist.

EXPERIMENTAL SECTION

Bisphenol A(1) and 4,4'-dihydroxydiphenyl(2) were purchased from Pfaltz and Bauer Company. All chemicals were of reagent grade and were used without further purification. The ethanol used in ultraviolet measurements was of "photrex" grade from Sargent-Welch. Melting points were taken on hot stage and are uncorrected. Infrared spectra were recorded on a Perkin Elmer model 710B spectrophotometer using both nujol and hexa-chloro-1,3-butadiene mulls. PMR spectra were obtained on a Varian EM 360A spectrometer with tetramethylsilane as an internal standard. Elemental analyses were performed by Galbraith Laboratories, Knoxville, Tenn. Ultraviolet measurements were done on a Bausch and Lomb spectronic 21 spectrophotometer.

Synthetic procedural details in each step, enroute to polymers 11A and 11B in Scheme 1 were essentially the same. Catalytic hydrogenation of aromatic nuclei to give cycloalkyl-diol(3) was achieved by the procedure of Meyer.[7]

Oxidation of diol to yield dione(4) was performed as described by Wilds[8] and subsequent oximation of dione to dioxime(5) done following the procedure of Campbell.[2] The dinitro compounds were prepared according to the procedure of Nielsen.[9]

4,4'-Dinitroisopropylidenebicyclohexane(6A)

The crude product obtained from peroxytrifluoroacetic acid oxidation[9] of 5A was recrystallized from 95% ethanol. A product (47%) melting in the range of 110-150°C was obtained. Such a wide melting range, the absence of a carbonyl band at 1720 cm^{-1} and oximino bands at 1720 cm^{-1} and oximino bands at 3300 and 1675 cm^{-1}, and the presence of nitro absorbance at 1570 cm^{-1} in the IR spectrum suggested that the product was a mixture of isomers of 6A. This was supported by PMR spectroscopy which was in very good agreement with the structure proposed for 6A. Several recrystallizations of this product gave a pure isomer, mp 151-53°C. Anal. Calcd. for $C_{15}H_{26}N_2O_4$: C, 60.40; H, 8.72; N, 9.39. Found: C, 60.66; H, 8.85; N, 9.17. υ max 2960s (doublet), 2880m, 1575s. 1525m, 1475w, 1455m, 1390s, 1375m, 1350m, 1320w, 1085w, 1050w, 1020w, 920m, 885w, 810w, 795w, cm^{-1}, absorbances - s = strong, m = medium, w = weak. δ(CDCl$_3$) 4.3 (2H, protons on C attached to - NO$_2$ group) 2.28-0.7 (18H broad cycloalkyl + 6H, methyl protons).

The Nef reaction[10] was used to characterize 6A by dissolving 0.1g (0.0005 mol) of 6A in 5 ml of 1N KOH solution (9:1,
water: ethanol) and after stirring at room temperature for 1
hour, the pH of the solution was adjusted to 7 with 50% HCl. A
yellowish white solid (65%) precipitated out and was filtered,
dried and recrystallized from ethanol-water. m.p. 140-46°C. 4A
obtained from oxidation of 3A (see Scheme 1) melted at 146-48°C.
There was no depression in the melting point of a mixture of the
two samples. The infrared spectrum was identical to that of 4A
prepared by oxidation of 3A.

2-(4-nitrocyclohexyl)-2-(4-oxocyclohexyl)propane(7A)

Residues obtained from the filtrates of multiple recrystallizations of 6A were combined to give a gummy solid. This was
subjected to column chromatographic separation on silica gel with
chloroform and ethanol as successive eluants.

After separation of residual quantities of 6A in the first few
fractions, 7A was recovered from the subsequent fractions (6%).
The mixture of isomers had m.p. 90-125°C; υ max 2970s, 2885s,
1735s, 1570s, 1455m, 1430m, 1390m, 1375m, 1340m, 1320w, 1300w,
1270w, 1220w, 1200w, 1180m, 1080w, 1060w, 1000w, 960m, 915m,
860w, 820w, 790w, 760w cm^{-1}; $\delta(CDCl_3)$ 4.20 (1H, proton on
carbon attached to $-NO_2$ gr), 2.2-0.67 (18H, broad cycloalkyl
protons, + 6H, methyl protons).

2-(4-nitrocyclohexyl)-2-(4-oximinocyclohexyl)propane(8A)

7A was oximated according to the procedure of Campbell.[2]
The nitrooxxime (50%), 8A, was recrystallized from 95% ethanol.
The mixture of isomers had m.p. 170-190°. Anal. Calcd. for:
$C_{15}H_{26}N_2O_3$; C, 63.82; H, 9.18; N, 9.89. Found: C, 63.56;
H, 9.30; N, 9.69. υ max 3300-3150m, 2950s, 2890m, 1680w,
1580s, 1560s, 1470m, 1450m, 1390m, 1305w, 1280w, 1250w, 1165w,
1005m, 980m, 960m, 925m cm^{-1}; $\delta(DMSOd_6)$ 4.4 (1H, proton on
carbon attached to $-NO_2$ group), 3.23 (1H, Oximino proton),
2.5-0.67 (18H, broad cycloalkyl protons + 6H, methyl protons).
8A was hydrolyzed to 7A in conc. HCl following the procedure of
Vogel.[3] The product of hydrolysis resembled 7A in all respects,
m.p. 88-110°C. IR spectra of the product showed the absence
of oximino peaks at 3300 and 1680 cm^{-1} and the presence of keto
and nitro frequencies at 1720 and 1570-1550 (doublet) cm^{-1}
respectively and was identical in all respects to the infrared
spectrum of 7A obtained as described above by column chromatography.

4,4'-Dinitrobicyclohexane(6B)

Two of the several possible isomers of this compound were obtained after oxidation of 5B following the procedure of Nielsen.[9]

Low melting isomer, m.p. 80-90°. Anal. Calcd. for: $C_{12}H_{20}$ N_2O_4: C, 56.25; H, 7.81; N, 10.93. Found: C, 56.23; H, 8.07; N, 10.78. υ max 2970m, 2875m, 1540m, 1450m, 1390s, 1350m, 1315s, 1265w, 1240w, 1214w, 1200w, 1190w, 1180w, 1080m, 1000m, 915m, 860m, 840w, 820w cm^{-1}; δ(CDCl$_3$): 4.43 (2H, protons on carbon attached to nitro group), 2.28-1.2 (18H, broad cycloalkyl protons).

High melting isomer m.p. 180-85°C. Anal. Found: C, 56.14; H, 7.89; N, 10.79. The IR spectrum resembled that of the low melting isomer except for the absence of peaks at 1200, 1195 and 860 cm^{-1}; δ(CDCl$_3$) 4.24 (2H, protons on carbon attached to nitro group), 2.15-1.0 (18H, broad cycloalkyl protons). The Nef reaction[10] was used to characterize both isomers of 6B by following the procedure described earlier. IR and PMR spectra of the reaction products (60%) were identical to an authentic sample of 4B, m.p. 118-119°C.

4-(4-nitrocyclohexyl)cyclohexanone oxime(8B)

Residues obtained from filtrates of multiple recrystallizations of 6B were combined and subjected to column chromatographic separation with silica gel as adsorbent and chloroform and ethanol, respectively, as eluants. Residual quantities of 6B separated out in the first fraction followed by a compound showing the presence of both carbonyl and nitro absorption frequencies in the IR spectrum. This was followed by 8B (0.5%) in the pure form. The mixture of isomers had m.p. 155-60°C. Anal. Calcd. for: $C_{12}H_{20}N_2O_3$ C, 60.00; H, 8.33; N, 11.66. Found: C, 60.08; H, 8.57; N, 11.45. υ max cm^{-1}: 3300-3250m, 2960m, 2900m, 1680w, 1550s, 1460m(doublet), 1440m, 1390m, 1345w, 1310w, 1280w, 1255w, 1215w, 1110w, 1000w, 960w, 945m, 920m cm^{-1}; δ(CDCl$_3$): 4.25 (1H, proton on carbon attached to nitro group) 3.55-3.13 (1H, broad, = N-OH), 2.12-1.12 (18H, broad cycloalkyl protons).

Polymer Preparation

In a typical experiment, 1.0g of dinitrocycloalkane(6) and 1.0g of ammonium chloride were stirred in 80 ml. of 50% aqueous ethanol and the mixture heated under reflux with stirring until all solid material had dissolved. If necessary, additional alcohol was used to dissolve the solids. While heating and stirring continued, 1.0g of zinc dust was added slowly over 4-5 minutes

and the mixture refluxed gently for 30 minutes. The zinc dust
was removed and the filtrate acidified with 20% HCl while ex-
ternally cooled with an ice bath. Bromine water (0.01g Br_2/ml)
was added with stirring until the bromine color no longer disappeared. A white precipitate formed, and the mixture was re-
frigerated overnight. The mixture was filtered, dried and
extracted in a Soxhlet extractor with 95% ethanol for 15-20
hours. The precipitate was mixed with 100 ml. of benzene and
half of the benzene was distilled off to remove water and ethanol
azeotropically. Filtration of the mixture gave polymer.

Yields: (from 6) 11A - 23-26%, 11B - 2-3%

m.p.:   11A - 170°C becomes yellow

             210-220°C melts to reddish brown liquid

             11B - 180°C turns brown

             215°C turns light yellow

             236-40°C decomposes giving black residue.

11A Calcd. for: $C_{1335}H_{2314}N_{178}O_{180}$: C, 67.54; H, 9.83; N,
10.50. Found: C, 67.03; H, 10.01; N, 10.03. 11B Anal. Calcd.
for $C_{144}H_{240}N_{24}O_{26}$: C, 63.50; H, 8.88; N, 12.34. Found C,
63.70 H, 8.41 N, 11.90. 11A υ max 3350w, 2950s, 2875s, 1720w,
(doublet) 1550s, 1480w, 1470w, 1450s, 1400s, 1340m, 1300w,
1280w, 1260w, 1200s, 1185w, 1040w, 1020w, 980w, 940w, 920w,
860w, 730s. 11B υ max 2950s, 2875s, 1720m, 1580s, 1550s,
1460m, 1438w, 1420m, 1390m, 1380w, 1340m, 1310m, 1270w,
1260w, 1240w, 1200s, 1080m, 1060m, 1000m, 920m, 820w, 840w,
790w.

The blue test[6] for the presence of nitro group was performed on both 11A and 11B by dissolving a small amount of
polymer in 1.0N aqueous NaOH. The basic solution was then
poured into 2.0N HCl and a transient blue color, indicative of
the presence of nitro group was observed.

Both isomers of 6B yielded the same polymer (11B).

Determination of Molecular Weights

Solutions of various concentrations of dioxime(5) and
Nitrooxime(8) in 0.1M NaOH in 15% aqueous ethanol were prepared. Ultraviolet spectra were determined for each of the
solutions and absorbance vs. concentration curves were established for each compound to determine their respective molar
extinction coefficients. The absorbances of anions(8 and 5) were

additive and non-interfering at 232nm and the solutions obeyed Beer's Law.   Calibration curves for analysis were prepared.

A polymer(11) solution in 0.1M 15% aqueous ethanolic NaOH was made such that the polymer concentration was 0.05 g/L.   To obtain a clear solution it was necessary to heat the mixture at 70-80°C or 2-3 hours.   The spectrum of the clear solution was recorded and the amount of each oxime, 8 and 5, determined. From this, the molecular weight of the polymer was calculated using the equations described above.   Yields of 8 and 5 were quantitative for each polymer.

ACKNOWLEDGMENT

We are indebted to the National Science Foundation for their support of this project under grant number DRM-7818803.

REFERENCES

1. W.L. Childress, and L.G. Donaruma, Macromolecules, 7, 427 (1974).
2. T.W. Campbell, V.S. Foldi, and R.G. Parrish, J. Appl. Polym. Sci., 2 (4), 81 (1959).
3. A.I. Vogel, Textbook of Practical Organic Chemistry, Langmann, London, 1972, p. 1075.
4. O. Kamm in "Organic Syntheses," H. Gilman and A.H. Blatt, Eds. Collect. Vol. I, 2nd Ed. Wiley, New York, 1964, p. 445; G.H. Coleman, C.M. McCloskey, and F.A. Stuart, ibid., E.C. Hornung, Ed., Collect. Vol. III, Wiley, New York, 1955, p. 668.
5. B.G. Gowenlock, H. Spedding, J. Trotman, and D.H. Whiffen, J. Chem. Soc., 3927 (1957).
6. V. Meyer, Ann, 175, 88 (1875); O. Piloty and A. Stock, Ber., 35, 3093 (1902); H.B. Hassand, E.F. Riley, Chem. Revs.; 33, 399 (1943).
7. A.I. Meyers, W.N. Beverung, and R. Gault, Organic Syntheses, 51, 103 (1971).
8. A.L. Wilds, C.H. Shunk, and C.H. Hoffman, J. Am. Chem. Soc., 76, 1733 (1954).
9. A.T. Nielsen, J. Org. Chem., 27, 1993 (1962).
10. J.U. Nef, Ann., 280, 266 (1894).

MACROMOLECULAR DYES - SYNTHETIC STRATEGIES

Smarajit Mitra

Central Research Laboratories
3M Company
St. Paul, MN 55144

INTRODUCTION

Colored macromolecular compounds abound in nature. Many colored resinous materials may be extracted from plant and animal sources, but a large majority of them are chemically complex mixtures, characterization of which remain serious analytical challenges. On the other hand, the syntheses of simpler and more clearly defined structurally colored polymers are of more recent origin, evolving largely from the need to impart hue to textile materials. Common dye molecules have low molecular weights and are soluble in aqueous or organic solvents and are, therefore, susceptible to continuous loss upon prolonged usage by diffusional and leaching processes. When such dye molecules are chemically bound to a macromolecule, be it the actual fibers of the textile or a secondary polymeric vehicle that strongly adheres to the fibers, significant enhancement of the fastness of the dyes result. Subsequent to this observation, many other applications of polymer bound dyes have been documented and some of these results will be described in the following pages.

From a synthetic chemistry standpoint, three general strategies may be seen to have been employed to make polymeric dyes. These are:

A) the synthesis of monomeric dyes which have polymerizable groups, which are subsequently combined by standard polymerization techniques to give macromolecules;

B) the reaction between preformed polymers and dyes, where the two components have mutually reactive functional groups, resulting in chemical bond formation between the polymer and the dye;

C) the synthesis of a polymeric dye precursor which, by appropriate transformations, generate the dye structures on the polymer chain.

Each of these general pathways may in turn have great latitude in terms of the actual chemistries involved in making the polymer bound dye systems. In the following section, these schemes will be examined in closer detail with examples from published literature. This will be followed by brief discussions on some preliminary results from our laboratory.

POLYMERIZABLE DYES

Monomeric dyes may be prepared bearing functional groups which can participate in polymer forming reactions. Such functional groups may either belong to vinyl types which can undergo free radical chain growth polymerization reactions or they may be of one or more classes of reactive sites on the dye which can undergo step growth polycondensation reactions. Several examples of both these polymerization pathways are available in the literature.

Chain Growth Polymerizations

Much of the work in this area has focussed on polymeric dyes which can participate in redox reactions. Early work by Manecke and Kossmehl on vinyl Malachite Green (VMG) is specially significant as it established a simple method of entry into the polymerizable triphenylmethane class of dyes.[1] Thus the carbinol base form (III) of VMG was synthesized[2] by a Grignard reaction between 4-vinyl-phenyl magnesium chloride (I) and Michler's ketone (II).

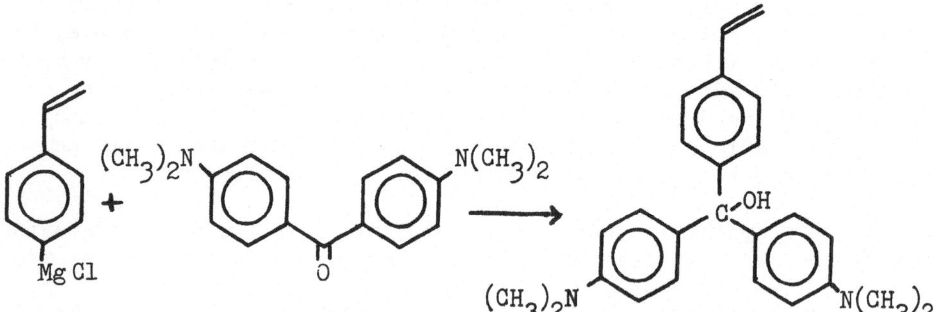

(III) polymerizes readily with free radical initiators like benzoyl peroxide and is also copolymerizable with other vinyl monomers like styrene and divinylbenzene. As anticipated, these materials exhibit properties of a redox polymer.

Phenothiazines are well known redox systems and poly(2-vinyl phenothiazine) (IV) is readily oxidized by agents like ferric chloride or bromine and by oxygen in presence of light.[3] Shigehara and coworkers synthesized the acrylamide derivatives of the thiazine dyes (V), Thionin and Azure B, and copolymerized them with N-vinyl pyrrolidone by free radical methods to give water soluble polymeric dyes.[4]

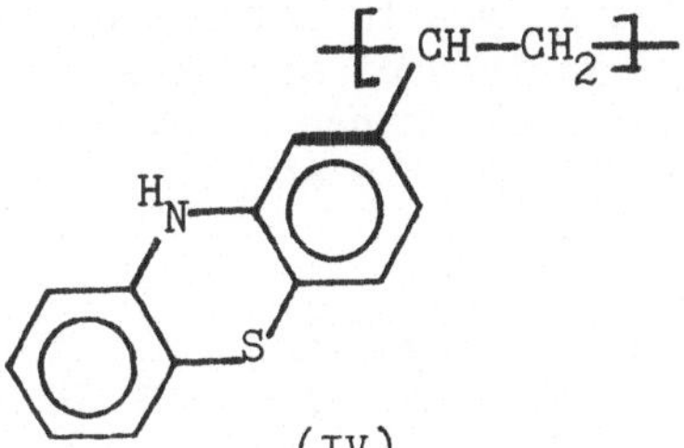

(IV)

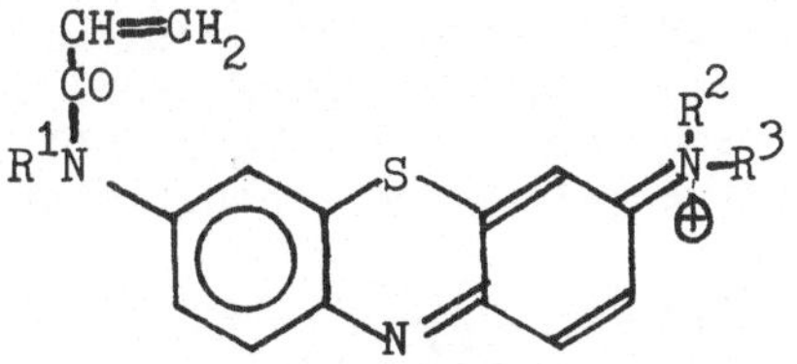

(V)

$$R^1 = R^2 = R^3 = H$$
$$R^1 = R^2 = R^3 = CH_3$$

However, these polymers are found to be photolabile and do not show photoredox reactivity with low valency metal ions, including ferrous salts.

Dye sensitization of photoconductivity has emerged as an area of intense research in the last few years. Okamoto and coworkers have examined the effect of polymeric Malachite Green on the photo-conductivity of polyvinylcarbazole (PVCz) by comparing three systems - PVCz mixed with normal monomeric Malachite Green, a co-polymer of VCz with VMG and mixtures of homopolymers of VCz and VMG.[5] The degree of sensitization in the copolymer was somewhat less than that in which the monomeric dye was present and the sensitization in the homopolymer blend was much poorer. The results led to some reasonable interpretations of the state of aggregation of the dyes in these polymeric forms.

Macromolecular azo dyes which behave as photochromic polyelec-trolytes have been reported[6] where 4-aminoazobenzene dyes (VI and VII) are converted to the corresponding acrylamide and methacryl-amide derivatives and are copolymerized with acrylic and methacrylic acids.

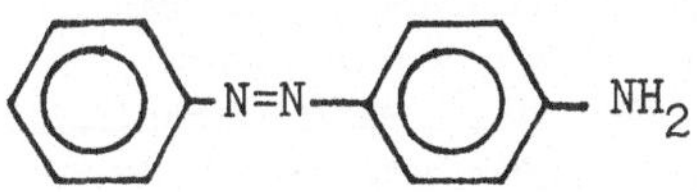

(VI)

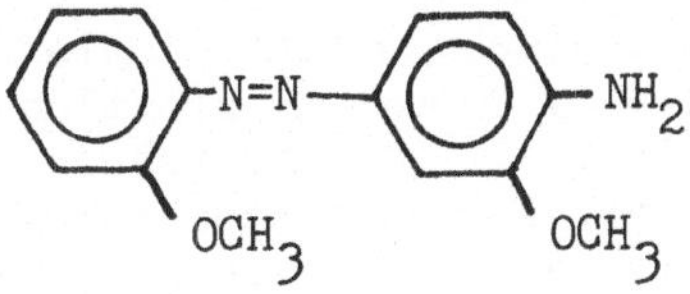

(VII)

Optical behavior of the isomerizing side chains, specially the cis-trans isomerization process of the azo linkage, was compared to those of nonpolymeric model compounds. The irradiation response and dark relaxation of the polymeric dyes are affected by the charge and conformation of the copolyelectrolyte.

Over the last decade, Marechal and coworkers in France have developed elaborate methods of functionalizing dyes to introduce polymerizable unsaturation sites on them. A variety of acrylate and acrylamide derivatives of dyes, covering the whole range of colors, have been reported.[7,8] Thus (VIII) is converted to the acrylate dervative (X) in two synthetic steps:

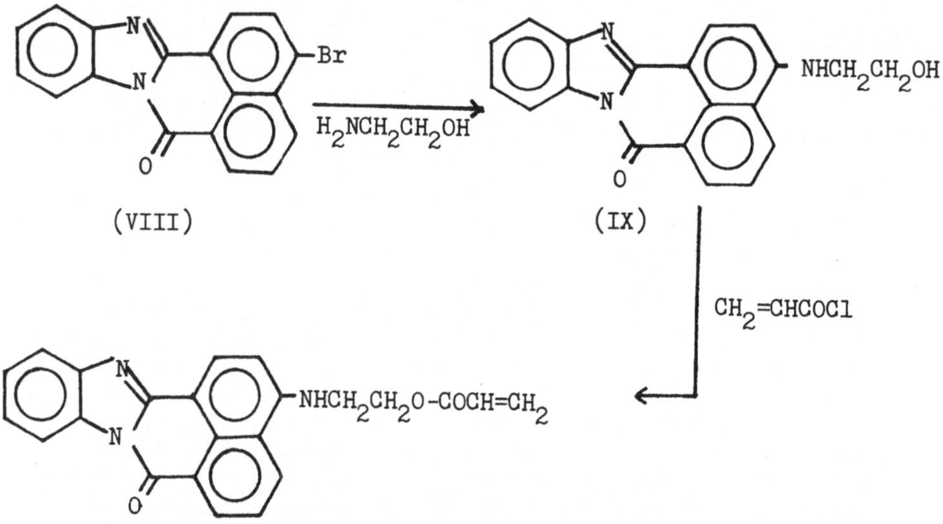

Homo- and copolymers of many of these dye monomers have been prepared followed by extensive patent coverage.

Step Growth Polymerization

Again, the redox properties of the phenazine, oxazine and thiazine dyes were the driving force for the preparation of some of the earliest examples of condensation polymerization of dyes.

Sansoni prepared several samples of polycondensates of resorcinol, formaldehyde and Methylene Blue and tested their redox properties by repeated oxidation reduction cycles,[9] with ammoniacal sodium dithionite as the reductant and acidic hydrogen peroxide as the oxidant. The best resins had a redox capacity of 4.5-5.8 meq of Fe(III) per gram. This novolac type of condensation was later used with azo dyes to give thermosetting colored coating compositions.[10] When dye molecules are available with two condensable functional groups such as amines, alcohols or carboxylic acids etc. the dye may be polycondensed with other difunctional monomers to give polymers. Thus polyesters have been prepared from dihydroxy-azo dyes (XI) to give structurally colored polymers.[11]

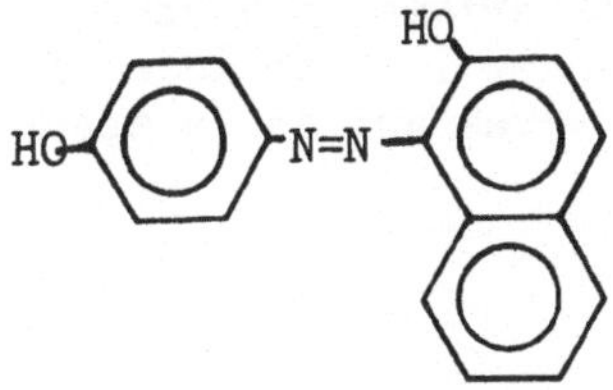

(XI) 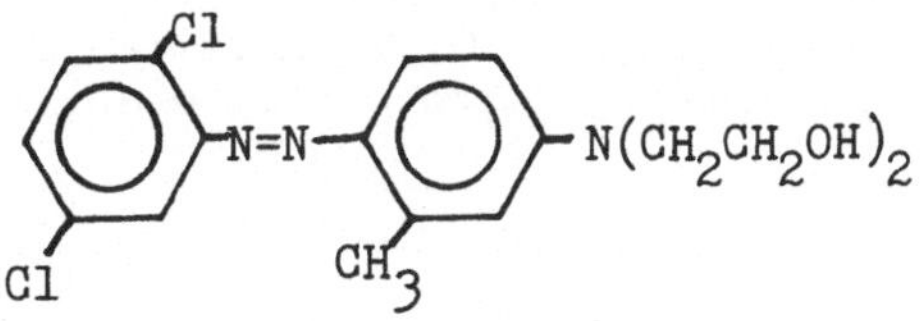   (XII)

Bonnet has described photocurable, colored urethane - acrylate resins from sequential copolymerization of (XII) with diisocyanates and multifunctional acrylates.[12] Polyurethanes (XIII) and poly-ureas (XIV) bearing triphenylmethane and anthraquinone dyes may be used for dyeing of natural and synthetic fibers or as electrophoto-graphic toner materials.[13]

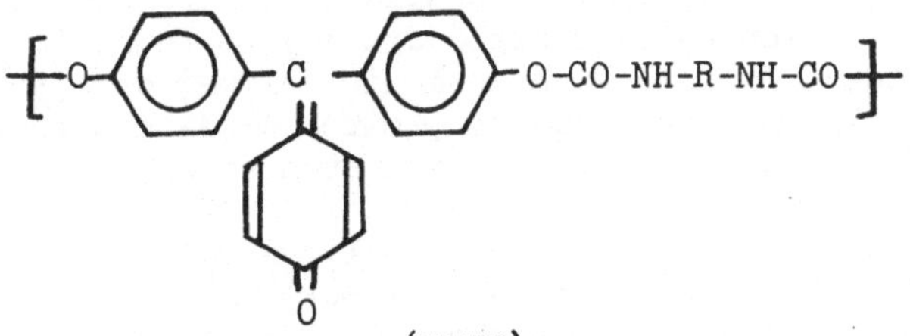

(XIII)

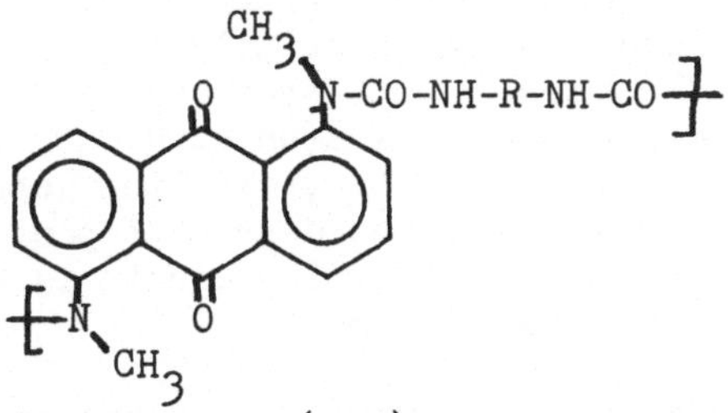

(XIV)

Marechal and coworkers have expanded on this technique of polycon-densation reactions to prepare a host of macromolecular dyes ranging through polyamides[14,15], polyesters[16,17] and polyurethanes.[18] Many of these dye monomers were synthesized in their laboratory introducing polycondensable groups on the dyes.

More recently Carraher et. al. have combined the syntheses of

organometallic polymers with those of polymeric dyes.[19] Using bis-
cyclopentadienyl-titanium dichloride, condensation reactions were
carried out with phenyl sulfonaphthalein dyes, Nigrosine, Indigo
Carmine, Eriochrome Black T, Congo Red etc. by interfacial techniques.
These polymeric dyes impart fluorescence to materials they are
combined with, and retain their fluorescence over long periods of
time. Moreover, the Cp$_2$Ti moieties in the polydyes act as uv absorbers and thereby improve weatherability.

DYE BINDING TO POLYMER

     One of the more common methods of binding dyes to polymers is
through ionic interactions. This may take the form of actual counterion binding between a charged polymer and an oppositely charged dye
or by strong dipolar interactions between polymer and dye, as is
often the case for many textile dyeing processes. Such physically
bound dyes may participate in redox reactions, as is the case for
Dowex 50/Methylene Blue systems where the polymeric dye can be used
for deoxygenating water, chemical reductions and precipitations of
finely divided metals.[20] Where the substrate polymer is not electrically charged, an ionically charged dye may be bound to it by
pretreatment of the substrate with an ionic polymer of opposite
charge. Thus cellulosic fibers may be pretreated with a polyamine
and subsequently dyed with an anionic dye such as Calcocid Blue 2G.[21]

     Covalent binding of dyes to functionalized polymers is also a
viable route. Suitable functionalized polymers that have been
described for such binding include those derived from p-chlomethyl-
styrene, allyl glycidyl ether or N-hydroxymethylacrylamide.[2,4]
Azine and thiazine dyes like Thionine, Azure Blue, Saframine-O and
Neutral Red have been reacted with these polymers through nucleophilic substitution with the amino groups available on the dyes. Photochromic and thermochromic behavior is seen with such polymer films
and the reversibility of these optical properties depend strongly
on the water content of the films. The degree of photobleaching of
the polymeric dyes in presence of reducing agents is generally
smaller than those of monomeric systems.

     A significant impact has been made by researchers at Dynapol
in the area of polymer bound dyes as nonabsorbable food colorants.
The high molecular weights of the colorants prevent their absorption from the gastrointestinal tract and toxicity problems are
avoided.[22,23] The general procedure for preparing many of these
macromolecular dyes is the reaction between a nucleophilic polymer,
preferably polyvinylamine, with bromoanthraquinones,[24] catalyzed by
Cu(I).

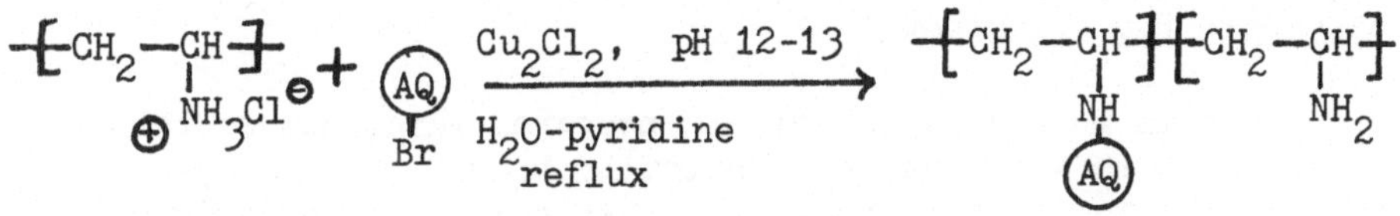

Sulfamation of some of the residual amines with trimethylamine-SO_3 complex extends the water solubility of the polymeric dyes to basic pH ranges. An alternate route to such water soluble polymeric dyes is through the binding of the anthraquinone dyes to copolymers of vinylamine and vinyl sulfonic acid.

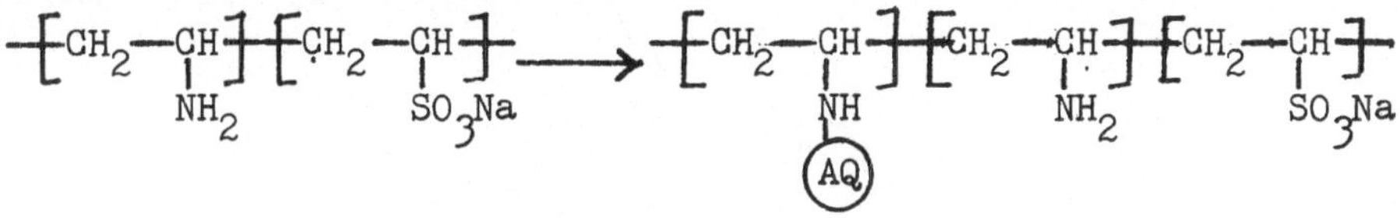

This general scheme of covalent binding of dyes to amino polymers has been extended to other polymer backbones and other dyes.[25]

The binding of dyes to chromatographic column material is finding widespread application in the purification of proteins and enzymes. Thus, Nimmo and Holms describe the use of Cibacron Blue (XV) bound to Sepharose in the purification of isocitrate dehydrogenase from E.coli.[26]

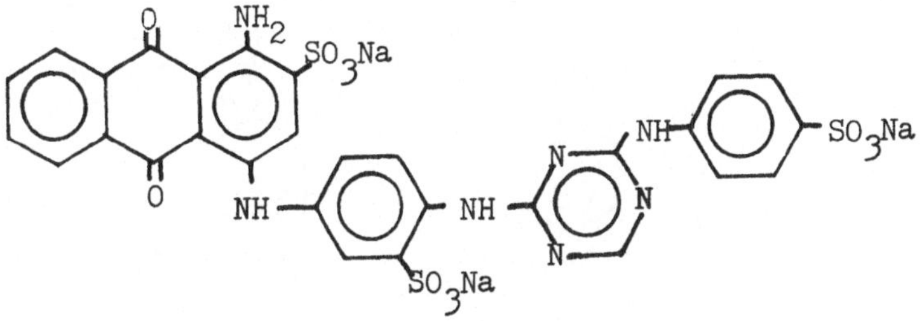

(XV)

The usefulness of these chlorotriazinyl dyes in binding to textiles had been recognized in the 1950's. However, optimal binding procedures of these dyes to polysaccharide matrices were developed more recently.[27] The application of 'dye - ligand chromatography' to protein purification has been reviewed recently.[28]

POLYMERIC DYE PRECURSORS

In this method, a polymer is prepared which has certain structural features which can be chemically transformed to dye structures
in one or more steps. Perhaps the most common class of dyes that
have been so synthesized on a polymer backbone is the azo dyes,
which are simply formed by the coupling of a polymeric aryl diazonium salt with an activated aromatic system. The former are prepared
from polymeric aromatic amines by diazotization. Thus polymers and
copolymers of 2,2'-divinylbenzidine (XVI) have been diazotized and
coupled with sodium naphthionate, sodium anthranilate and sulfanilic
acid to give polymeric azo dyes which act as pH indicators.[29]

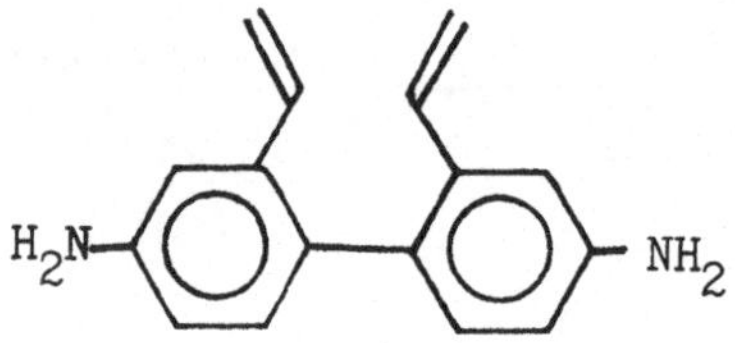

(XVI)

The polyvinylamine prepared by the Dynapol group has been functionalized with p-acetamidobenzene sulfonyl chloride and the sulfonamide
adduct has been diazotized after deacetylation.[30]

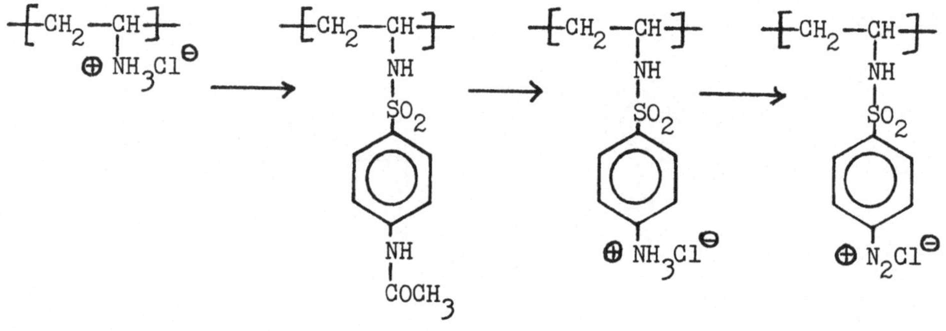

Coupling with various sulfonate salts produced a host of polymeric
azo dyes. Some of these dyes may be used as food colorants.[31] The
polymer may in turn be used as the electron rich aromatic moiety
and coupled with diazonium salts. Polymers of N-ethyl-N-(2-methac-
ryloylethyl)aniline have been so converted and the resulting dyes
act as color pH indicators which may be used in flow through column
processes.[32] Poly(m-N,N-dimethylaminostyrene) has been coupled with

N,N-dimethyl-p-phenylenediamine by the method of Bernthsen[33] to give poly(4-vinyl Methylene Blue).[1] This polymer acts as a hydrogen acceptor in the enzymatic dehydrogenation of ethanol. The carbinol form (XVII) of polymeric Malachite Green may be prepared by the nucleophilic attack of poly(p-lithium styrene) on Michler's ketone;[34] acidification gives the colored dye (XVIII).

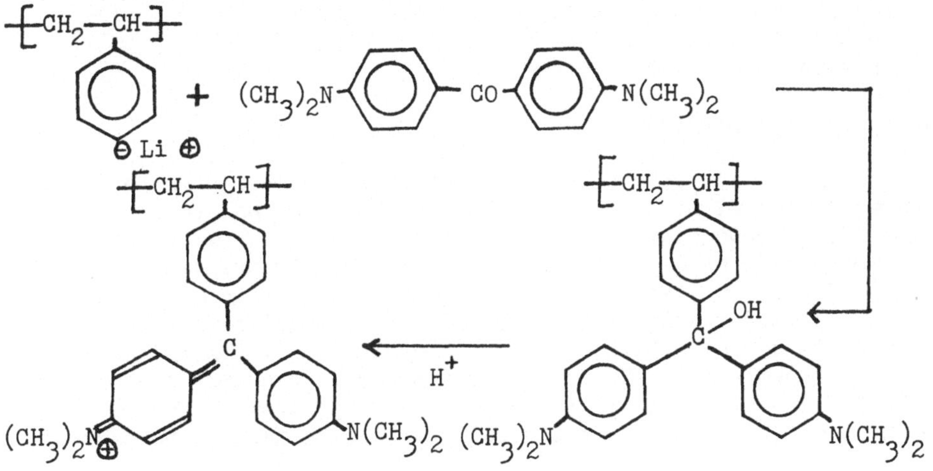

(XVIII)   (XVII)

Spectroscopically the polymer is very similar to p-ethyl Malachite Green and this non-diffusing dye behaves as a polyelectrolyte in methanol.

We have used this third strategy to prepare polymeric triaryl-methane dyes chemically bound to condensation polymer backbones. The experimental details and a discussion of some of our results follow.

EXPERIMENTAL

2,2'-Phenyliminodiethanol (PIE) was obtained from MCB and 2,2'-m-tolyliminodiethanol (TIE) was obtained from Eastman and they were recrystallized from a benzene/hexane mixture. Terephthaloyl chloride (TC), isophthaloyl chloride (IC), toluene-2,4-diisocyanate (TDI), oxalyl chloride (OC) and ethylene glycol (EG) were obtained from Aldrich and purified by distillation. Polyethylene glycol (av. MW 3000-3700) was obtained from MCB. All the aromatic ketones were commercially available and used without purification. All solvents used were purified by distillation over appropriate drying agents.

Synthesis of Polyesters

The polyesters were typically prepared by mixing the diol and the diacid chloride in equivalent amounts (0.2-0.3 M conc. of monomer) in 1,2-dichloroethane and adding an organic tertiary amine (pyridine or triethylamine) as a base catalyst. The mixture is heated under nitrogen on a constant temperature oil bath and after the appropriate time the polymer was recovered by precipitating into methanol, and purifying by reprecipitation from halogenated solvents into methanol or ether.

Synthesis of Copolyesters

A mixture of 3.1 g (50·mmoles) of EG and 9.0g (50 mmoles) of PIE was dissolved in 200 ml of 1,2-dichloroethane and to it was added 20.3g (100mmoles) of TC. The mixture was stirred under nitrogen and 30 ml of pyridine was added. The mixture was heated at 80°C for 6 hrs., and then concentrated under reduced pressure and the polymer precipitated into 1200 ml of methanol. It was purified by reprecipitation from chloroform into methanol and dried under a vacuum. The yield of polymer was 85%. From NMR, the ratio of EG/PIE incorporation in the copolyester appeared to be 3/4. The M_n= 2500 and M_w = 3500 from GPC calibrated with respect to polystyrene.

When a 3/1 ratio of EG/PIE was used in the feed for polymerization, 80% yield of a copolymer with a 2/1 ratio of incorporation of EG/PIE was observed.

Synthesis of Poly(ether-ester) Block Copolymer

The same procedure for the polymerization was used except that polyethylene glycol was used instead of EG. The block copolymer was precipitated in diethyl ether instead of in methanol and was thoroughly extracted with water to remove all pyridinium hydrochloride and unreacted polyethylene glycol. Using a 1:5 molar ratio of the polyethylene glycol to PIE, a polymer was obtained which showed a 1:7 ratio of incorporation of polyethylene glycol/PIE by NMR.

Synthesis of Polyurethane

Equivalent amounts of TIE and TDI (50 mmoles each) were dissolved in an organic solvent like 2-butanone (1.1M conc.) and to this solution was added 0.1g of dibutyltin dilaurate as a catalyst. The mixture was heated under nitrogen at 75°C for 2 hrs. and the polymer was recovered by precipitation into acetone followed by reprecipitation into diethyl ether from chloroform. It was dried under a vacuum to yield a quantitative conversion to polymer. From GPC the M_n = 7,500 and M_w = 46,900, calibrated with respect to polystyrene.

Synthesis of Polymeric Dyes

The polyesters or polyurethanes were dissolved in halogenated solvents like chloroform, chlorobenzene or 1,2-dichloroethane and to the solutions were added equivalent amounts of the appropriate diarylketones and phosphorous oxychloride. The mixtures were heated to reflux for several hours and the solvents were thereafter removed under reduced pressurs. The highly colored solids were purified by precipitation from N-methyl pyrrolidone solutions into acetone. They were filtered, washed and dried under a vacuum. The visible spectra of the polymeric dyes were measured in methanol solutions.

RESULTS AND DISCUSSION

The low temperature polyesterification between diacid chloride and diols in the presence of a tertiary amine has been extensively utilized. The process has been variously called acceptor catalytic[35] or non-equilibrium[36] polyesterification. The case with PIE or TIE is of special interest as the diols themselves have tertiary amino functionalities. The reaction rate of polycondensation between PIE and TC was measured at 80°C, using four equivalents of pyridine as the added base. The results are shown in Figure 1. It is readily observed that under these conditions, polymer yields of 64% are obtained in 15 mins., thereafter which, prolonged heating for 4 hrs. does not increase the yield significantly. This suggests that either chain ends are deactivated by some process to inhibit further growth or an equilibrium is established. Sulzberg and Cotter have noted that under high temperature conditions, intramolecular ring cyclization of the hydroxy chain terminal may cause chain termination through the formation of free carboxylic acid and N-aryl morpholine.[37]

The reduced viscosity for the polymer prepared in 15 mins. was found to be 0.06 dl/g in N-methyl pyrrolidone at 25°C (0.5 g/dl).

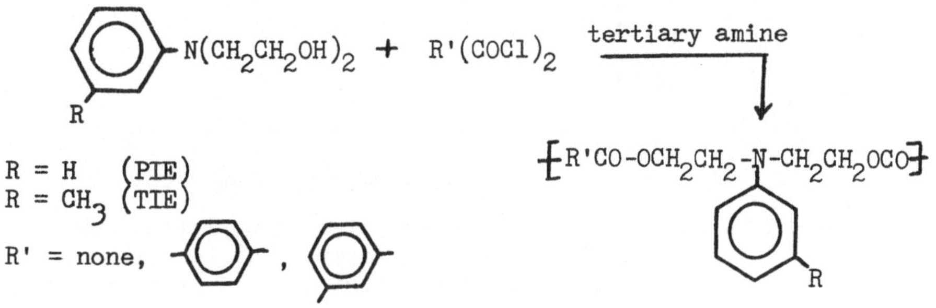

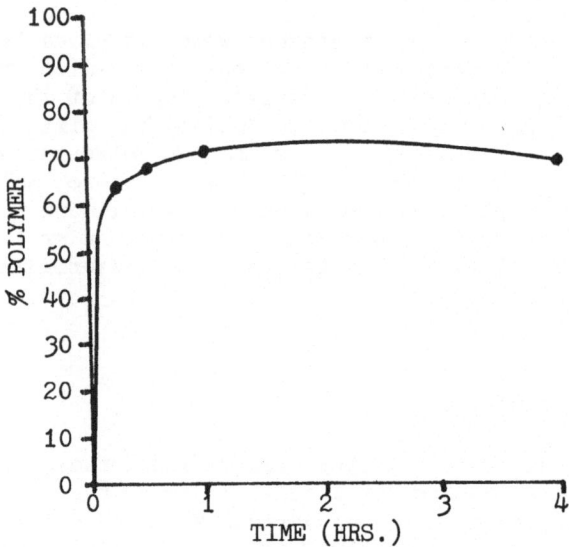

Figure 1.   The rate of polymerization of equivalent amounts of
2,2'-phenyliminodiethanol and terephthaloyl chloride in presence of
4 equivalents of pyridine at 80°C.

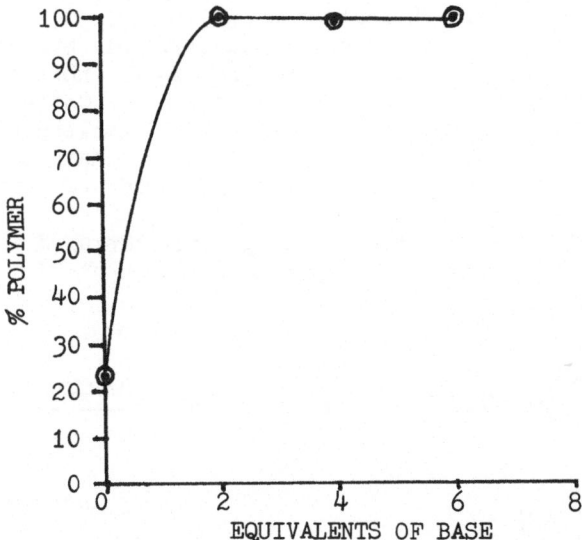

Figure 2.   The effect of triethylamine concentration on the
amount of the polymer formed in 1 hr. at 40°C from
equivalent amounts of 2,2'-phenyliminodiethanol and
terephthaloyl chloride.

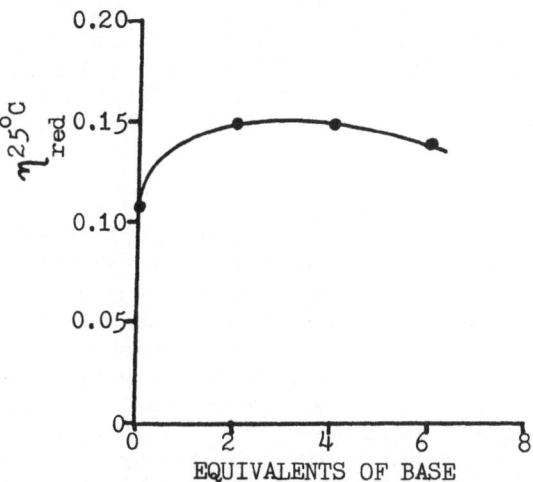

Figure 3. The effect of triethylamine concentration on the molecular weight of the polymer formed in 1 hr. at 40°C from equivalent amounts of 2,2'-phenyliminodiethanol and terephthaloyl chloride. Reduced viscosity in $C_2H_2Cl_4$ (0.2 g/dl)

Table 1. Absorption characteristics of some representative polymeric dyes from 2,2'-phenyliminodiethanol terephthalate polyester

Diaryl ketone	λ_{max}(nm)
Me$_2$N–⬡–CO–⬡	616.4
Et$_2$N–⬡–CO–⬡–NEt$_2$	602.3
Me$_2$N–⬡–CO–⬡–NMe$_2$	594.8
CH$_3$O–⬡–CO–⬡–OCH$_3$	553.1 & 519.8
HO–⬡–CO–⬡ / OH	552.3 & 513.1

Korshak et.al. had observed that in the polyesterification
with aliphatic diols, like hexamethylene diol, stronger organic
bases like triethylamine were more effective than weaker bases like
pyridine.[38] The polycondensation of PIE with TC was therefore
examined with triethylamine as the base. Even at low temperatures
of 40°C, polymerizations for 1-4 hrs. resulted in 97-98% polymer
formation using four equivalents of the base. The reduced viscosity
of the polymer formed in 1 hr. was 0.15 dl/g in tetrachloroethane
at 25°C (0.2 g/dl).

The effect of concentration of base employed in the polyesterification was also studied. Variation of the amount of triethylamine
introduced into the reaction shows that there is no effect of excess
base on the polymer yield as seen in Figure 2. Where no external
base was used, 24% polymer was still formed, presumably aided by
the participation of PIE as a tertiary base. Korshak et.al. have
noted similar independence of polymer yields from base concentrations
in the formation of poly(hexamethylene terephthalate), though in
their case the yields were much lower.[39] The molecular weights of
the polymers formed were found to be minimally affected by the
triethylamine concentration (Figure 3) and not as dramatic as in
the case of hexamethylene diol.

In the copolyesterification with EG, it was found that PIE was
incorporated in the recovered polymer in greater relative quantities
than was EG. This was also the case with polyethylene glycol. The
poly(ether-ester) block copolymer had better solubility properties
in common organic solvents than the homopolyesters. The polyurethanes
gave higher molecular weights than did the polyesters.

Polymeric dyes were prepared by coupling the p-positions of
the N-aryl groups of the polymer backbones with variously substituted diarylketones using phosphorous oxychloride as an acid catalyst.
The λ_{max} for the absorption of the polymeric dyes vary considerably
with the nature of the substituents on the diarylketones used for
the coupling. Some representative examples of the ketones employed
and the absorption characteristics of the polymer dyes are shown
in Table 1.

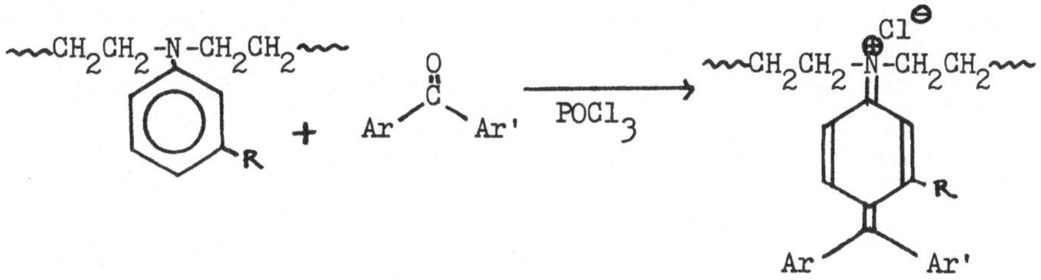

Fabrics and fibers colored with these macromolecular dyes showed good resistance to solvent wash.

CONCLUSION

As briefly described in the previous pages, polymer bound dyes are interesting and desirable materials for a multitude of applications. Apart from the numerous benefits of non-leachable and non-diffusive coloration of articles, they generate stimulating possibilities in electrical and optical materials. Since many dyes possess rather attractive electrochemical properties, binding of these to macromolecules should create new resources conducive to many industrial applications. The biologically non-absorbable food colorants area is an exciting breakthrough, the general plan of which should lead to the syntheses of other non-toxic additivies which take advantage of the "anatomical compartmentalization" concept. As analytical separation media, though research activity has spanned the better part of a decade, the full scope of this powerful tool has hardly been realized.

For the synthetic polymer chemist the challenges are countless. A vast array of new dyes are prepared every year, some of them with rather unique properties. From the wealth of information available on polymer synthetic routes, the chemical combination of these dyes with macromolecules should prove to be an open field for exploration, the pursuit of which will have many beneficial results.

REFERENCES

1. G. Manecke and G. Kossmehl, Makromol. Chem. 80 22 (1964)
2. G. Manecke and G. Kossmehl, Chem. Ber. 93 1899 (1960)
3. H. Kamogawa, J. M. Larkin, K. Toel and H. G. Cassidy, J. Polym. Sci. A 2 3603 (1964)
4. K. Shigehara, H. Matsunaga and E. Tsuchida, J. Polym. Sci. Polym. Chem. Ed. 16 1853 (1978)
5. K. Okamoto, Y. Hasegawa, S. Kusabayashi and H. Mikawa, Bull. Chem. Soc. Jap. 41 2563 (1968)

6. R. Lovrien and J. C. B. Waddington, J. Amer. Chem. Soc. 86
    2315 (1964)
7. Th. Dreyfus and E. Marechal, Bull. Soc. Chim. Fr. 1196, 1646
    (1975)
8. M. Champenois and E. Marechal, Bull. Soc. Chim. Fr. 2217, 2220,
    2223 (1975)
9. B. Sansoni, Naturwissenschaften 41 212 (1954)
10. A. Ravve, C.W. Fitka and J. C. Brichta, U.S. Patent 3,267,064
    (1966)
11. R. L. Meek, C. E. Feazel, P. M. Daugherty, F. C. Mallory and
    E. P. Colfield,Jr., U. S. Patent 3,278,486 (1966)
12. E. J. M. Bonnet, Ger. Offen. 2,946,965 (1980)
13. S. Nagashima, K. Tsuchiya and T. Tsuneda, Japan 73 08,562 (1973)
14. Ph. Gangneux and E. Marechal, Bull. Soc. Chim. Fr. 1466, 1483
    (1973)
15. E. Bonnet, Ph. Gangneux and E. Marechal, Bull. Soc. Chim. Fr.
    504, 507 (1976)
16. A. LePape and E. Marechal, C. R. Acad. Sci. C284 517, 561, 619,
    659 (1977)
17. A. LePape and E. Marechal, Bull. Soc. Chim. Fr. Chimie Molecul.
    263 (1978)
18. B. Petir and E. Marechal, Bull. Soc. Chim. Fr. 1591, 1597, 1602
    (1974)
19. C. E. Carraher,Jr., R.A. Schwarz, J. A. Schroeder and M. Schwarz,
    J. Macromol. Sci. Chem. A15 773 (1981)
20. B. Sansoni, Naturwissenschaften 39 281 (1952)
21. J. L. Keen, U. S. Patent 3,619,356 (1971)
22. D. Dawson, R. Gless and R. E. Wingard,Jr., Chemtech 6 724 (1976)
23. D. J. Dawson, Aldrichimica Acta 14 23 (1981)
24. D. J. Dawson, K. M. Otteson, P. C. Wang and R. E. Wingard,Jr.,
    Macromolecules 11 320 (1978)
25. K. M. Otteson and D. J. Dawson, U. S. Patent 4,178,422 (1979)
26. H. G. Nimmo and W. H. Holms, Biochemical Soc. Trans. 8 390 (1980)
27. P. D. G. Dean and D. H. Watson, J. Chromatography 165 301, 319
    (1979)
28. A. J. Turner, Trends in Biochem. Sci. July 171 (1981)
29. M. Tahan, D. Perez and A. Zilkha, Isr. J. Chem. 9 191 (1971)
30. D. J. Dawson, R. D. Gless and R. E. Wingard, J. Amer. Chem. Soc.
    98 5996 (1976)
31. N. Bellanca, U. S. Patent 3,940,503 (1976)
32. M. Bleha, Z. Plichta, E. Votavova and J. Kalal, U. S. Patent
    4,166,804 (1979)
33. A. Bernthsen, Liebigs Am. Chem. 251 1 (1889)
34. D. Braun, Makromol. Chem. 33 181 (1959)
35. V. A. Vasner and S. V. Vinogradova, Russ. Chem. Rev. 48 16 (1979)
36. S. V. Vinogradova, Vys. Soed. Ser. A 19 667 (1977)
37. T. Sulzberg and R. J. Cotter, Macromolecules 2 146 (1969)
38. V. V. Korshak, V. A. Vasner, S. V. Vinogradova and A. V. Vasilev
    Vys. Soed. Ser. A 16 502 (1974)
39. V. V. Korshak, S.V. Vinogradova, A. V. Vasilev and V. A. Vasnev
    Vys. Soed. Ser. A 14 56 (1972)

QUATERNIZATION OF CONDENSATION POLYMERS

William H. Daly and Shih-Jen Wu

Department of Chemistry
Louisiana State University
Baton Rouge, LA 70803

INTRODUCTION

Although the preparation of quaternary ammonium containing resins from appropriately substituted vinyl monomers has been studied extensively[1] elaboration of condensation polymers has not received comparable attention. Condensation polymers cannot be prepared with sufficiently high molecular weights or satisfactory chemical stability for most applications involving flocculation or viscosity enhancement. However, the excellent mechanical properties exhibited by many aromatic condensation polymers suggest that speciality applications, in particular, permselective membranes, could be served by suitably modified resins. We have observed that polymers containing oxy-phenylene repeat units are subject to facile electrophilic substitution including chloromethylation; the reactive derivatives can be cast into tough flexible films.[2]

Quaternary ammonium ion exchange resins were produced initially by chloromethylating crosslinked polystyrene beads with chloromethyl methyl ether, followed by quaternization with tertiary amines.[3] We have circumvented exposure to the highly carcinogenic bis(chloromethyl)ether, a common contaminent of commercial chloromethyl methyl ether, by employing 1,4-bis(chloromethoxy)butane or 1-chloromethoxy-4-chlorobutane and have produced chloromethylated poly(oxy-2,6-dimethyl-1,4-phenylene) and polysulfone.[4] Alternatively, chloromethyl methyl ether can be generated from acetyl chloride and methylal[5], and the reaction mixture utilized directly in chloromethylation of activated aromatic repeat units.

Since the quaternization of triethylamine with ethyl iodide was
first studied by Menschutkin[6], extensive kinetic investigations have
established the basic conditions governing an Sn2 reaction between
two neutral species forming a charged product. Further, quaternization of low molecular weight compounds was found irreversible by
Harman and coworkers;[7] upon reaction of excess trimethylamine or
equivalent amounts of dimethylaniline with tritium-labelled methyl
iodide in either ethanol or benzene, no evidence for exchange of
methyl groups could be detected. Since quaternization is irreversible under normal experimental conditions, only the kinetically
favored products are obtained and the reaction affords a relatively
sensitive probe into the influence of the polymer backbone on rates
of polymer reactions. The formation of a charged transition state
is favored by more polar solvents, which solvate the developing
charge. Polymer backbones may be considered as part of the solvent
matrix, thus reactions on more polar condensation polymers should be
more facile than those occuring in a hydrocarbon matrix like
polystyrene.

Synthesis of quaternary ammonium polymers can be achieved
either by treatment of chloromethylated polymers with various
amines, or by adding alkyl halides to polymeric amines. Although
the former approach has the greater commercial significance, the
latter technique was the subject of most early kinetic studies. The
reactivity of functional groups attached to a polymer chain was
generally assumed to be equivalent to that of low molecular weight
analogs by Flory.[8] However, in many instances, the quaternization
rates of reactions involving polymers are significantly different
from those of the corresponding low molecular weight analogs;
moreover, distinct differences in the rate profiles are often
observed when reactions on polymer substrates are conducted. Both
rate accelerations and decelerations relative to reaction rates of
appropriate model systems are observed. This "polymer effect" can
be attributed to many factors including: steric effects, neighboring
group effects, the influence of an electrostatic field, and
conformational and configurational changes of the polymer chain
during the reaction. We will assess the relative contributions of
each of these factors to the relative rates of polymeric
quaternizations.

Quaternization of Polymeric Amines with Alkyl Halides

During preparation of polyelectrolytes from poly(4-vinyl-
pyridine) (P-4-VP) and alkyl halides, some puzzling phenomena were
observed by Coleman and Fuoss.[9] These authors found quaternization
of pyridine, 4-picoline and 4-isopropylpyridine with n-butyl bromide
in tetramethylene sulfone followed normal second-order kinetics.
Similar results were reported by Hinshelwood and coworkers[10] for the
reaction of pyridine and alkyl halides. However, on quaternization
of poly(4-vinylpyridine) under the same conditions, they observed

that the reaction begins at about the same rate as the simple 4-
alkylpyridine, but, as the reaction proceeds, the rate begins to
decrease and a negative deviation from conventional second-order
kinetics is observed. An extensive series of model compounds were
quaternized with n-butyl bromide[11] in an effort to ascertain the
significant factors leading to rate retardation in the polymer
system.[12] The reactivity of the second nitrogen in diamines was
significantly less than that of the first, i.e., $k_2/k_1 = 0.5 \sim 0.7$.
If an intramolecular inductive effect were responsible for the
diminished nucleophilicity of the second nitrogen atom, a significant
effect should be observed in the conjugated model compound,
PyCH=CHPy; however, only a slight effect was observed. Further, no
difference between the rates of the second quater-nization step of
1,3-(bis-4-pyridyl)propane and 1,2-(bis-4-pyridyl)-ethane was
detected. These results indicate that inductive effects do not
contribute significantly to the relative nucleophilicity of the
second amino function. Steric effects are also eliminated by the
failure of the additional methylene spacer to influence the second
quaternization rate. Fuoss et al.[13] concluded that the rate
decreases observed between the first and second quaternization steps
in the model compounds could be attributed to a volume field effect,
i.e.,an electrostatic effect produced at the site of the second
nitrogen by the positive charge on the initially quaternized
nitrogeh.

   Arcus and Hall[14] reported long range field effects in a kinetic
study of quaternization of poly(4-N,N-dimethylamino-styrene) with
methyl iodide in DMF. Two homopolymers with different molecular
weights, and one copolymer composed of a 2:1 styrene:4-N,N-dimethyl-
aminostyrene monomer mixture were quaternized along with p-isopropyl
N,N-dimethylaniline and N,N-dimethylaniline as low molecular weight
analogues. The two model compounds exhibited normal S_N2 kinetics.
However, a rate deceleration was observed when the extent of quaternization of each of the three polymers reached 20 ~ 25%. The initial
rate constants of the homopolymers and copolymers were nearly
equal. Changes in the molecular weight and in the spacing of amino-
groups by the insertion of inert units had little effect on the
initial rate constant. Plots of the instantaneous rate constants
against percentage quaternization indicated that, at 50%
quaternization, the rate constants fall to about 60%, and 75% of the
initial value observed for the homopolymers and copolymer
respectively (Figure 1). Apparently, the accummulation of positive
charge on the macromolecules as reaction proceeded, increases the
energy required to introduce additional charge. This effect was more
enhanced for the homopolymers than for the copolymer due to the
accumulation of a greater charge-density at a given degree of
quaternization.

   Arcus and Hall also studied the reaction of P-4-VP and n-butyl
bromide in tetramethylene sulfone, DMF and their mixtures. In

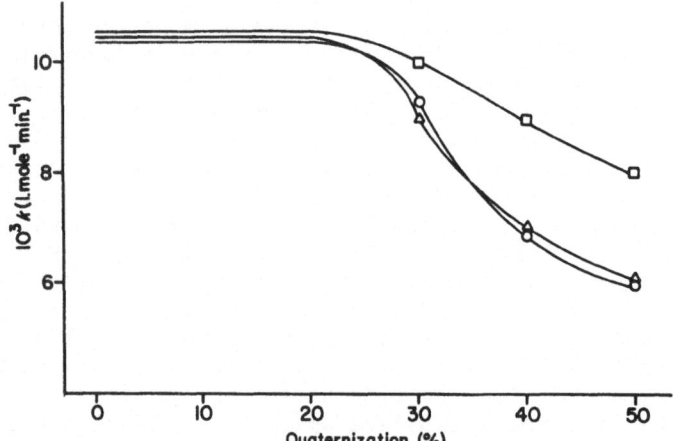

Figure 1.   Changes in second order rate constants, k, with extent
             of quaternization: o, poly-(4-N,N-dimethylamino-styrene),
             MW = 65,000; Δ, poly-(4-N,N-dimethylamino-styrene,
             MW = 97,000;   □, 2:1 styrene:4-N,N-dimethylaminostyrene
             copolymer.

tetramethylene sulfone, they observed rate retardations similar to
those reported by Coleman and Fuoss.[9] In tetramethylenesulfone-DMF
mixtures, the initial rate constant and extent of deceleration
decreased as the proportion of DMF was increased, until, in pure DMF,
there was no appreciable deviation from second order kinetics.
Quaternization of pyridine and P-4-VP with n-butyl bromide in the
presence of N-butylpyridinium bromide and poly(N-butyl-4-vinyl-
pyridinium bromide, respectively, was also studied. The results
show no appreciable alteration in the course of reaction; the
initial rate was increased 5% due to a small accelerative salt
effect.

    Morcellet-Saurage and Loucheux[15] investigated quaternization of
homopolymers and copolymers with styrene of 2-vinyl-, 2-methyl-5-
vinyl- and 4-vinylpyridine with butyl bromide. Both the homo- and
co-polymers of 2-vinylpyridine, exhibited a rate retardation at
about 4% conversion. The steric hindrance exerted by the polymer
backbone appears to be the major factor causing deceleration.
Normal second order kinetics were observed for 2-methyl-5-vinyl-
pyridine and 4-vinylpyridine-styrene copolymers containing up to
20 mole % pyridine units, moreover, within experimental error, the

quaternization rate constants of these copolymers are almost equal
to those of the corresponding model compounds. These results
indicate that the polymer bound pyridine group does not alter its
reactivity if it has at least two unreactive neighbors. In
contrast, the reactions of homopolymers, 2-methyl-5-vinyl- and 4-
vinylpyridine were retarded at conversions of about 38% and 50%,
respectively. This phenomena can also be attributed to the
accumulation of positive charge along the polymer chain.

Retardation during quaternization of P-4-VP with a series of
alkyl halides was also reported by Boucher and coworkers.[16]
Following a detailed neighboring group analysis,including kinetic,
and viscometric studies during reaction, they concluded that the
factors leading to rate retardation included a short range neighboring group effect imposed by steric hindrance, and a global
electrostatic effect as the polymer molecules initially expand and
then contract as the polymer-solvent interaction becomes less
favorable.

Quaternization of Chloromethylated Polystyrene

Although the production of quaternary ammonium salt polyelectrolytes on a polystyrene matrix was reported initially by Jones[3],
the kinetics of the quaternization process were investigated first
by Lloyd and Durocher.[17] Substrate polymers were prepared by
emulsion polymerization of vinylbenzyl chloride with 0.05-0.1 %
divinylbenzene, and the quaternization of these crosslinked resins
with aqueous trimethylamine was studied. The rate of quaternization
of poly(vinylbenzyl chloride) latex, PVBC, composed of about 60%
para and 40% ortho-isomers, decelerated markedly at about 50% conversion. This deceleration was attributed to the different reactivities of o- and p-chloromethyl substituents on the aromatic
rings. In order to ascertain the impact of positional isomers of
the "copolymer" rates, two "homopolymer" latexes were prepared from
99% p-vinylbenzyl chloride and 95% o-vinylbenzyl chloride respectively. The p-PVBC reacted very fast, while the high-ortho
analogue, after an initial surge, reacted much more slugishly.
Furthermore, for p-PBVC, the rate constant at 20 ~ 50% conversion
is significantly higher than that of the initial 0-20% conversion;
deceleration was observed in the case of o-PVBC. The accelerating
rate with increasing conversion of p-PVBC suggested that the
quaternized group provided a more polar environment for unreacted
neighboring sites, consequently, the formation of a charged
transition state was favored. Charges introduced close to the
polymer backbone in the ortho position were buried in a hydrocarbon
matrix and failed to influence the environment of the residual
chloromethyl groups. Thus the steric hindrance imposed by the
backbone exerted the dominating influence on quaternization rate.
The rates of both "homopolymers" and the "copolymer" were retarded
at very high conversions(>90%) due to an electrostatic field
effect.

In a systemic kinetic study of the reaction of various amines with chloromethylated polystyrene (CMPS), Kawabe and coworkers[18-20] found the rate profiles to be significantly influenced by the nature of amine and the reaction media. They observed that some cases conformed to simple second order reaction rate laws, but slight variations in the reaction conditions produced either deceleration or acceleration as quaternization proceeded (Table I). The deceleration was observed in the reactions of bulky amines i.e., diethylamine or dibutylamine, in DMF, DMSO and dioxane. This rate retardation could be attributed to the steric hindrance imposed by the quaternized neighboring group. The nature of the solvent also played an important role in the quaternization process; for instance, a normal second order reaction was observed between triethylamine and CMPS in DMF. In DMSO, a deceleration attributed to an electrostatic effect was observed[20].

Accelerations occurred commonly when hydroxyethyl amines were allowed to react with CMPS in dioxane or dimethylacetamide, indicating that a hydrophilic effect or the formation of intramolecular hydrogen bonds may play an important role in the promoting quaternization[21]. Autoacceleration occurred at about 33% conversion suggesting that when a chloromethyl group has at least one quaternized neighboring group, the presence of hydroxyl in that quaternary group contributes to the solvation of adjacent transition states. Rate comparisons at different temperature regimes demonstrated that the hydrophilic effect was stronger in dioxane than that in dimethylacetamide. Quaternizations of CMPS with triethylamine, N-2-hydroxyethyl-dimethylamine or N,N-bis-hydroxyethylmethylamine conducted in DMF, also exhibited stepwise acceleration[22].

Different kinetic behavior was observed when secondary hydroxyalkylic amines, methyl-2-hydroxyethylamine and butyl-2-hydroxyethylamine, were employed as nucleophiles. Autoacceleration appeared in dioxane for both secondary amines however, normal second order kinetics were followed in DMF when the nucleophile is methyl-2-hydroxyethylamine which has less bulky substituents. In the reaction of butyl-2-hydroxyethylamine with CMPS in DMF, rate retardation began when the conversion reached about 75% owing to the steric hindance of the bulky butyl group[23]. Thus the sensitivity of the rate profiles to reaction media and nucleophile structure complicates assessment of "polymeric effects".

The data in the literature is based on functional polymers prepared from vinyl monomers, namely styrene and vinylpyridine isomers. These polymers have flexible hydrocarbon backbones which provide a non-polar environment for the active sites. The major factors causing negative or positive deviation from normal second order kinetics in quaterniztion in polymeric systems are limited to steric, electrostatic field and hydrophilic effects. The influence of backbone rigidity or polarity could not be ascertained.

Table I. Variations in Kinetic Rate Profiles for Quaternization of Chloromethylated Polystyrene.

Amine	Solvent	Kinetic Behavior	Factor	Ref.
n-butyl-	dioxane	deceleration	steric	18a
	DMF	normal		18c
	MEK	deceleration	steric	18d
t-butyl-	DMF	deceleration	steric	18d
di-n-butyl-	dioxane	deceleration	steric	18a
	DMF	deceleration	steric	19a
diethyl-	dioxane	deceleration	steric	18d
	DMF	deceleration	steric	18d
triethyl-	DMF	normal		19d,21
	DMSO	normal		19d
2-amino-ethanol	dioxane	acceleration	*	19e
	DMF	normal		19e
	DMSO	normal		19e
diethanol-	dioxane	acceleration	*	19a
	DMF	deceleration	steric	19a
	DMSO	deceleration	steric	19a
triethanol-	DMF	acceleration	*	19e
	DMSO	normal		19e
1-amino-2-propanol	dioxane	acceleration	*	19e
	DMF	normal		19e
	DMSO	normal		19e
2-amino-2-methyl-propanol	dioxane	acceleration	*	19e
	DMF	normal		19e
	DMSO	normal		19e
trihydroxyl-methanamine	DMSO	normal		19e
2-amino-butanol	dioxane	acceleration	*	19d
	DMF	normal		18c
	DMSO	normal		18c

* hydrogen bond formation in the transition state.

EXPERIMENTAL

Chloromethylation of Polysulfone. A solution of chloromethyl
methyl ether (6 mmole/ml) in methyl acetate was prepared by adding
acetyl chloride (141.2 g, 1.96 mol) to a mixture of dimethoxymethane
(180 ml, 2.02 mol) and anhydrous methanol (5.0 ml, 0.12 mol)[5]. The
solution was diluted with 300 ml of 1,1,2,2,-tetrachloroethane and
SnCl$_4$ (1.05 ml, 0.009 mol) was added. After the mixture was brought
to reflux, a solution of polysulfone (40 g, 0.09 eq) in 500 ml
tetrachloroethane was added slowly. Refluxing was maintained for 3
hours before the catalyst was deactivated by injecting 5 ml of water
into the reaction mixture. The reaction volume was reduced to 400 ml
before precipitating the chloromethylated polysulfone, 3, in
methanol. After reprecipitating 3 from chloroform in methanol,
46.5 g of chloromethylated resin was obtained. The elemental
analysis was consistant with 1.89 chloromethyl groups per repeat
unit; Found: C, 64.97; H, 4.72; Cl, 12.55; S, 6.20; O, 11.9.

Preparation of Polysulfone Model. Treatment of the disodium
salt of bisphenol-A (20 g, 0.088 mol) with chlorophenyl phenyl
sulfone (44.3 g, 0.175 mol) in a 2:1 v:v toluene:DMSO (100 ml)
according to the procedure of Johnson et al.[24] afforded 51.2 g of
crude adduct. Recrystallization from benzene yielded the pure
adduct 1, m.p. 182–183°; mol. wt. (mass spec) 660.1. Analysis:
Calc'd. for C$_{39}$H$_{32}$O$_6$S$_2$: C, 70.91; H, 4.84; O, 14.53; S, 9.72.
Found: C, 70.83; H, 4.99; O, 14.54; S, 9.66.

Chloromethylation of 1. A solution of 1 (40 g, 0.06 mol) in
200 mol of 1,1,2,2-tetrachloroethane was blended with 200 ml of the
chloromethyl methyl ether-methyl acetate mixture and 0.7 ml of
SnCl$_4$. The reaction mixture was refluxed (80–85°) for 8 hr under
argon. After deactivating the catalyst with 2 ml of water, the low
boiling components were distilled. The remaining tetrachloroethane
was evaporated in vacuo and the residual oil crystallized in
ethanol. Recrystallization from ethanol afforded 40.2 g, 87% yield
of pure 3, m.p. 90–91°C. Analysis: Calc'd. for C$_{41}$H$_{34}$O$_6$S$_2$Cl$_2$:
C, 65.01; H, 4.49; O, 12.67; S, 8.47; Cl, 9.36. Found: C, 64.86;
H, 4.69; O, 12.30; S, 8.62; Cl, 9.47.

Kinetic Investigations. Stock solutions of 1.42 g of 3 and
1.92 g of 2 in 100 ml of either DMSO or 1:1 v:v dioxane:DMSO were
equilibrated at the reaction temperature. Aliquots (10 ml) of the
substrate solutions were mixed with similarly equilibrated solutions
of triethylamine, quinuclidine or tributylamine at a ratio which
assured a 2:1 excess of amine to chloromethyl substituent. The
reaction was terminated at the desired reaction time by adding 15 ml
of 0.1 N HNO$_3$. The ionic chloride content was assayed potentio-
metrically with 0.025 N AgNO$_3$ using a chloride selective electrode
to detect the end point.

Quaternization of amines with alkyl halides exhibit second-order kinetics and for such reactions, the integrated form of the rate expression is:

$$f(x) = \frac{1}{(a-b)} \ln \left[\frac{b(b-x)}{a(b-x)}\right] = kt$$

where x is concentration of halogen ions at time t, a is initial concentration of base, b is initial concentration of alkyl halide, and k is the rate constant. A plot of f(x) vs. time should be linear, with a slope equal to k.

RESULTS AND DISCUSSION

The second order nature of quaternization reactions in our system was confirmed by rate studies on model compounds. Although benzyl chloride is usually selected as the model for chloromethylated polymers[20], we chose to synthesize a difunctional model that would be sensitive to neighboring group effects. Condensation of 4-chlorophenyl phenyl sulfone with the disodium salt of bisphenol-A yielded an excellent model for the polysulfone segment, 1. Quantitative chloromethylation of 1 with a chloromethyl methyl ether/ methyl acetate mixture in the presence of stannic chloride afforded the corresponding bischloromethyl adduct, 2.

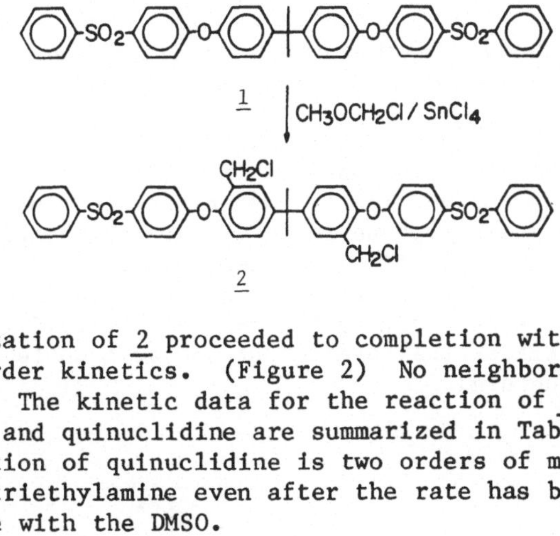

Quaternization of 2 proceeded to completion with no deviation from second order kinetics. (Figure 2) No neighboring group effect was observed. The kinetic data for the reaction of 2 with triethylamine and quinuclidine are summarized in Table II. Note that the reaction of quinuclidine is two orders of magnitude faster than that of triethylamine even after the rate has been moderated by mixing dioxane with the DMSO.

The equivalent reactivity of the two chloromethyl groups on 2 is contrary to the observations of Chow and Fuoss[11], who reported that the quaternization of the second nitrogen in the bis-pyridylalkanes was much slower. This negative deviation was attributed to the extramolecular electrostatic field effect produced by the positive charge on the first nitrogen. It is obvious that the electrostatic effect did not appear in the polysulfone model system,

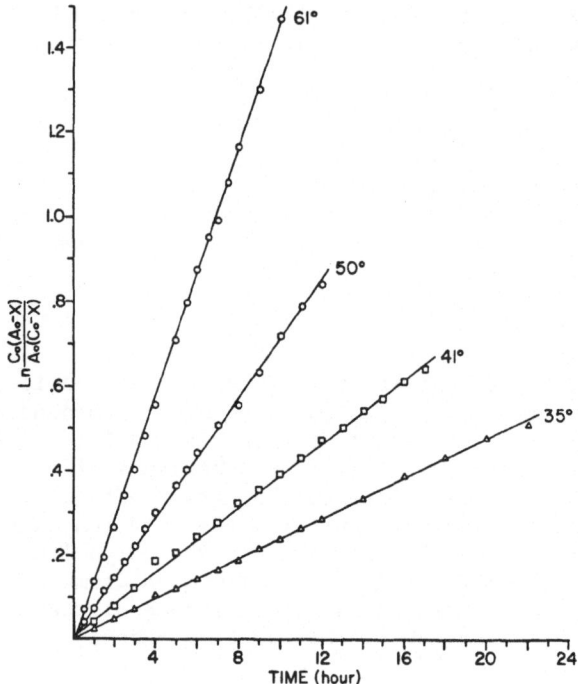

Figure 2.   Quaternization of 2 with triethylamine in DMSO;
             $[-CH_2Cl]$, 0.0248 meq/ml; [TEA], 0.053 mmol/ml.

furthermore, steric hindrance can also be ruled out. It may be
argued that the highly polar aprotic DMSO (dielectric constant =
46.7) solvated the charged species and diminished the electrostatic
effect. However, this hypothesis is immediately contradicted by the
results obtained in a significantly less polar DMSO/dioxane (50:50)
mixture. Apparently the chloromethyl sites in the model compound
are shielded from each other by the relatively rigid isopropylidine
bridge which restricts the rotation of the aryl units in the
oxyphenylene structure.

Table II. Quaternization of the Polysulfone Model, 2.

run	T, °C	a mmol/ml	b meq/ml	a/b	$k \times 10^3$ $l.mol^{-1}sec^{-1}$	% conv.
With Triethylamine in DMSO.						
1	61	0.0530	0.0248	2.14	1.46	86.3
2	50	0.0530	0.0248	2.14	0.691	71.4
3	41	0.0510	0.0249	2.05	0.402	63.9
4	35	0.0500	0.0250	2.00	0.267	55.2
	Ea= 13.5 kcal/mol			log A=5.99		
In DMSO/Dioxane						
5	60	0.0535	0.0246	2.17	0.335	60
With Quinuclidine in DMSO/Dioxane						
6	35	0.0505	0.0254	1.99	101.	82.3
7	30	0.0520	0.0251	2.07	80.5	81.3
8	26	0.0520	0.0258	2.02	64.3	74.8
9	20	0.0535	0.0257	2.08	47.1	75.1
	Ea=9.6 kcal/mol			log A=5.8		

Quaternization of Chloromethylated Polymers

A survey of the reaction conditions required to quaternize chloromethylated condensation polymers in a homogeneous media revealed that mixed solvent systems would be required to handle poly(oxy-2,6-dimethyl-3-chloromethyl-1,4-phenylene), 6. The reaction of triethylamine with chloromethylated polysulfone proceeded cleanly in pure DMSO, and a model compound was easy to synthesize. Therefore, we focused our initial attention on polysulfone derivatives.

Chloromethylated polysulfone containing an average of 1.9 chloromethyl groups per repeating unit, 3, was treated with triethylamine in DMSO. We expected the isolation of active sites demonstrated with the model 2 would prevail in the polysulfone system. This was not the case, as is evident in Figure 3. The kinetic plots are concaved downward because the quaternization of the polysulfone proceeds less and less rapidly as the degree of conversion increases. The quaternization of 3 with TEA can be modeled by two reaction rate constants, k_0 and k_2. Normally, three individual rate constants, k_0, k_1 and k_2 are defined for polymer modification. These constants depend on the distribution of reacted sequences along the chain. In the initial stage of quaternization, the reaction is undoubtedly a random process along a given polymer chain, and probability of quaternization of a site which has reacted neighbor is small, but as the reaction proceeds beyond 1/3 conversion, a deviation should appear if $k_0 \neq k_1$. On the other hand, when the degree of quaternization is 50%, the majority of the

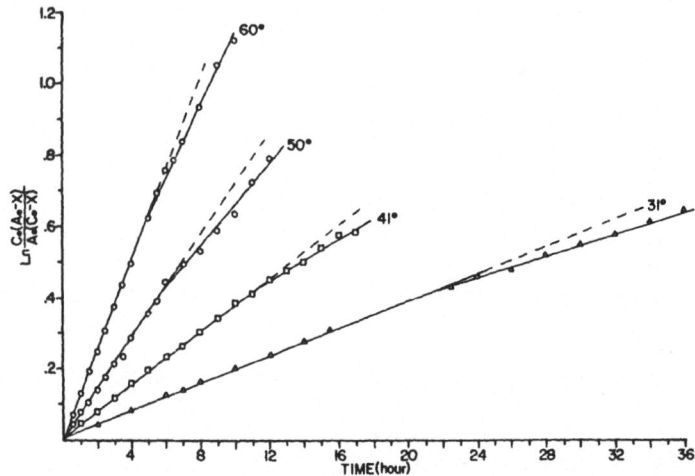

Figure 3.   Quaternization of 3 with triethylamine in DMSO;
             [-CH₂Cl], 0.024 meq/ml; [TEA], 0.053 mmol/ml.

chloromethyl groups have two quaternized neighbors. However, the
kinetic behavior of the model compound indicated that the chemical
reactivity of a given $-CH_2Cl$ group is independent of whether its
nearest neighbor has reacted or not. Therefore, it is reasonable to
assume that for quaternization of 3, the relation between these rate
constants is $k_0 \simeq k_1 > k_2$, where k_2 pertains to reactions on an
extended chain with a high charge density.

The values for k_0 and k_2 estimated from the slopes of
conventional second-order plots are summarized in Table III. The
ratios, k_0/k_2, are also reported. These ratios remain quite
constant for all runs except runs 14 and 15, in which the concentration ratio of amine:$-CH_2Cl$ is about 10. Note also that the rate
constants for these two runs, 14 and 15 are slower than runs 11 and
12 where the concentration ratio of amine : $-CH_2Cl$ is 2. A similar
observation was made by Kawabe[20] in the amination of chloromethylated polystyrene with 2-aminobutanol in DMSO. This abnormal
phenomena is not fully understood, but the possibility of solvent
polarity change due to the high concentration of amine required can
not be ruled out.

Copolymers of high content styrene and 4-vinyl pyridine has
been reported to follow normal second order kinetics by Morcellet-
Saurage and Loucheux[15]; they concluded that even the presence of the
macromolecular backbone does not change the reactivity of the
functional group if its two neighbors are inert. One can extend
this idea to a polysulfone system, only if one can uniformly distri-
bute a small number of chloromethyl groups along the polysulfone
chain. An attempt to use a polysulfone with an average of
$0.5-CH_2Cl$/repeat unit was thwarted by its insolubility in DMSO. The
lowest degree of substitution at which the derivative remained

Table III. Quaternization of Chloromethylated Polysulfone.

run	T, °C	a mmol/ml	b meq/ml	a/b	k_0 x 10^3 $1.mol^{-1}sec^{-1}$	k_2 x 10^3 $1.mol^{-1}sec^{-1}$	k_0/k_2
Treatment of 3 with triethylamine in DMSO.							
10	60.5	0.0530	0.0252	2.11	1.22	1.03	1.18
11	50.5	0.0530	0.0248	2.14	0.709	0.585	1.21
12	41	0.0510	0.0250	2.04	0.393	0.311	1.26
13	31.2	0.0505	0.0238	2.12	0.196	0.169	1.16
		Ea = 12.4 kcal/mole			log A = 5.26		
14	51	0.249	0.0251	9.92	0.464	0.309	1.50
15	41	0.251	0.0249	10.1	0.263	0.173	1.52
In the presence of 1% NH_4NO_3							
16	51	0.0510	0.0251	2.03	0.815	0.653	1.25
In DMSO/Dioxane							
17	60	0.0535	0.025	2.15	0.336		
Treatment of 4 with triethylamine in DMSO.							
18	60	0.0545	0.0250	2.18	1.23	1.13	1.09
19	50	0.0505	0.0250	2.02	0.703	0.565	1.24
20	41	0.0530	0.0250	2.12	0.375	0.341	1.10
21	31	0.0505	0.0243	2.08	0.195	0.173	1.13
		Ea = 12.4 kcal/mole			log A = 5.37		
Treatment of 3 with quinuclidine in DMSO/dioxane.							
22	31.5	0.0498	0.0247	2.02	8.70	58.9	1.48
23	26.2	0.0510	0.0255	2.00	6.60	45.5	1.45
24	20	0.0484	0.0249	1.94	4.91	32.9	1.49
		Ea = 8.8 kcal/mole			log A = 5.2		
Treatment of 3 with tri-N-butylamine in DMSO/Dioxane							
25	66	0.052	0.025	2.07	0.280		
26	61	0.052	0.025	2.08	0.210		
27	55	0.051	0.025	2.04	0.123		
		Ea = 16.4 kcal/mole			log A = 7.05		

soluble was an average of 1.2-CH_2Cl/repeat unit, 4. The experimental results from quaternization of 4 are included in Table 3,
runs 17-20. Comparison of the results from polysulfones substituted
to two different degrees shows that there is no significant difference between them, moreover, the kinetic parameters, Ea and A, are
equal within experimental error. Furthermore, the break points
where deviation from the initial rate occur are at the same percentage conversion, that is 51 ~ 52%.

A possible explanation for the failure of the isolated reactive site hypothesis on polysulfone could be nonuniform distribution of chloromethyl groups along the polymer backbone. Although the average degree of chloromethylation is 1.2-CH_2Cl per unit, it is difficult, even impossible, to regulate the micro distribution of these groups on the polymer. The chloromethylation of polymer is a random process, but during the reaction some segment of polymer chain may develop a high density of functional groups. Up to this point, one can conclude that the average degree of chloromethylation has little influence on the kinetic behavior.

Impact of Steric Hindrance

Most observations of rate retardation in polymer modifications have been attributed to steric hindrance. In order to estimate the steric influence of the relatively bulky triethylbenzylammonium substituent on unreacted site during quaternization, quinuclidine was chosen as nucleophile. It is well known that nucleophilicity of quinuclidine in displacement reactions is greater than that of triethylamine, since bicyclic amines are less sterically hindered. Preliminary experiments on the quaternization of chloromethylated polysulfone with quinuclidine in DMSO showed that the reaction velocity was too rapid to investigate using our experimental techniques, i.e., 85% conversion was obtained with three minutes. Therefore, we were forced to add a less polar solvent to DMSO in order to reduce the reaction rate. It was found that a 50:50 (v/v) mixture of dioxane and DMSO dissolved both chloromethylated and quaternized polysulfone so the rate could be measured in a homogeneous system. The introduction of a nonpolar solvent reduced the initial rate of triethylamine substitution fourfold (Table III, run 17).

The initial velocity of quinuclidine substitution is significantly faster than that of triethylamine at the same temperature (Table III, runs 21-23), even though the former was investigated in a mixed solvent. Similar results were found in the quaternization of the model compound. If a steric effect were considered to be the sole factor producing the decrease in k_2 with respect to k_0, one would expect that: (1) k_0/k_2 for quinuclidine substitution should be smaller than k_0/k_2 for TEA substitution and/or (2) the initial linearity in the second order plot would extend beyond 52% conversion where deviation occurs in the triethylamine system. Experimental results refute these expectations; rate retardation is enhanced in quinuclidine reactions, furtherfore, the break point is almost the same for both cases.

We anticipated a more pronounced steric effect to appear in substitutions with tri-n-butylamine. Since this amine is not soluble in DMSO, we were forced to use the DMSO/dioxane mixture. The corresponding initial rate for tributylamine was slightly less,

(runs 25-27), than the TEA rate, and no break point was detected.
The rate reduction can be attributed to the reduced nucleophilicity
of tributylamine and the nonpolar solvent character of long chain
aliphatic amines. The absence of a break point indicates that no
significant neighboring group steric effect is contributing to the
retardation. These observations lead us believe that a steric effect
is not the sole factor responsible for the deceleration.

Impact of Added Electrolyte

Whenever an uncharged polymer chain is converted to a ionic
polymer either by titration or by chemical modification, the mutual
repulsion of the ionized species may lead to chain expansion. The
consequence of chain expansion will indeed affect the conformation
of polymer in solution, consequently, it could be in turn change the
chemical reactivity of the polymer. In order to investigate whether
the conformational change of polymer during reaction influence the
kinetics of quaternization, Boucher and co-workers[16] measured the
reduced viscosity during the quaternization of poly(4-vinylpyridine)
with n-butyl bromide in propylene carbonate; these measurements were
conducted in the presence or absence of an added electrolyte, N-ethyl
pyridinium bromide. The reduced viscosity was found increase to a
maximum value at 15% conversion, and then gradually decrease. Upon
addition of electrolyte, the reduced viscosity maintain constant
throughout the reaction. They found no significant effect of added
salt on kinetic measurements, except in runs which exhibited a cloud-
point as the reaction proceeded. In these cases an enhanced retardation was observed. The retardation in rate produced by added salt
was proposed to be due to a global effect, i.e., a decrease in the
overall dimension of the macromolecules as the cloud-point was
approached.

In order to ascertain the influence of added salts on the
quaternization of polysulfone, the kinetics were measured in the
presence of 1% ammonium nitrate (run 16 in Table III). Salt concentrations up to 1% could be employed before the chloromethylated
polysulfone began to salt-out, and no interference in titration
process was detected. As shown in Figure 4 there was no appreciable
alteration in the rate profile but the initial rate increased by 15%;
this rate increase is consistent with the anticipated accelerative
salt effect for reactions involving polar transition states. The
reduced viscosities measured as the reaction progressed are shown in
Figure 5. A sharp increase in the viscosity was observed in the
initial stage of reaction, if no electrolyte was added. Upon addition of 1% NH_4NO_3, no significant change in the reduced viscosity was
observed. These results imply that quaternized polysulfone tends to
expand to maximum extension if no salt is present, but maintains an
equilibrium dimension when electrolytes are present to shield the
developing charges along the chain. However, no change is observed
in the rate profile in the presence of added salts so the chain

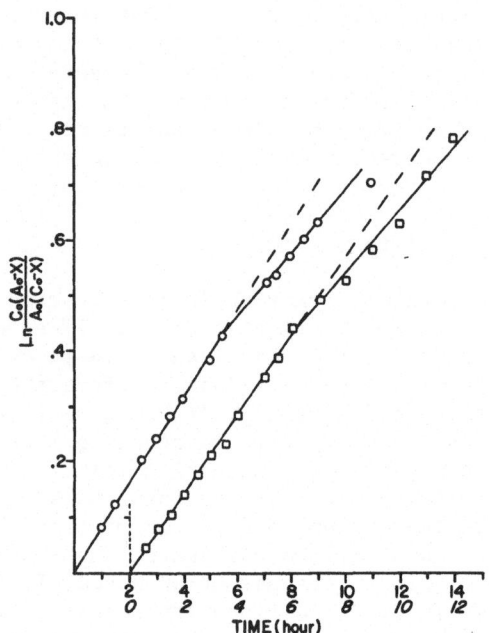

Figure 4.   Quaternization of 3 with triethylamine in DMSO in the
             presence of NH_4NO_3 at 51°C: o, $[-CH_2CL]$, 0.025 meq/ml;
             [TEA], 0.051 mmol/ml, 1 wt%. NH_4NO_3; □, $[-CH_2Cl]$, 0.025
             meq/ml; [TEA], 0.053 mmol/ml.

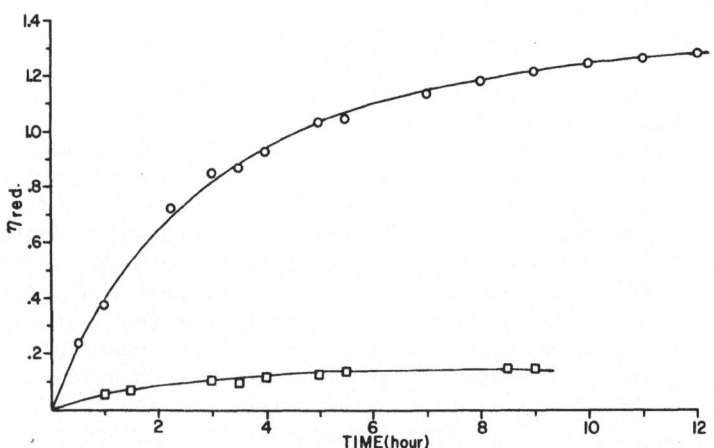

Figure 5.   Relative viscosity of 3 during quaternization with
             triethylamine in DMSO at 50°C: o, $[-CH_2Cl]$, 0.024 meq/ml;
             [TEA], 0.0505 mmol/ml; □, $[CH_2Cl]$, 0.024 meq/ml; [TEA],
             0.0505 mmol/ml, 1 wt% NH_4NO_3.

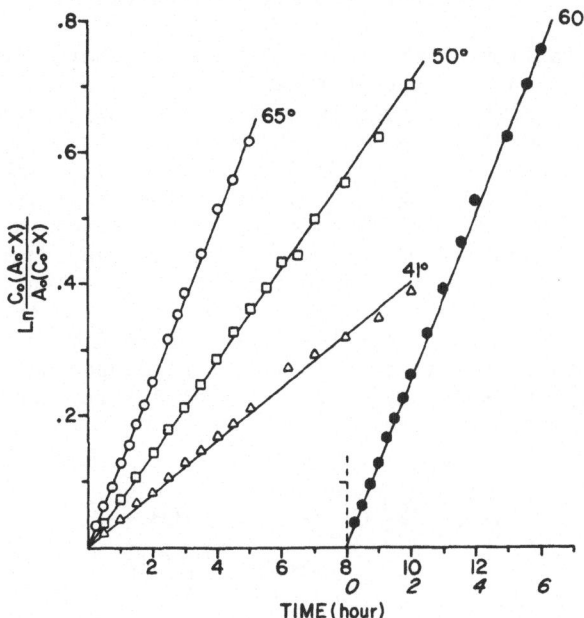

Figure 6.   Quaternization of poly(vinylbenzylchloride) with
             triethylamine in DMSO: o, [-CH$_2$Cl], 0.023 meq/ml, [TEA],
             0.048 mmol/ml; •, [-CH$_2$Cl], 0.024 meq/ml,
             [TEA], 0.055 mmol/ml; □,[-CH$_2$Cl], 0.024 meq/ml,
             [TEA], 0.051 mmol/ml; Δ [-CH$_2$Cl], 0.025 meq/ml, [TEA],
             0.053 mmol/ml.

expansion can not be responsible for the rate retardation at high
degrees of conversion.

Impact of Chain Flexibility

Chloromethylated polysulfone indeed exhibits different kinetic
behavior in the quaternization with TEA than its corresponding model
compound. From experimental results, it is clear that the rate
retardation is not due to steric hindrance, the degree of chloromethylation on the polymer chain, or a salt effect. Stereoisomeric
effects are not a potential factor, since chloromethylated polysulfone consists of only one detectable isomer. In spite of these
results, we knew that the polymer backbone must play an important
role in this reaction. Under the same experimental conditions used
to quaternize chloromethylated polysulfone, poly(vinylbenzyl
chloride) exhibited normal second-order kinetics with an Ea of 10.4
kcal/mole, as shown in Figure 6. Noda and Kagawa also observed the
same phenomenon in the quaternization of chloromethylated polystyrene with TEA in DMF (Ea = 10.5 kcal/mole)[21]. The major difference between these two systems is the polymer backbone; polysulfone

being composed of a sequence of stiff aryl units.

Bond flexibility which in turn determines chain flexibility, has significant influence on the physical properties of a polymer. The flexibility of a given polymer is derived from the freedom of internal rotation of bonds in the main chain. In solution, if rotations and folding of the chain backbone involve relatively small changes of energy, the entropy gain from increasing number of accessible configurations may favor the formation of charged products that require different environments. On the other hand, if the required energy for rotation of a polymer chain bond is high, rotation to form more stable product configurations will be inhibited, and the accessibility of the remaining reactive sites may be limited. In order to investigate this factor in more detail, a condensation polymer with a flexible connecting segment, chloromethylated acetylated phenoxy resin, 5, was synthesized.

Phenoxy resin was prepared from bisphenol-A and epichlorohydrin and the secondary hydroxyl groups produced by the condensation were protected by acetylation. After chloromethylation, the functional groups are located on the biphenol-A unit ortho to the ether linkage. Functional groups on polysulfone are located in the same relative position so the activity of $-CH_2Cl$ on either polymer should be identical. In fact chloromethylated phenoxy resin is more active than 3, (see Figure 7). No appreciable change in the rate profile was observed, but the break point shifted to a higher extent of conversion. The functional groups are more accessible and the charged groups can be stabilized by formation of polar domains. Introduction of a flexible glyceryl ether repeat unit reduces the stiffness of the polymer backbone; for example, the glass transition for bulk phenoxy resin is 100° C and the T_g of the chloromethylated acetylated derivative drops to 60°. In contrast, chloromethylated polysulfone exhibits a glass transition of 175°; this rigidity must persist in solution and limit free rotation of the polymer chain. The rigid aromatic groups extend the chain and make it difficult for chain folding to create small domains.[24]

Poly(oxy-2,6-dimethyl-3-chloromethyl-1,4-phenylene), 6 contains approximately the same concentration of reactive sites as poly-(vinylbenzyl chloride), but it is not a mixture of stereoisomers. Treatment of 6 with a 25% methanolic trimethylamine solution produces a solution of quaternized derivative suitable for film casting. Approximately 60% of the active chloride can be quaternized at room temperature; complete quaternization required an aftertreatment at

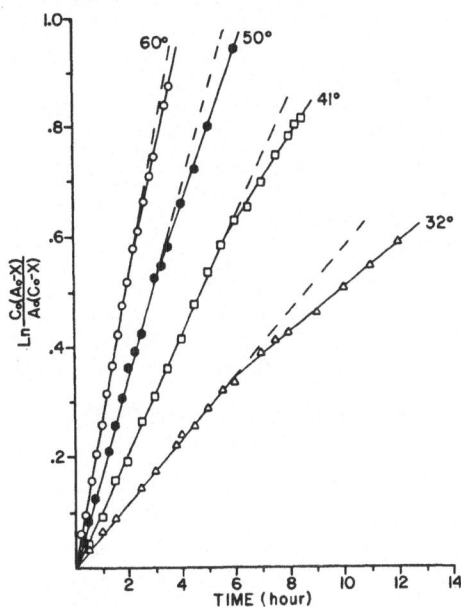

Figure 7.   Quaternization of $\underline{5}$ with triethylamine in DMSO;
             o, [-CH$_2$Cl], 0.026 meq/ml; [TEA], 0.051 mmol/ml;
             ●, [-CH$_2$Cl], 0.025 meq/ml; [TEA], 0.052 mmol/ml;
             □, [-CH$_2$Cl], 0.026 meq/ml; [TEA], 0.052 mmol/ml;
             Δ, [-CH$_2$Cl], 0.025 meq/ml; [TEA], 0.051 mmol/ml.

90° for several hours. The rate retardation noted qualitatively was
confirmed by a careful evaluation of the reaction of triethylamine
with $\underline{6}$ in 1:1 DMSO-THF. As Figure 8 shows, a pronounced retardation
occurs when 33% of the chloromethyl substituents had reacted.
Although it would appear that the bulky triethylammonium substitutent
is creating a steric hindrance to further reaction, the orthogonal
conformation of adjacent aromatic rings should allow ready access to
residual chloromethyl sites. The bulk glass transition of $\underline{6}$ occurs
around 150°C and the segment rigidity is augmented by localized
restrictions to chain rotation. Thus, quaternized $\underline{6}$ can not fold
into the configurations which stabilize high charge density.

     During quaternization of chloromethylated polymers, the positive
charge density gradually increases on the polymer chain. The
polyelectrolyte fails to dissociate completely; an equilibrium is
established between the macroion and variously bound counterions.[26]
The counterions can be classified into three categories: counterions
freely moving outside the region occupied by macroions, those bound
but mobile within small domains in the macroion, and those bound to
individually charged groups on the macroion. The relative
concentration or freely moving counterions is very small and can be

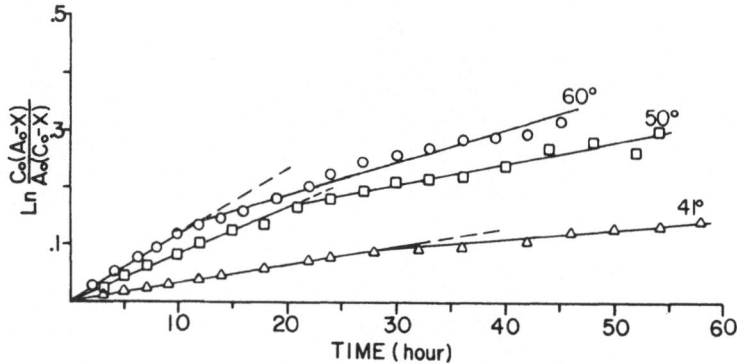

Figure 8.   Quaternization of 6 in 1:1 DMSO-THF: [-CH$_2$Cl], 0.024
meq/ml; [TEA], 0.0525 mmol/ml.

surpressed completely by adding salts. Thus, the most significant
factor controlling introduction of additional charge will be the
distribution between bound but mobile and individually bound counterions. In polar solvents, flexible polymers fold into configurations which contain polar domains with high dielectric constants
and most of the ionic functional groups reside within these
domains. The number of localized counterions is estimated to be
small in strongly dissociative polyelectrolytes.

If the flexibility of the polymer chain is reduced, the tendancy
to fold in to domains must decrease. Each charge must be localized
within the polymer backbone, which is a low dielectric region.
Solvation of individual counterions will become the primary mode of
product stabilization. If the counterions are tightly bound to
individual groups, the solvent molecules must be highly ordered to
effect solvation. Further, the extent of solvation will have a
pronounced impact upon the stability of the transition state, and the
reaction rate will be very sensitive to solvent variations. We have
noted that DMSO is very difficult to remove quantitatively from
quaternized polymers and that the polyquats are extremely hydro-
scopic. These observations indicate that solvent molecules are very
strongly bound to the ionic sites.

Our results to date suggest that the flexibility of the polymer
backbone plays the most important role in polymer reactivity, particularly if there is a significant change in the polarity of the
polymeric derivative. Rigid polymer chains can not contribute significantly to solvation. Retardation of substitution rates occurs

because the resultant product must be formed in a non-polar environment and stabilization thru solvation is restricted. Thus, accessibility of active sites is not the dominent factor in determining
reactivity.

REFERENCES

1. F. Hoover, J. Macromol. Sci. Chem., A4 1327 (1970).
2. W. H. Daly, Supon Chotiwana, and Yih-Ching Liou, in "Polymeric
    Amines and Ammonium Salts" ed. by E. Goethals, Pergamon Press
    1980, p 37.
3. G. W. Jones, Ind. Eng. Chem., 44, 2686 (1952); K. W. Pepper, H.
    W. Paisley and M. A Young, J. Chem Soc., 4097 (1953).
4. W. H. Daly, Supon Chotiwana, and R. A. Nielsen, Polymer
    Preprints, 20, 835 (1979).
5. J. S. Amato, S. Karady, M. Sletzinger and L. M. Weinstock,
    Synthesis, 970 (1979).
6. N. Menschutkin, Z. Phys. Chem., 6, 41 (1890).
7. D. Harman, T. D. Stewart, and S. Ruben, J. Am. Chem. Soc., 64,
    2294 (1942).
8. P. J. Flory, J. Am. Chem. Soc., 61, 3334 (1939).
9. B. D. Coleman, and R. M. Fuoss, J. Am. Chem. Soc., 77, 5472
    (1955).
10. a. C. A. Winkler, and C. N. Hinshelwood, J. Chem. Soc., 1147
       (1935);
    b. N. J. Pickels, and C. N. Hinshelwood, Ibid., 1353 (1936);
    c. K. J. Laidler, and C. N. Hinshelwood, Ibid., 858 (1938);
    d. H. C. Raine, and C. N. Hinshelwood, Ibid., 1378 (1939).
11. a. P. L. Kronick, and R. M. Fuoss, J. Am. Chem. Soc.,77, 6114
       (1955); b. E. Hirsch, and R. M. Fuoss, Ibid., 77, 6115 (1955);
    c. M. Watanabe, and R. M. Fuoss, Ibid., 78, 527 (1956); L.-Y.
    Chow, and R. M. Fuoss, Ibid., 80, 1095 (1958).
12. The model compounds included: pyridine, 4-picoline, 4-
    isopropylpyridine, N,N-dimethylaniline, p,p-methylene-bis-(N,N-
    dimethylaniline), 1,3-di-(4-pyridyl)propane, 1,2-di-(4-
    pyridyl)ethane, and 1,2-di-(4-pyridyl)ethylene.
13. R. M. Fuoss, M. Watanabe, and B. D. Coleman, J. Polym. Sci. 48,
    5.(1960).
14. C. L. Arcus, and W. A. Hall, J. Chem. Soc., 4199 (1963); 5995
    (1964).
15. J. Morcellet-Sauvage, and C. Loucheux, Makromol. Chem., 176, 315
    (1975).
16. a. E. A. Boucher, and C. C. Mollett, J. Polym. Sci. Polym. Phys.
       Ed. 15, 283 (1977); b. E. A. Boucher, J. A. Groves, C. C.
    Mollett, and P. W. Fletcher, J. Chem. Soc. Faraday Trans. 1, 73,
    1629 (1977); c. E. A. Boucher, E. Khosravi-Babadi, and C. C.
    Mollett, Ibid., 75, 1728 (1979).
17. W. G. Lloyd, and T. E. Durocher, J. Appl. Polym. Sci., 7
    2025.(1963); 8, 953 (1964).

18. a. H. Kawabe, and M. Yanagita, Bull. Chem. Soc. Jpn., 41, 1518 (1968); b. 44, 896 (1971); c. 46, 38 (1973); d. 46, 3627 (1973).

19. a. H. Kawabe, Bull. Chem. Soc. Jpn. 47, 2936 (1974); b. 48, 163 (1975); c. 49, 2043 (1976); d. 54, 1914 (1981); e. 54, 2886 (1981).

20. H. Kawabe, Sci. Pap. Inst. Phys. Chem. Res. (Jpn.) 76, 43 (1982).

21. I. Noda, and I. Kagawa, Kogyo Kagaku Zasshi (J. Chem. Soc. Japan, Ind. Chem. Sec.) 66, 857 (1963).

22. a. S. Dragan, I. Petrariu, and M. Dima, J. Polym. Sci. Polym. Chem. Ed. 10, 3077 (1972); b. 18, 2333 (1980); c. C. Luca, I. Petrariu, and M. Dima, Ibid., 17, 3879 (1979).

23. E. Tsuchida, and S. Irie, J. Polym. Sci. Polym. Chem. Ed., 11, 789 (1973).

24. R. N. Johnson, A. G. Farnham, R. A. Clendinning, W. F. Hale and C. N. Merriam, J. Polym. Sci. Part A-1, 5, 2375 (1967).

25. a. G. Allen, J. McAinsh and C. Strazielle, Eur. Polym. J., 5, 319 (1969); b. J. S. Ham, Polymer Preprints, 23 13 (1982).

26. F. Oosawa, "Polyelectrolytes" Marcel Dekker, Inc. 1971.

PARA-METHYLSTYRENE: A NEW COMMERCIAL MONOMER FOR THE STYRENICS

INDUSTRY

Warren W. Kaeding and George C. Barile

Mobil Chemical Company
P. O. Box 240
Edison, New Jersey 08818

INTRODUCTION

Styrene was first discovered in 1839 as a distillation component of balsam oil by E. Simon (1), who recorded the name "styrol"
for this substance. It was accompanied by a sticky, viscous
material which we can now safely assume was the corresponding polystyrene. This discovery began a field of chemistry that has evoked
intense academic and industrial interest for over a century and
whose fascinating early history is recorded in Boundy and Boyer's
treatise on styrene (2).

Serious commercial interest in styrene goes back to pre-World
War II years, when companies like I. G. Farbenindustrie and Dow
Chemical began developing technology. This work was rapidly accelerated when natural rubber from the Far East was abruptly cut off.
The GR-S rubber program was established in the United States to
develop synthetic substitutes (3). Copolymers of styrene and
butadiene were identified as suitable rubbers. Nine companies
banded together to design and build production in a crash program.
By 1944, 200 million pounds per year of styrene production was
achieved (4) and today U.S. capacity is greater than seven billion
pounds.

Many other vinyl aromatic monomers and polymers were synthesized and investigated during this postwar period of discovery
following styrene's early commercial success. Polymers of vinyl
toluene, based on a more plentiful supply of toluene, had properties similar to those of polystyrene and were proposed as an
attractive alternative (5). During this time, there was also great
interest in α-methylstyrene (6), divinylbenzene (7) and various
halogenated styrene monomers (8).

However, styrene is the simplest vinyl aromatic with the
fewest production and purification problems and an adequate supply
of benzene was soon assured by the development of a new process to
convert toluene to benzene (9). Construction of larger scale
plants and growing production experience accelerated the cost advantage of styrene over its rivals, and, as a result, styrene/poly-
styrene became the dominant vinyl aromatic monomer/polymer in the
marketplace. For three decades that situation has remained
virtually unchallenged by a substituted styrene.

MOBIL PMS MONOMER

Mobil PMS is a new styrenic monomer which is composed of 97%
para-methylstyrene and 3% meta isomer (0% ortho). It is a
different mixture of methylstyrene isomers than commercially
available vinyltoluene, which contains approximately 35% para and
65% meta isomers. It is important to compare the processes and
properties for PMS and vinyltoluene in order to understand the
significance of the new material. A summary of the typical properties of styrene, vinyltoluene and PMS is shown in Table 1. PMS
and vinyltoluene have higher boiling points and significantly lower
vapor pressure than styrene, which reduces the inhalation exposure
in the work environment and decreases the potential fire hazard.

The manufacture of vinyltoluene is analogous to that of styrene where toluene is substituted for benzene (Equations 1 and 2)
in a conventional acid catalyzed alkylation with ethylene. The
process gives rise to three isomers during alkylation (Equation
1). The close boiling points (Table 2) of the meta and para
isomers make it impractical to accomplish a separation by distillation (10). The ortho isomer, however, is removed and recycled by a
careful and costly distillation. This step is necessary because
some of the ortho isomer undergoes cyclization (Equation 3) during
the dehydrogenation step to produce indan and indene (11).

After removal of the ortho-ethyltoluene, which is recycled,
subsequent dehydrogenation produces the same isomer distribution as
in the alkylate. Although vinyltoluene has not approached styrene
in commercial significance in the polymer (discussed later), it has
found a place in specialty polymer and copolymer applications.

MOBIL PARA-METHYLSTYRENE (PMS): ALKYLATION REACTION

Mobil PMS is prepared by an acid-catalyzed alkylation of
toluene with ethylene to make ethyltoluene, which is subsequently
dehydrogenated to methylstyrene. The key technological innovation
which makes Mobil PMS different from vinyltoluene is the ability to
produce the para isomer in high selectivity (> 97%) during the.

Table 1

Mobil PMS Monomer Properties

Structural Formula

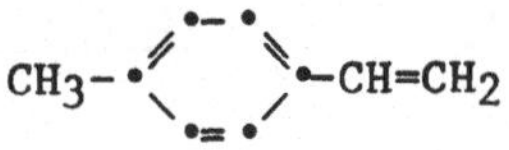

CH_3- ⬡ $-CH=CH_2$

Molecular Weight   118.18

Typical Isomer Distribution   97% para/3% meta/0% ortho

Chemical Analysis	Test Method	Typical Values		
		Mobil PMS Monomer	Styrene	Vinyl-Toluene
Purity, wt% vinyl	GC	99.7	99.7	99.5
Isomer Distribution	GC			
para		97	–	33
meta		3	–	66.7
ortho		–	–	0.3
Solvent Compatibility				
a. Acetone		∞	∞	∞
b. Carbon Tetrachloride		∞	∞	∞
c. Benzene		∞	∞	∞
d. Ether		∞	∞	∞
e. N-Heptane		∞	∞	∞
f. Ethanol		∞	∞	∞
ΔH vap. @ 25°C, cal/g		98.2	102.4	101.8
Specific Heat, @ 25°C, cal/g/°C		0.413	0.408	0.414
Refractive Index, n_D^{25}		1.5408	1.5440	1.5395
Density, g·cm⁻³ @ 25°C (77°F) 0.892		0.902	0.892	
Viscosity, cps @ 25°C		0.79	0.72	0.78
Surface Tension, dynes–cm⁻¹ @ 25°C		34	32	31
Boiling Point, °C @ 760 mmHg		170	145	168
Freezing Point, °C		– 34	– 31	– 77
Vapor Pressure, mmHg at 0°C (32°F)		–	1	–
20 (68°F)		–	4	1
40 (104°F)		4	14	4
60 (140°F)		14	38	13
80 (176°F)		35	90	35
100 (212°F)		80	191	83
120 (248°F)		170	363	176
140 (284°F)		325	661	341
160 (320°F)		580	1114	616

Refractive Index uses n_D^{25}.

Table 2

Some Significant Boiling Points (°C)

Isomer	Ethyltoluene[a]	Methylstyrene
Para	162	170
Meta	161	168
Ortho	165	171
Indane	177	
Indene	181	

(a) Reference 10

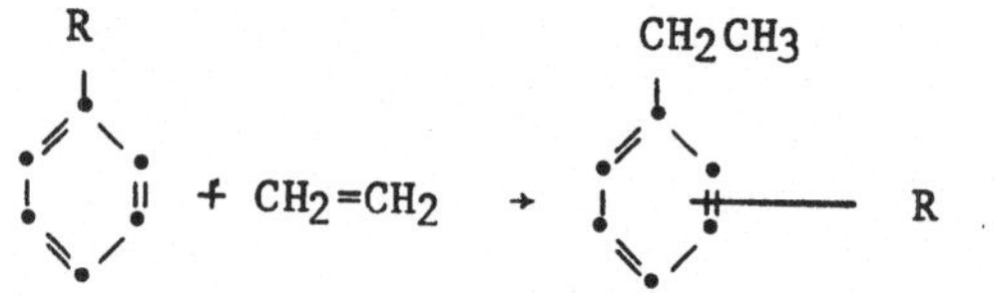

$$
\begin{array}{c}
\text{R} \\
\bigcirc + CH_2{=}CH_2 \rightarrow \overset{CH_2CH_3}{\bigcirc} - R
\end{array}
\tag{1}
$$

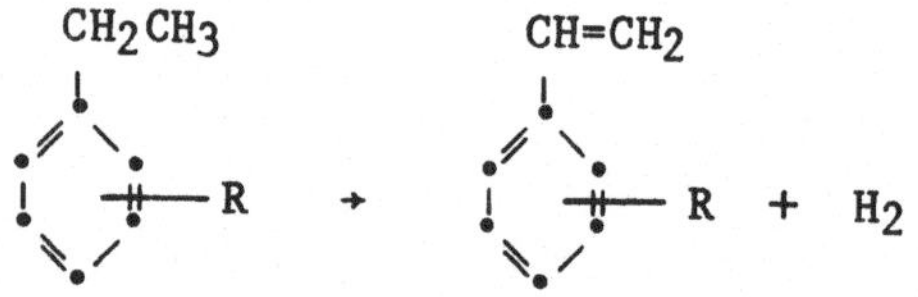

$$
\overset{CH_2CH_3}{\bigcirc} - R \rightarrow \overset{CH{=}CH_2}{\bigcirc} - R + H_2
\tag{2}
$$

$$
\overset{CH_2CH_3}{\bigcirc}\!\!\!^{CH_3} \xrightarrow{-H_2} \overset{CH_2-CH_2}{\bigcirc}\!\!\!^{CH_2} \xrightarrow{-H_2} \bigcirc\!\!\!\bigcirc
\tag{3}
$$

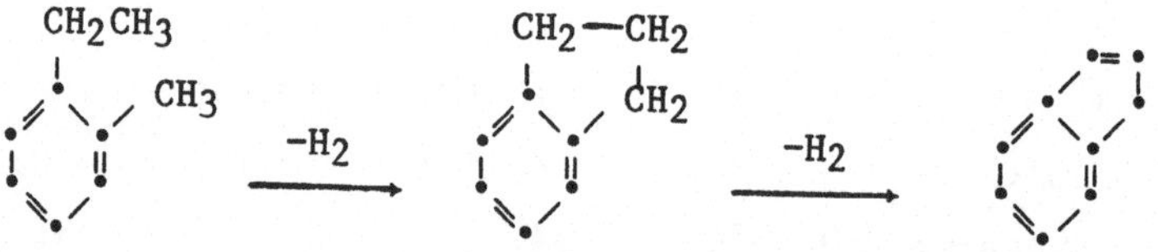

Indan   Indene

R = H (styrene) R=CH$_3$ (vinyltoluene)

alkylation by using a modified Mobil ZSM-5 zeolite catalyst. A comparison of typical product streams produced by alkylation of toluene with ethylene using conventional aluminum chloride, Mobil ZSM-5, and modified ZSM-5 is shown in Table 3. Aluminum chloride and unmodified ZSM-5 gives a near equilibrium ratio of meta/para isomers in the product ethyltoluene. The alkylation over modified ZSM-5 shows a dramatic increase in the yield of the para isomer. Furthermore, formation of the undesired ortho isomer is eliminated, Table 3.

Table 3

Alkylation of Toluene with Ethylene;
Composition of Typical Product Streams

Compound (wt%)	HCl–AlCl$_3$ [a]	Catalysts	
		ZSM-5 Class Zeolite	
		Unmodified	Modified
Light gas and benzene	0.2	1.0	0.9
Toluene [b]	48.3	74.4	86.2
Ethylbenzene and xylenes	1.2	1.2	0.5
p–Ethyltoluene	11.9	7.0	12.0
m–Ethyltoluene	19.3	14.7	0.3
o–Ethyltoluene	3.8	.3	0
Aromatic C$_{10}$+	14.4	1.4	0.1
Tar	0.9	0	0
Total	100.0	100.0	100.0
Ethyltoluene isomers (%)			
Para	34.0	31.8	97.5
Meta	55.1	66.8	2.5
Ortho	10.9	1.4	0
	100.0	100.0	100.0

(a) Reference 11.
(b) Excess toluene is used to prevent polyalkylation and
    resultant build-up of C$_{10}$+ and tars.

Zeolite catalysts are composed of silicon and aluminum oxides
in a crystal structure that is permeated by intracrystalline pores
and cavities of precise and uniform dimensions. Chemical reactions
occur primarily within these pores. If the intracrystalline structure is chosen to have certain precise dimensions, the ease of
accommodation of reactant and product molecules will depend critically on the shape of the molecules. It is thus possible to generate molecular shape selective catalysts (12-16).

A distinguishing feature of the new class of synthetic zeolites used in selective alkylation is a high silica/alumina mole
ratio, i.e., greater than 12. Since each aluminum has a net -1
charge, the presence of a cation is required to maintain ionic
balance. This cation is an acidic proton, which is the catalytic
site for alkylation to produce ethyltoluene (Equation 1).

The number of oxygen atoms bound to silicon and aluminum atoms in a zeolite pore opening determines its size. A ring of eight oxygen atoms characteristic of zeolite A, for example, is sufficiently large to admit and expel straight-chain aliphatic hydrocarbons, but is so small as to exclude aromatic and branched chain compounds.

Certain Mobil ZSM-5 type zeolites have pore openings with rings of 10 oxygen atoms (17). This structure permits access to reactant or product molecules with larger dimensions, such as substituted aromatic compounds, which can diffuse in and out of the catalyst. A mixture of toluene and alkylating agents, such as methanol or ethylene can easily enter the pores and react at an acidic site to produce the corresponding xylenes or ethyltoluenes. In previously reported work, thermodynamic equilibrium mixtures of isomers were produced (18). Furthermore, individual isomers were isomerized to the equilibrium mixture under alkylation conditions over a zeolite with similar properties (19).

More recently, methods have been developed for reducing the effective pore and channel dimensions. These techniques employ both physical treatments and chemical reagents. They have provided the basis for para-selective alkylation catalysts (18). These modified zeolites permit discrimination between molecules of slightly different dimensions. As a result, the para-isomers of the xylene or ethyltoluene products with the smallest minimum dimensions (Table 4) are able to diffuse out of the catalyst pores at rates about three orders of magnitude greater than those for the corresponding ortho- and meta-isomers (20). This discrimination capability is schematically represented in Figure 1, where the effective size of a para-selective catalyst pore is shown by the dashed line.

DEHYDROGENATION REACTION

The second step of the PMS process is a catalytic dehydrogenation of the ethyltoluene alkylate and is analogous to the conventional technology developed for the conversion of ethylbenzene to styrene (Equation 2). The hydrocarbon is mixed with steam in the vapor phase and passed over the catalyst at about 600°C to yield methylstyrenes and hydrogen. Many commercial catalysts contain mixtures of iron oxide, potassium carbonate and small amounts of other metals (22) as promotors and inhibitors. At these temperatures, side reactions such as step-wise cleavage of the side chain or even rupture of the aromatic ring itself are observed. With a methyl and ethyl group on the ring in para-ethyltoluene, side reactions occur on both groups to give xylenes and ethylbenzene in addition to the toluene and benzene by-products observed in styrene production. Modification of process conditions and catalysts is required to maximize yields of para-methylstyrene.

Table 4

Minimum Relative Dimensions of Alkyl
Aromatic Molecules [a]

Hydrocarbon	Minimum Cross Section (Å)
Benzene	7.0
Toluene	7.0
Ethylbenzene	7.0
o-Xylene	7.6
m-Xylene	7.6
p-Xylene	7.0
o-Ethyltoluene	7.7
m-Ethyltoluene	7.6
p-Ethyltoluene	7.0

(a) From Fischer-Hirschfelder-Taylor hard sphere molecular
     models.

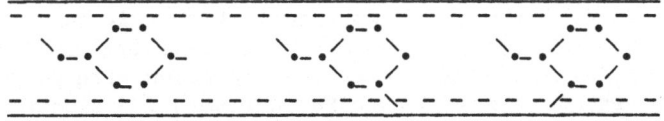

Figure 1

This need was recognized by the producers of vinyltoluene and a
number of patents have appeared that describe various promoter
oxides such as zinc or copper (23), mixtures of different iron
oxides (24), and procedures for operation and regeneration (25).

We have identified dehydrogenation catalysts with high activity, stability for prolonged steady-state operation, excellent
selectivity to product and operability at practical steam/ethyl-
toluene feed ratios.

This research on alkylation and dehydrogenation has culminated
in the construction and operation of a 35 MM lb. PMS monomer production unit jointly owned and operated by Mobil and the American
Hoechst Corporation at the latter's facility in Baton Rouge,
Louisiana.

POLYMERIZATION BEHAVIOR

Since our ambitious goal from the outset was to replace significant volumes of polystyrene in present markets and to develop new
uses based on special properties of this new monomer, the polymerization behavior and the physical and processing characteristics of
poly-para-methylstyrene were extensively studied. The early literature contains considerable information on polymers of the individual methylstyrene isomers and documents the effort expended to
develop reliable characterization methods (5, 8, 11). Attention
was also focused on the subtleties of the polymerization process
itself. A consequence of this early work was the establishment of
increasingly more rigorous specifications for monomers and the
corresponding polymers and copolymers. But more definitive results
had to await development of modern chromatographic instruments and
analytical techniques. Recent work suggests that the isomeric
composition of early monomer samples was not accurately known.
Consequently, the reported properties of the corresponding polymers
may not be accurate.

In order to establish the preferred para isomer content for
this new monomer, the properties of the polymers produced from
methylstyrenes over a wide range of isomer compositions were investigated. The measurements clearly indicated that the higher the
para content, the higher the glass transition temperature (Figure
2). Similar trends were observed for the Vicat softening temperature (26). The monomer containing 97% para-methylstyrene and 3%
meta isomer was chosen for commercialization. It is a mixture
which can be efficiently produced and gives polymers with superior
thermal and processing qualities in comparison with polystyrene.
This 97/3 mixture has been used for our detailed process development, product evaluations and commercial production.

The storage stability of PMS is similar to that of other styrenic monomers over a broad range of conditions. The heat of
polymerization is comparable to that of styrene and vinyltoluene.
The copolymerization Q and E values are also similar (27). The
volume contraction on polymerization, calculated at 25°C, is lower
than for styrene. These data are summarized in Table 5.

The kinetic behavior of PMS in thermal and peroxide-initiated
polymerizations was compared to that of styrene. The isothermal
polymerization rates for PMS and styrene were measured over the
conventional thermal polymerization temperature range of 110°C to
150°C. The conversions for inhibitor-free monomers were monitored
as a function of time and were very similar (Figure 3). The
effects of solvent on polymerization rates were also measured and
found to be similar. In 10% solvent, either ethylbenzene or para-
ethyltoluene, the conversion rates are about 12% lower than without
solvent for both styrene and PMS. The molecular weights and

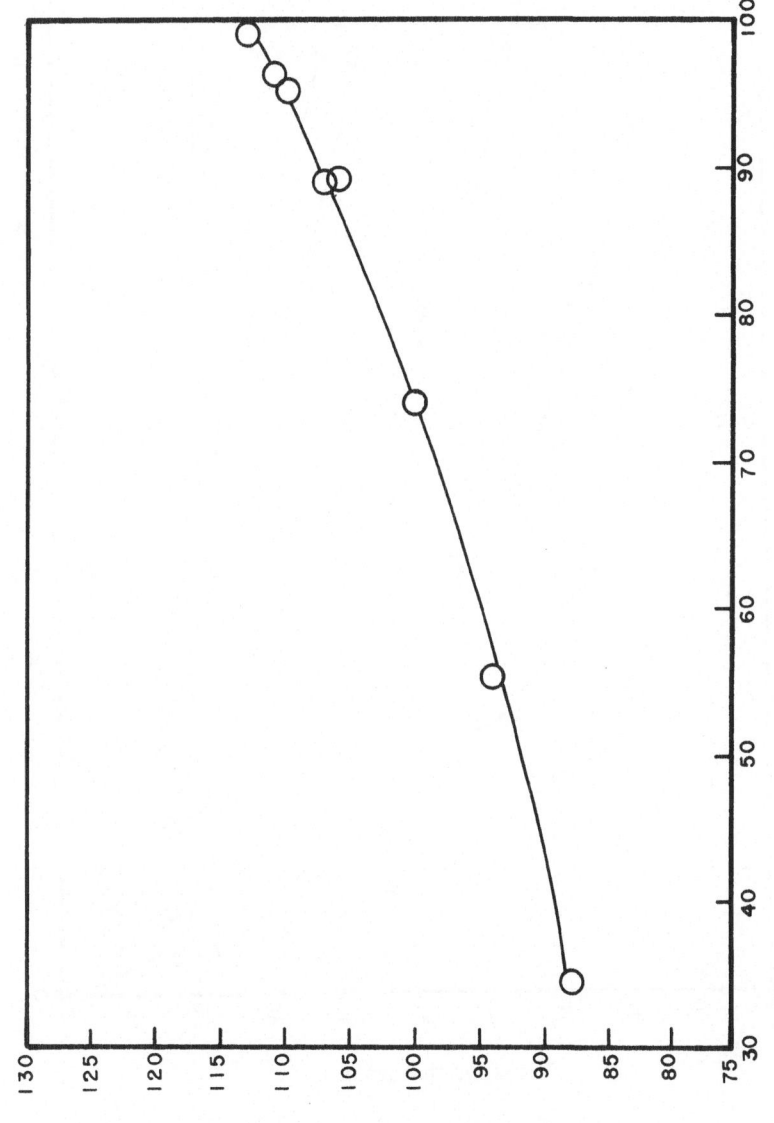

DEPENDANCE OF THE GLASS TRANSITION TEMPERATURE
ON THE PARA ISOMER CONTENT OF PMS POLYMERS

PARA ISOMER, WT PC

Figure 2

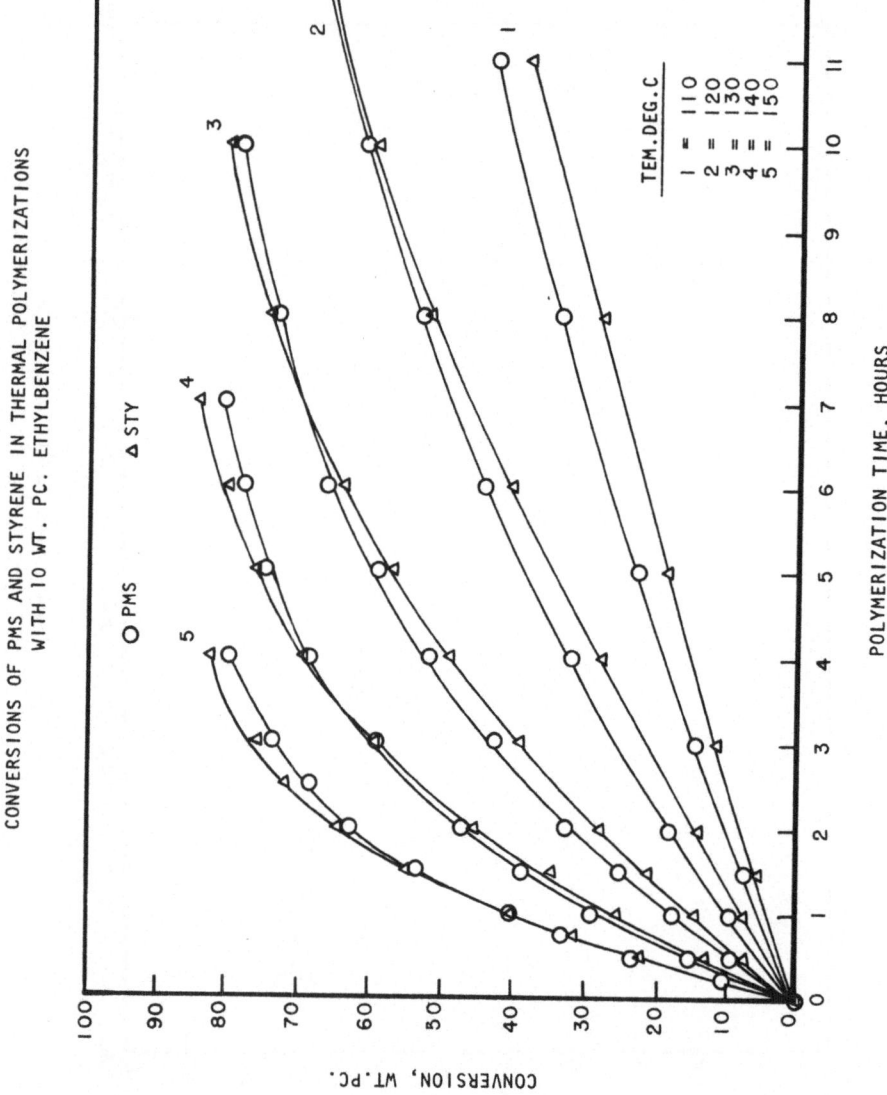

CONVERSIONS OF PMS AND STYRENE IN THERMAL POLYMERIZATIONS
WITH 10 WT. PC. ETHYLBENZENE

O PMS   △ STY

TEM. DEG. C		
1	=	110
2	=	120
3	=	130
4	=	140
5	=	150

POLYMERIZATION TIME, HOURS

CONVERSION, WT. PC.

Figure 3

molecular weight distributions of these thermal polymers were very similar.

The solution or process viscosities of poly-PMS are also similar to those of polystyrene in bulk-solvent systems. In broad terms, the polymerization kinetics, molecular weight, heat transfer and viscosities of PMS are comparable to those observed in the homopolymerization of styrene.

A wide variety of polystyrene-like polymers and copolymers (crystal, impact modified, ABPMS, PMS-AN, PMS-BR, PMS-MA and PMS-MMA)(28) have been prepared from PMS using bulk, solvent and suspension polymerization techniques in our laboratories and pilot plants using thermal, anionic and chemical initiation. From a resin manufacturing point of view, PMS monomer can be processed in existing styrene polymerization equipment to produce poly-PMS analogues. However, process development must be done to optimize conditions for each resin type.

POLYMER PROPERTIES

The large variety of PMS polymers and copolymers being prepared in our investigations are compared to the polystyrene analogs. The basic property differences seen in homopolymer are also observed in other derivatives. Poly-PMS has a greater free volume than the polystyrene equivalent, as indicated by its 4% lower specific gravity. The glass transition temperature is 113°C, 10°C higher than for polystyrene. Similar increases in Vicat softening and heat distortion temperatures were also observed. A comparison of the densities and the Vicat softening temperatures of four resin types is shown in Table 6. From the table, it can be seen that the extent of the difference is a function of the PMS concentration in the polymer.

A comparison of the physical properties of a high heat homopolymer and a rubber modified resin is shown in Tables 7 and 8.

The other properties of the homopolymer of PMS are similar to those of polystyrene, with the exception of the flexural modulus, where poly-PMS is 8% less stiff and the hardness is slightly higher. Both homopolymers are highly transparent, the haze being under 5%.

POLYMER FABRICATION

Techniques for compounding, fabrication, pigmentation, pelletizing, inhibiting and reinforcing of the two polymer types are also very similar. PMS resins have two processing differences. The higher heat properties of poly-PMS decrease molding cycle times for crystal and impact resins by shortening the cooling

Table 5

Properties	Mobil PMS	Styrene	Vinyltoluene
Heat of Polymerization Kcal/Mole.)	16 ± 1	17 ± 1	16 ± 1
Volume Contraction on Polymerization (% Calculated at 25°C)	12	14	12
Copolymerization Q & E Values			
Q	1.27	1.00	1.06
E	−0.98	−0.80	−0.78

Table 6

	Specific Gravity		Vicat Softening Temperatures	
Resin Type	Mobil PMS	Styrene	Mobil PMS	Styrene
Homopolymer	1.01	1.05	116	108
Rubber Modified	1.01	1.05	108	100
Acrylonitrile Copolymer	1.04	1.07	112	108
ABS Polymer	1.01	1.03	107	103

time in the mold. In-house and field evaluations have demonstrated 5-20% cycle time advantages in molding thin and thick wall parts, depending on molding conditions.

PMS resins have also shown better mold fill properties than polystyrene analogs in spiral flow and commercial molding studies because of poly-PMS' lower melt viscosities at typical molding temperatures (Figure 4). This advantage reduces the need for mercaptan modifiers to control molecular weight or for lubricants to improve flow. Use of these modifiers can adversely affect thermal and/or physical properties of polystyrene. Foam extrusion and thermoforming properties of poly-PMS appear to be similar to those of polystyrene, with the PMS polymers generally exhibiting higher melt strength.

Table 7

Comparison of Physical Properties of
High Heat Crystal Poly-PMS and Polystyrene

Typical Properties	ASTM Test Method	Unit	Typical Value	
			Poly-PMS	Polystyrene
Nominal Melt Flow Rate(200°C, 5000g, Condition G)	D-1238	g/10 min.	4.1	4.0
Vicat Softening Temp.	D-1525	°C	116	108
Deflection Temp. Under Load at 264 psi (Injection Molded, Annealed)	D-648	°C	92	88
Tensile Strength at Break	D-638	psi	7,200	7,200
Tensile Elongation at Break	D-638	%	3.0	3.0
Tensile Modulus	D-638	psi	320,000	360,000
Flexural Modulus	D-790	psi	434,000	465,000
Flexural Strength at Break	D-790	psi	11,500	13,400
Izod Impact Strength at 73°F (1/8" thick injection molded specimen, notched)	D-256	ft-lbs/in.	0.3	0.3
Hardness - Rockwell	D-785	M Scale	80	70

Table 8

Comparison of the Physical Properties of
High Impact Poly-PMS and Polystyrene

Typical Properties	ASTM Test Method	Unit	Typical Value	
			HIPPMS	HIPS
Nominal Melt Flow Rate (200°C, 5,000g, Condition G)	D-1238	g/10 min.	3.5	4.0
Izod Impact Strength at 73°F (1/8" Shick Injection Molded Specimen, notched)	D-256	ft-lb/in	1.6	1.8
Vicat Softening Temperature	D-1525	°C	108	100
Tensile Yield	D-638	psi	4,000	3,800
Tensile Strength at Break	D-638	psi	4,100	3,900
Tensile Elongation	D-638	%	20	35
Tensile Modulus	D-638	psi	220,000	240,000
Flexural Strength at Break	D-790	psi	6,000	5,400
Flexural Modulus		psi	240,000	290,000
Hardness, Rockwell	D-785	L Scale	65	60

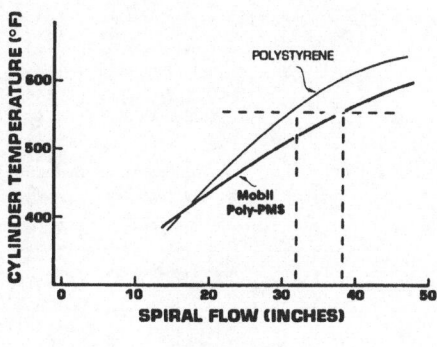

Figure 4

Much work remains to be done in defining the fabrication profile for each specific resin type. However, indications are that
poly-PMS can be used interchangeably with polystyrene resins in
many applications with some processing advantages.

POLYMER PRODUCT

In addition to a lower density than polystyrene, poly-PMS
shows several other product differences which are advantageous in
certain uses. Poly-PMS has potential for those applications where
polystyrene has borderline thermal performance characteristics. An
extra safety margin is provided for higher temperature use and
storage. Poly-PMS tumblers can withstand boiling water in a micro-
wave oven, whereas polystyrene parts become severely distorted in a
few minutes under these conditions (29). Similar behavior is observed in dishes placed in a forced hot air oven. The poly-PMS
dish survives 104°C for 10 minutes without distortion, whereas the
polystyrene part shrinks and distorts.

A potential area of application in which poly-PMS can give a
significant improvement in cost and performance is flame retar-
dance (FR) or ignition resistance. PMS-based impact resins require
lower loadings of costly and heavy conventional FR additives to

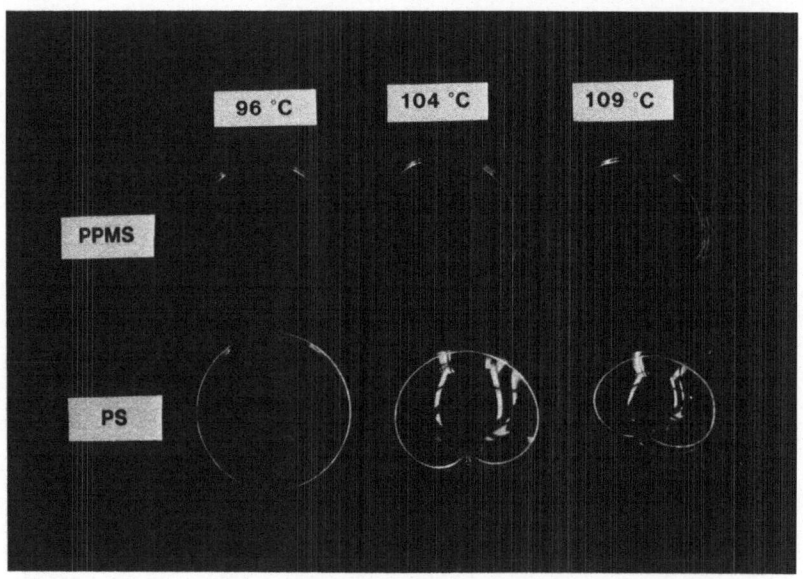

Figure 5

Left to right: Dishes exposed to 96°C, 104°C and 109°C for 10
minutes show the superior thermal properties of poly-PMS.

make products that meet the Underwriters Laboratory V-0 flammability rating (30). The antimony oxide and decabromodiphenyl-
oxide flame retardant system normally is added to impact polystyrene in 15-20% by weight in order to meet the requirement. With
poly-PMS, however, the same effect is achieved with 20-30% less
additives. The differences between the poly-PMS and polystyrene
based resins in their flame retardant formulation is thought to be
related to the differences in the mechanism of polymer degradation
in the flame. Poly-PMS can undergo chain extension whereas polystyrene undergoes chain scission. This difference reduces the need
for additives to flame retard the PMS-based system.

Polystyrene is very difficult to cross-link by electron beam
radiation or by peroxide without destroying its properties. As a
result, applications have not been developed. The methyl groups of
poly-PMS, by contrast, provide susceptible positions for such a reaction. PMS polymers can be cross-linked effectively by electron-
beam radiation. This unique property of ·poly-PMS could open
several new markets. Crosslinking turns a PMS-based thermoplastic
into a thermoset resin, significantly improving its grease resistance and flammability behavior.

Another area where the differences between PMS and styrene may
prove advantageous is in unsaturated polyester formulations. Sheet

and bulk molding polymers (thermoset, fiberglass reinforced, unsaturated polyester resins) contain about 30-50% styrene or vinyltoluene. Recently, the latter has been used more frequently for
this application because of growing concern for exposure of workers
to monomer vapors in the atmospheres of factories, warehouses or
confined spaces. This consideration is especially important in the
fabrication of large objects such as storage tanks or boat hulls,
production of automobile or appliance parts on assembly lines and
underground uses of products in mines where ventilation is difficult. The higher boiling point and lower vapor pressure of PMS
(and vinyltoluene), compared with styrene (Table 1) reduce this
potential hazard significantly. At 140°C, for example, the vapor
pressure of PMS is 60% lower than that of styrene.

In addition to having the vapor pressure advantage of vinyltoluene, PMS also provides an improvement in thermal properties
(probably associated with the high amount of para-isomer) when used
in the preparation of unsaturated polyester moldings.

TOXICOLOGY

Extensive toxicological testing of PMS monomer, polymers derived from PMS and of p-ethyltoluene has been conducted to determine the safety of these materials. The program has included tests
for acute oral, dermal and inhalation toxicity, skin and eye irri-
tation, mutagenicity and delayed neurotoxicity on several test
animal species. Chronic 90-day tests have been completed and two-
year chronic tests, including oncogenicity, are underway. An
extensive metabolism study has been initiated. Results thus far
indicate that PMS monomer and its polymers are similar in toxicology to styrene and polystyrene analogues. These observations
support the view that these materials have excellent potential for
a broad range of applications in industrial and consumer-oriented
products.

SUMMARY

Styrene has enjoyed a special position among vinyl aromatic
monomers because of its lack of isomers and consequent ease and
simplicity of production. Additional steps are normally required
to make substituted styrene derivatives, and separation and purification of isomers are usually difficult. Our discovery of novel
technology to produce p-ethyltoluene is a unique departure from
this situation. A highly selective catalyst enables the manufacture of a substituted vinyl aromatic monomer without coproducing
undesirable and difficult-to-separate isomers. Furthermore, direct
use of toluene eliminates the need for hydrodemethylation by which
much of the benzene starting material for styrene is produced.

Mobil PMS monomer is potentially a broad-scale replacement for styrene or vinyltoluene in polymer applications. In many cases, distinct and valuable property advantages have been demonstrated. These encouraging results from laboratory and pilot plant scale operations led to the design and construction of an alkylation unit based on Mobil's para-selective catalyst. In cooperation with American Hoechst Corporation, this unit was installed and integrated into an existing styrene plant in Baton Rouge, Louisiana, nominally producing 35 million pounds per year of PMS monomer. In the summer of 1982, the first large quantities of PMS monomer were produced. Supplies are now available for large-scale evaluation and commercial development. With continuing positive results from these evaluations, Mobil PMS will add a new dimension to the styrenics industry.

REFERENCES

(1)   Simon, E. Ann. 1839, 31, 265.

(2)   Boundy, R. H., Boyer, R. F., "Styrene--Its Polymers, Copolymers, and Derivatives"; Reinhold; New York, 1952.

(3)   Baruch Committee Report. "The Rubber Situation," House Document No. 836, U.S. Printing Office, Sept. 12, 1942; p. 28.

(4)   Boundy, R. H.; Stoesser, S. M., "Styrene--Its Polymers, Copolymers, and Derivatives"; Reinhold; New York, 1952; p. 11.

(5)   Amos, J. L.; Coulter, K. E.; Wilcox, A. C.; Soderquist, F. J., "Styrene--Its Polymers, Copolymers, and Derivatives"; Reinhold: New York, 1952; p. 1232.

(6)   Amos, J. L.; Coulter, K. E.; Tennant, F. M., "Styrene--Its Polymers, Copolymers, and Derivatives"; Reinhold; New York, 1952; Chapter 15.

(7)   Amos. J. L.; Rubens, L. C.; Horbacher, "Styrene--Its Polymers, Copolymers, and Derivatives"; Reinhold; New York, 1952; Chapter 16.

(8)   Amos, J. L.; Everson, J. W., "Styrene--Its Polymers, Copolymers, and Derivatives"; Reinhold; New York, 1952; Chapter 17.

(9)   Oliver, E. D. Stanford Res. Inst. Process Economics Report No. 30, 1967.

(10) Ferris, S. W. "Handbook of Hydrocarbons"; Academic Press; New York, 1955.

(11) a. Coulter, K. E.; Kehde, H.; Hiscock, B. F. "High Polymers"; Leonard, E. C., Ed.; Wiley-Interscience; New York, 1971; Vol. 24, pp. 548-549; b. Dixon, J. K.; Saunders, K. W. Ind. Eng. Chem.1954, 46, 652; c. Melchore, J. A. Modern Plastics 1956, 33, 163; d. Melchore, J. A. SPE J. 1957, 12, 33; E. Hennessy, B. J. Plastic Inst. (London), Trans. J. 1959, 27, 126.

(12) Weisz, P. B.; Frilette, V. J. J. Phys. Chem. 1960, 64, 382.

(13) Plank, C. J.; Rosinsky, E. J.; Hawthorne, W. P. Ind. Eng. Chem. Prod. Res. Dev. 1964, 3, 165.

(14) Chen, N. Y.; Weisz, P. B. Chem. Eng. Prog. Symp. Ser. 1967, 63, 86.

(15) a. Meisel, S. L.; McCullough, J. P.; Lechthaler, C. H.; Weisz, P. B. CHEMTECH 1976, 6, 86; b. Meisel, S. L. Philos. Trans. R. Soc. London 1981, A300, 157; c. Csicsery, S. M. "Zeolite Chemistry and Catalysis"; ACS Monograph No. 171, Rabo, J. A., Ed.; American Chemical Society: Washington, D.C., 1976; d. Weisz, P. B. Pure Appl. Chem. 1980, 52, 2091.

(16) Robson, H. CHEMTECH 1978, 176.

(17) a. Kokotailo, G. T.; Lawton, S. L.; Olson, D. H.; Meier, W. H. Nature 1978, 272, 437; b. Kokotailo, G. T.; Chu, P.; Lawton, S. L.; Meier, W. H. Nature 1978, 275, 120.

(18) a. Kaeding, W. W.; Chu, C.; Young, L. B.; Weinstein, B.; Butter, S. A. J. Catal. 1981, 67, 159; b. Kaeding, W. W., Chu, C., Young, L. B., Butler, S. A., J. Catal. 1981, 69, 392.

(19) Grandio, P.; Schneider, F. H.; Schwartz, A. B.; Wise, J. J. Hydrocarbon Process, 1972, 8, 85.

(20) Chen, N.Y.; Kaeding, W. W.; Dwyer, F. G. J. Am. Chem. Soc. 1979, 101, 6783.

(21) Csicsery, S. M. J. Catal. 1970, 19, 394

(22) Lee, E. H. Catal. Rev. 1973, 8(2), 285.

(23) Soderquist, F. J.; Boyce, H. D.; Butts, R., U.S. Patent 4,064,187, 1977.

(24) Turley, R. R.; Castor, W. M.; Nunnally, K. R., U.S. Patent 3,849,329, 1974.

(25) Soderquist, F. J.; Wazbinski, T. T.; Waldman, N., U.S. Patent 3,907,916, 1975.

(26) Polymer Preprints, ACS; Vol 23, Number 2, 93, 1982.

(27) Polymer Handbook, Second Ed., II – 387, 1975.

(28) (a) Murray, J. G.; U.S. Patent 4,367,320, 1982.
     (b) Prapas, A. G.; U.S. Patent 4,362,854, 1982.
     (c) Schwab, F. C.; U.S. Patent 4,260,694, 1981.
     (d) Prapas, A. G.; U.S. Patent 4,242,485, 1980.
     (e) Schwab, F. C.; U.S. Patent 4,176,144, 1979.
     (f) Schwab, F. C.; European Patent Appl.; EP 3405; 790808.
     (g) Myers, G. L., Org. Coating & Applied Polymer Science Proceedings, ACS, Vol 46, 302, 1981.
     (h) Sherman, A. M.

(29) Kaeding, W. W.; Young, L. B.; Prapas, A. G.; CHEMTECH 1982, 556.

(30) Tests for Flammability of Plastics Materials for Parts in Devices and Appliances, UL-94, 1978.

NEW VINYL ORGANOMETALLIC MONOMERS: SYNTHESIS AND POLYMERIZATION BEHAVIOR

Marvin D. Rausch,[*] David W. Macomber, and Francis G. Fang

Department of Chemistry
University of Massachusetts
Amherst, Massachusetts   01003

Charles U. Pittman, Jr.,[*] T.V. Jayaraman
and Ralph D. Priester, Jr.

Department of Chemistry
University of Alabama
University, Alabama   35486

INTRODUCTION

Organometallic polymers are useful in a variety of applications such as catalysts, UV absorbers, semiconductors, and antifouling agents.[1-3] The construction of organometallic polymers can proceed by derivatizing preformed polymers with organometallic units[3] or by preparing monomers with organometallic functions and polymerizing these monomers.[2] Monomers 1-3 represent three examples of organometallic monomers which have been polymerized.[4-7] In 2[6] and 3,[7] the organometallic unit is far removed from the polymerizable vinyl group and does not directly effect the polymerization behavior via electronic effects. However, in vinylferrocene, 1, the organometallic moiety is directly attached to the vinyl group which leads, in turn, to some unusual effects in polymerization. For example, the rate of polymerization of vinylferrocene is first order in both monomer and initiator when initiated by AIBN in benzene (V_p = 5.64 x $10^{-4}[1]^{1.12}$ $[AIBN]^{1.11}$).[5] This is the result of a monomolecular termination mechanism caused by an electron transfer from iron to the radical center, followed by termination and subsequent decomposition to a paramagnetic Fe(III) center in the polymer. The presence of the transition metal, Fe, in the monomer permits such an unusual redox behavior which is

243

impossible for 'normal' organic monomers (equation 1).

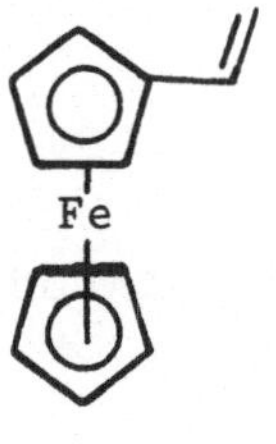

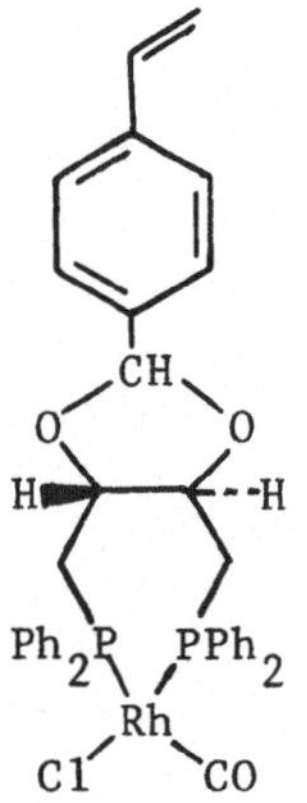

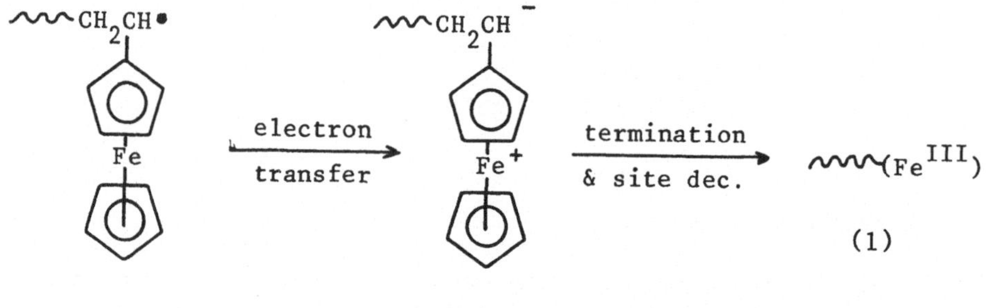

Vinylferrocene has been polymerized using radical,[4,5,8] cationic,[9] and Ziegler-Natta[10] initiators, but is inert to anionic initiation. Copolymerization studies have demonstrated that vinylferrocene has an exceptionally electron-rich vinyl group. An "e" value of -2.1 was established from copolymerizations with styrene.[11] Thus, 1 has a more electron-rich vinyl group in radical addition polymerization than does 1,1'-dianisylethylene (e = -1.96). The η^5-C_5H_5Fe moiety is therefore a strong electron donor to radical centers, a feature which is not unexpected in view of the known stabilizing effect of ferrocene on α-ferrocenyl cations. Finally, unlike most vinyl monomers, the molecular weight of poly(vinyl-ferrocene) does not increase with a decrease in initiator concentration, but it does increase with increasing monomer concentration.[4] This anomalous behavior is the result of vinylferrocene's high chain-transfer constant (C_m = 8 x 10^{-3} versus C_m = 6 x 10^{-5} for styrene at 60°C).[5]

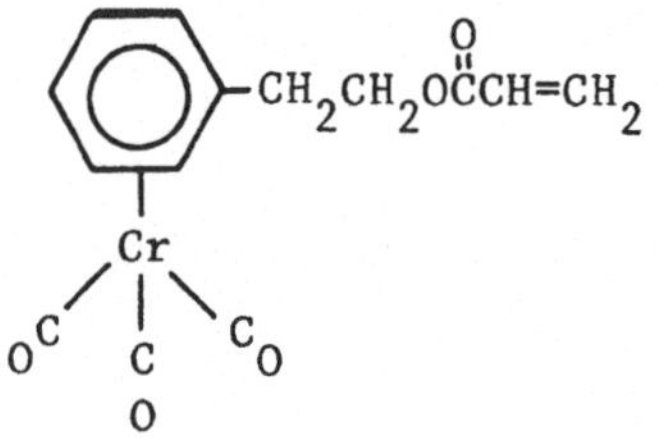

All of these interesting features of the model organometallic monomer $\underline{1}$ raised questions about other η^5-vinylcyclopentadienyl-transition metal monomers. Could they be synthesized, polymerized, or copolymerized? To that end, (η^5-vinylcyclopentadienyl)tricarbonyl-manganese, $\underline{4}$,[12,13] (η^5-vinylcyclopentadienyl)dicarbonylnitrosyl-chromium, $\underline{5}$,[14,15] and 3-vinylbisfulvalenediiron, $\underline{6}$,[16] were prepared.

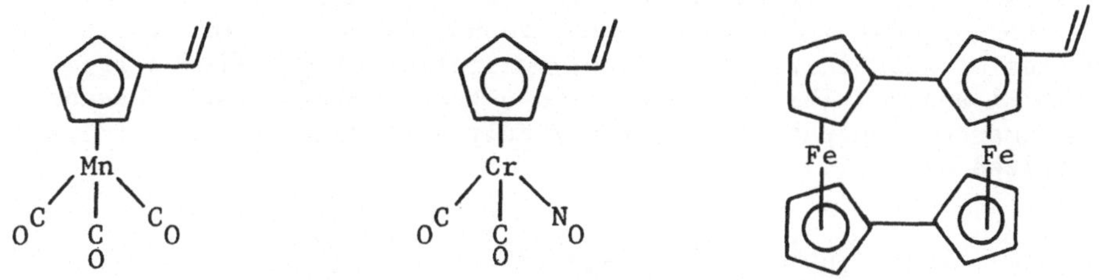

$\underline{4}$ $\qquad$ $\underline{5}$ $\qquad$ $\underline{6}$

In each case, the synthetic route involved Friedel-Crafts acylation of the Cp-metal complex followed by hydride reduction of the acetyl carbonyl function to an alcohol, and dehydration. This approach is illustrated in equation (2) for monomer $\underline{5}$.

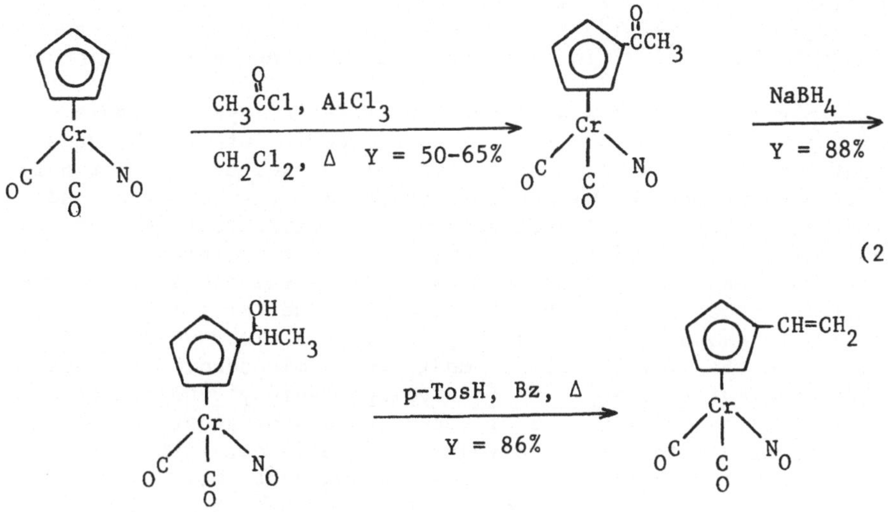

(2)

$\underline{5}$

Both 4^{12} and 5^{15} could be homopolymerized and copolymerized
with a variety of electron-donating or -attracting organic comonomers
(i.e., N-vinylpyrrolidone, styrene, methyl acrylate, acrylonitrile).
Remarkably, both 4^{12} and 5^{15} also exhibited large negative Alfrey-
Price "e" values (-1.99 and -1.98, respectively), almost identical
to that of vinylferrocene. Thus, both monomers have exceptionally
electron-rich vinyl groups. Moreover, replacing the second η^5-C_5H_5
in 1 by 3 CO groups, or by 2 CO and an NO group in 4 and 5, respectively, did not significantly change the "e" value. Unlike 1, the
molecular weight of the homopolymer of 5 increased as the 5/AIBN
ratio increased, suggesting that the rate was not first order in
initiator concentration as had been found previously for 1. The
kinetics of AIBN-initiated homopolymerization of 4 differed from
that of 1 and 5. The rate exhibited a three-halves order dependence
on monomer concentration in three widely different solvents,[13] as
summarized below.

$$V_p = 1.313 \times 10^{-4}[4]^{1.45}[AIBN]^{0.48} \quad \text{(in benzene)}$$

$$V_p = 1.980 \times 10^{-4}[4]^{1.58}[AIBN]^{0.47} \quad \text{(in benzonitrile)}$$

$$V_p = 1.50 \times 10^{-4}[4]^{1.54}[AIBN]^{0.47} \quad \text{(in acetone)}$$

$$\text{where } V_p \text{ is in molL}^{-1}s^{-1} \text{ and k in Lmol}^{-1}s^{-1}.$$

A rate equation of the form $V_p = k[M]^{1.5}[I]^{0.5}$ is derived, assuming
that the initiator efficiency is low and therefore initiation is
proportional to [M]. When this is the case, then $DP = V_p/V_i = [k_p/(2f'k_tk_d)^{0.5}]([M]/[I])^{0.5}$. Molecular weight measurements confirmed $DP \sim ([M]/[I])^{0.5}$ in agreement with this mechanistic picture.

At this stage, it was apparent that (η^5-vinylcyclopentadienyl)-
transition metal monomers were of great academic interest because of
the profound effect the organometallic moiety exerted on their polymerization behavior. We wished to further vary the structure and
examine, quantitatively, the effects of these structural variations
on polymerization behavior. However, many other features of organometallic polymers remained of interest. Mixed valence state polymers
of vinylferrocene (i.e., with both ferrocene and ferricinium units
present) were shown to be charge-hopping type semiconductors[17-19]
and more recently of interest for modifying semiconductor surfaces
for photochemical reactions.[20] (η^5-Cyclopentadienyl)rhodium di-
carbonyl sites on polymers are of interest as hydrogenation and
hydroformylation catalysts. Polymers of 5 could catalyze hydrogen-
ations of methyl sorbate.[15] Both thermal and photochemical stim-
ulation of polymers containing units of 7^{21} or 8^{22} resulted in
decomposition of the organometallic moiety. The metal carbonyl

stretching bands in the infrared spectra steadily disappeared and
metal oxide particles of tiny sizes were generated within the polymer
matrix.   Transition metal-containing polymers should have possible
applications as novel lithographic resists and in a myriad of other
areas.

$$\underline{7} \qquad\qquad\qquad \underline{8}$$

MONOMER SYNTHESIS

In order to further explore the field of organometallic polymers, one of the major needs has been for facile synthetic routes,
of general nature, to prepare a variety of new organometallic monomers.   The electrophilic aromatic substitution route used for the
preparation of $\underline{1}$, $\underline{4}$, $\underline{5}$, and $\underline{6}$ is severely limited in scope.   Only
a few η^5-cyclopentadienyl metal complexes other than ferrocene
undergo electrophilic aromatic substitution.[23]   These include ruthenocene and osmocene,[24] $(\eta^5-C_5H_5)M(CO)_3$,[25-29] where M = Mn, Tc
and Re, $(\eta^5-C_5H_5)V(CO)_4$,[30-32] $(\eta^5-C_5H_5)Cr(CO)_2(NO)$,[33] $(\eta^5-C_5H_5)$-
$Co(CO)_2$,[34] and $(\eta^5-C_5H_5)(\eta^4-C_4Ph_4)Co$.[35] As a consequence, alternate methods for the preparation of functionally substituted cyclopentadienyl metal compounds became an important goal of our
research.   Sodium cyclopentadienide containing aldehyde, ketone,
and ester substituents can be made by the reactions of C_5H_5Na with
ethyl formate, methyl acetate, methyl chloroformate, and dimethyl
carbonate (Scheme 1).[36-39]

By reacting reagents $\underline{9}$, $\underline{10}$, or $\underline{11}$ with $Co_2(CO)_8$ and I_2 in THF,
the corresponding formyl, $\underline{12}$, acetyl, $\underline{13}$, and carbomethoxy, $\underline{14}$,
$(\eta^5$-cyclopentadienyl)cobalt derivatives were prepared.[38] Their
rhodium counterparts, $\underline{15}$-$\underline{17}$, were made from reactions of $\underline{9}$, $\underline{10}$, and
$\underline{11}$ in THF with $[Rh(CO)_2Cl]_2$.[38] It appeared likely that this route
would be useful in the formation of vinyl-containing organometallic
monomers.   We therefore applied this approach to the synthesis of a
desired monomer, $(\eta^5$-vinylcyclopentadienyl)tricarbonylmethyltungsten,
$\underline{18}$.[40] The overall synthesis of $\underline{18}$ is shown in Scheme 2.   The yield
in the synthesis of $\underline{18}$ was good; however, this method requires a
Wittig reaction in the final step.   Although the latter reaction
was successful in the synthesis of $\underline{18}$ when phase-transfer conditions

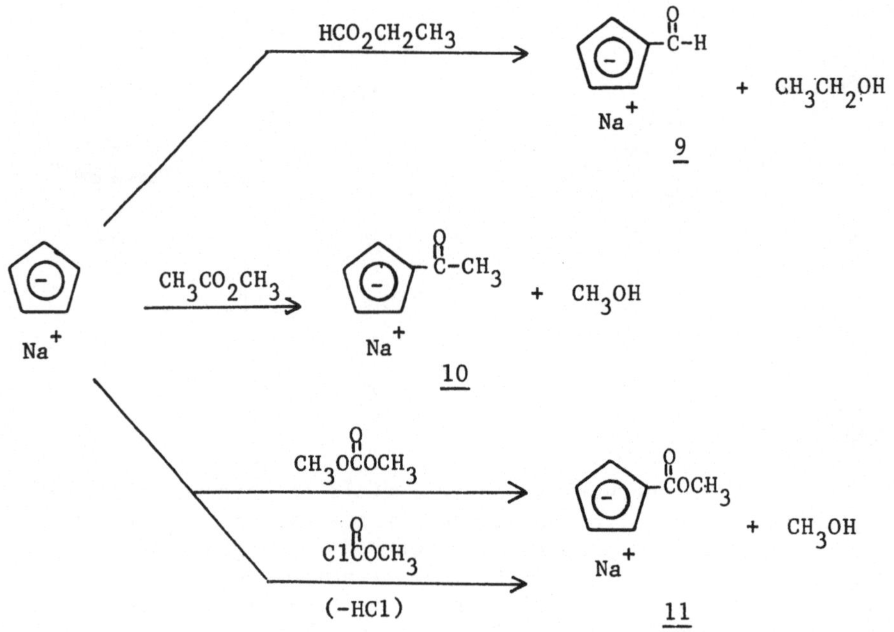

Scheme 1

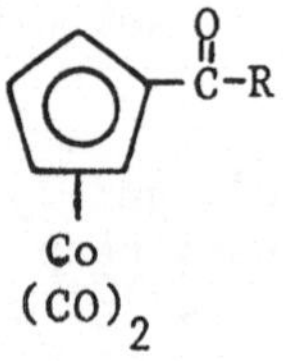

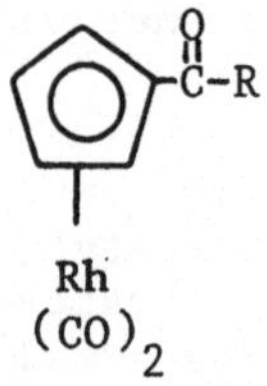

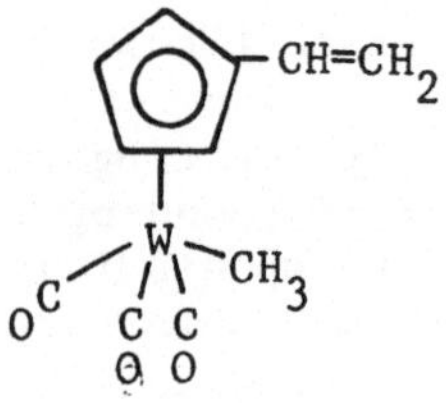

12 R = H	15 R = H
13 R = CH₃	16 R = CH₃
14 R = OCH₃	17 R = OCH₃

18

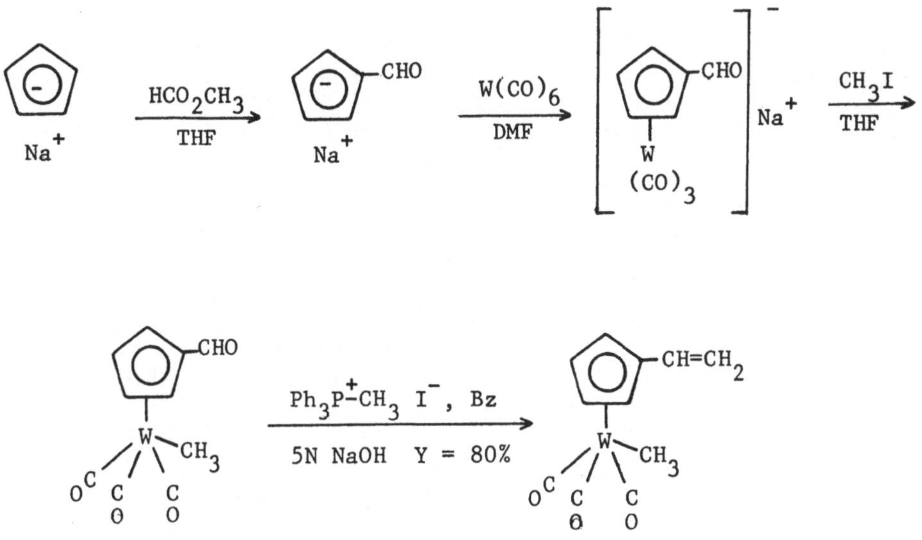

Scheme 2

were used, it has not proved to be generally applicable for the conversion of other formylcyclopentadienyl-transition metal compounds
to their corresponding vinyl analogs.[34,41]

Therefore, more direct methods for the synthesis of (η^5-vinyl-
cyclopentadienyl)metal complexes were desired. Knox and Pauson had
earlier prepared 1,1'-diisopropenylferrocene in ca. 60% yield from
6,6-dimethylfulvene, sodium amide, and ferrous chloride.[42] Also,
potassium isopropenylcyclopentadienide had been prepared from 6,6-
dimethylfulvene and potassium t-butoxide in diglyme.[43]

We now report an extension of this chemistry to a new and apparently general route for the synthesis of a wide variety of η^5-vinyl-
cyclopentadienyl organometallic monomers.[44] This method provides a
convenient means of introducing vinyl substituents onto η^5-cyclo-
pentadienyl rings in systems which are incapable of undergoing
electrophilic aromatic substitution.

6,6-Dimethylfulvene and 6-methylfulvene are reacted with lithium
diisopropylamide in THF at 25°C to produce isopropenylcyclopenta-
dienyllithium (19) and vinylcyclopentadienyllithium (20), respectively, in yields of 80-90% (equations 3 and 4). The solid products
are air-sensitive, as is cyclopentadienyllithium itself. The [1]NMR

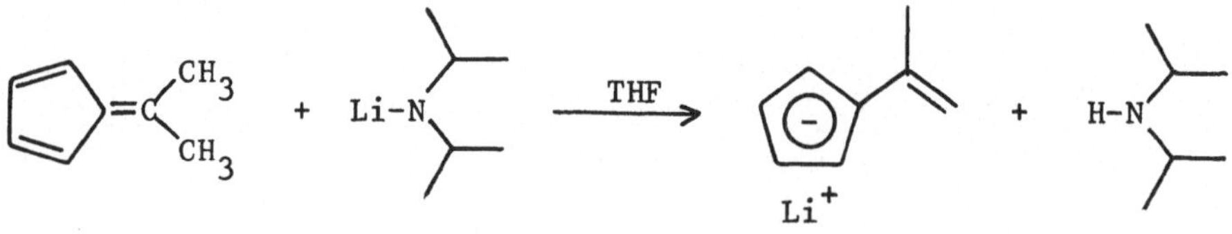

$$\underline{19} \quad Y = 80\text{-}90\% \qquad (3)$$

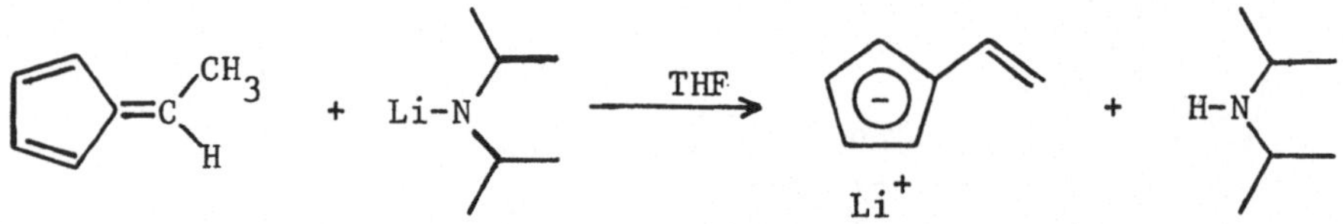

$$\underline{20} \quad Y = 80\text{-}90\% \qquad (4)$$

spectrum of $\underline{19}$ in THF-d_8 is in agreement with a recent report by Schore et al.[45] The ^1H NMR spectrum of the previously unknown compound $\underline{20}$ exhibits two apparent triplets due to the cyclopenta-dienyl protons, as well as the typical ABX pattern for a terminal vinyl group (Figure 1).

Both 19 and 20 react readily with a variety of organometallic compounds to afford the corresponding η^5-vinylcyclopentadienyl mo-nomers. The reaction between organolithium reagent $\underline{20}$ and (η^5-cyclo-pentadienyl)trichlorotitanium on ethyl ether produces (η^5-vinylcyclo-pentadienyl)(η^5-cyclopentadienyl)dichlorotitanium, $\underline{21}$ (mp 154-157°C, 16%) (Equation 5). Polymers of $\underline{21}$ after reduction could function

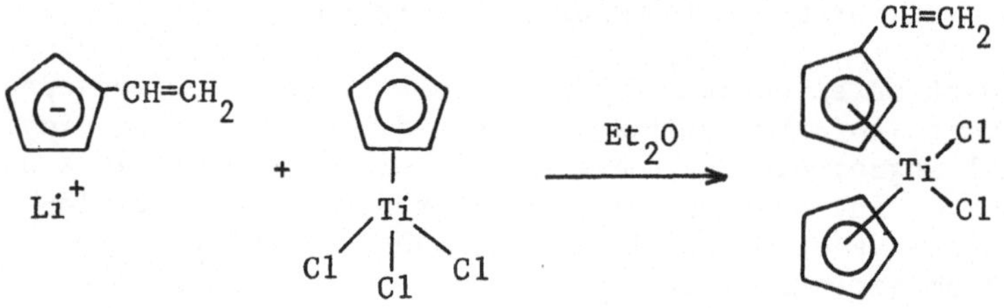

$$(5)$$

$$\underline{20} \qquad\qquad\qquad \underline{21} \quad Y = 16\%$$

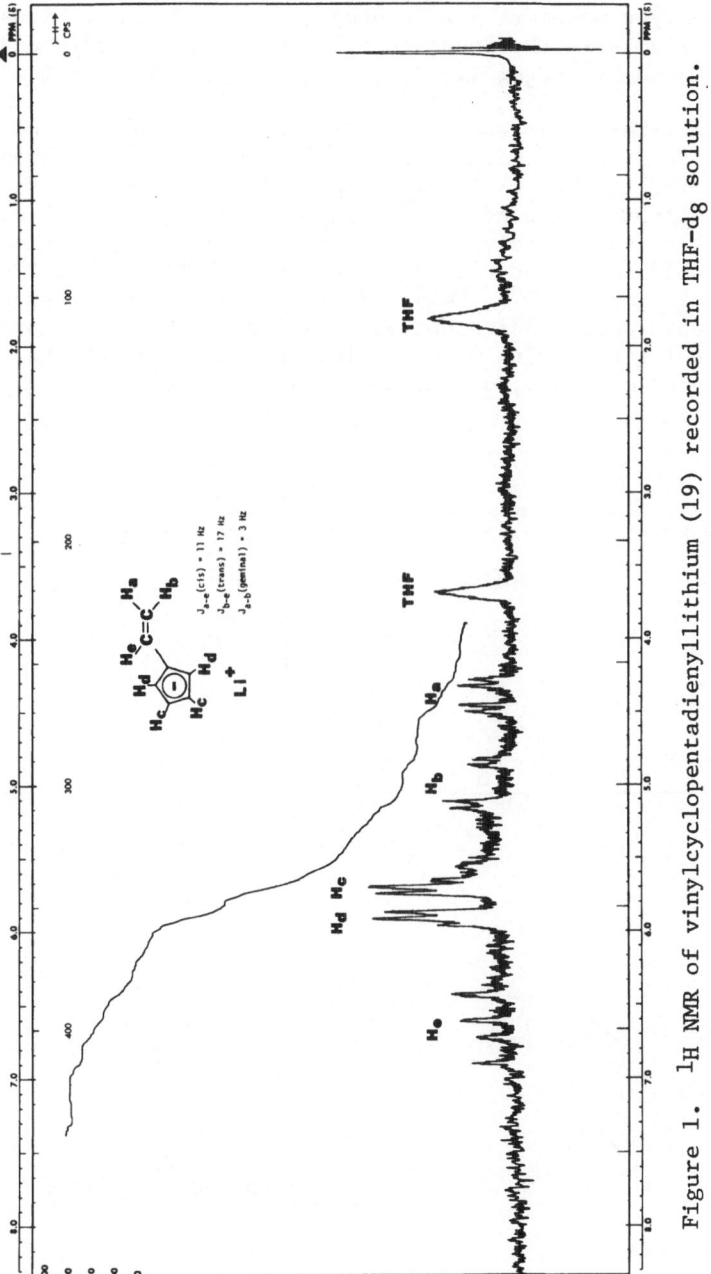

Figure 1. ^{1}H NMR of vinylcyclopentadienyllithium (19) recorded in THF-d_8 solution.

as hydrogenation catalysts, similarly to polystyrene-bound titanocene dichloride.[46] Further investigations into the synthesis of other titanocene-type monomers are in progress. For example, treatment of 21 with $Al/HgCl_2$ in the presence of carbon monoxide afforded $[(\eta^5\text{-}CH_2\text{=}CHC_5H_4)(\eta^5\text{-}C_5H_5)Ti(CO)_2]$: ν_{CO} 1980, 1910 cm^{-1} (in benzene).

Organolithium reagents 19 and 20 react with molybdenum hexacarbonyl and with tris(dimethylformamide)tricarbonyltungsten in refluxing THF to give η^5-vinylcyclopentadienyl monomers of these metals. Nitrosylation of the intermediate metal carbonyl anions with N-methyl-N-nitroso-p-toluenesulfonamide has afforded monomers 22 (liquid, 58%) (equation 6), 23 (mp 27-29°C, 93%), 24 (mp 22-23°C, 36%), and 25 (mp 38-40°C, 24%). This completes the series of (η^5-vinylcyclopentadienyl)dicarbonylnitrosyl group 6B metal monomers, when taken together with 5, and allows a comparison of the vinyl reactivity of monomers, 5, 22, and 24 as a function of the metal.

(6)

22 Y = 58%

23 Y = 93%

24 R = H Y = 36%

25 R = CH_3 Y = 24%

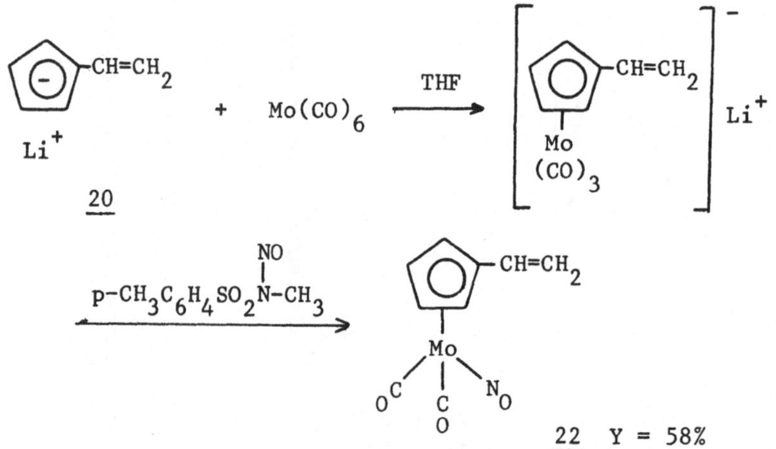

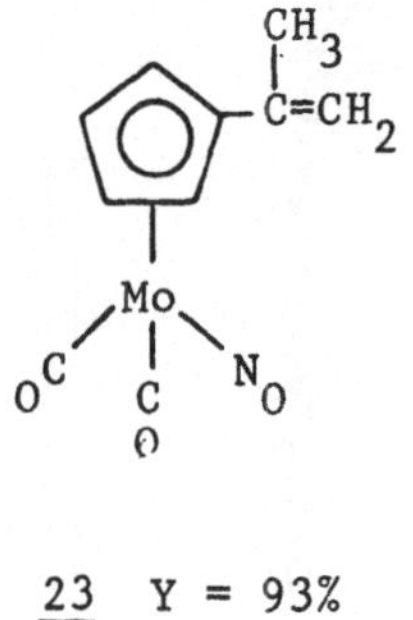

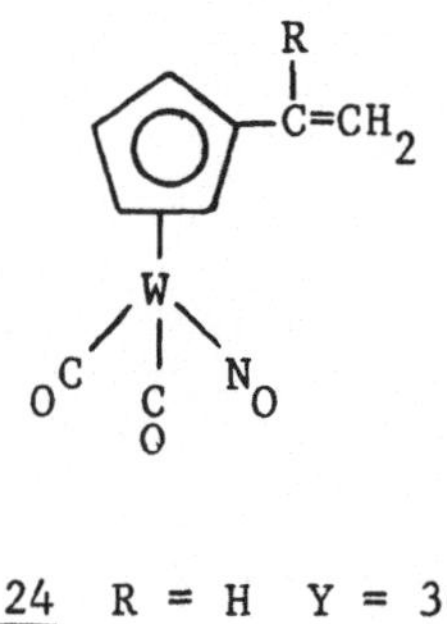

The use of polystyrene-bound (η^5-cyclopentadienyl)dicarbonyl-
cobalt as a catalyst[47-49] prompted us to synthesize monomers containing this metal. Reactions of 19 or 20 with an equimolar mixture of
$Co_2(CO)_8$ and I_2 in THF at 25°C give the cobalt monomers 26 (liquid,
15%) and 27 (liquid, 17%), respectively. Organolithium reagent 20

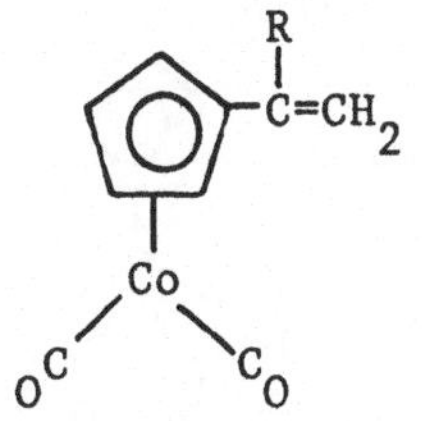

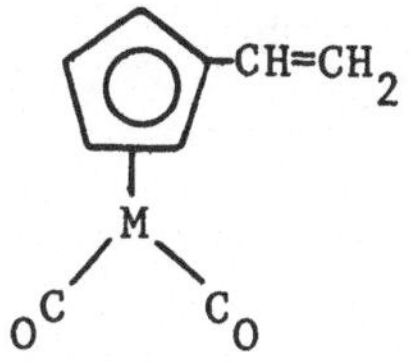

| 26 R = H   Y = 15% | 28 M = Rh   Y = 47% |
| 27 R = CH_3 Y = 17% | 29 M = Ir   Y = 91% |

is also a convenient precursor to the analogous rhodium and iridium
monomers 28 and 29. For example, the reaction of 20 with
$[Rh(CO)_2Cl]_2$ in THF at 25°C or with $Ir(CO)_3Cl$ in refluxing hexane
affords 28 (liquid, 47%) and 29 (liquid, 91%), respectively.

Using reagents 19 and 20, we have also been able to synthesize
the first organometallic monomers of copper. Thus, reactions of 19
or 20 with $(ClCuPEt_3)_4$ in ethyl ether give monomers 30 [mp 45°C
(dec), 66%] and 31 (mp 42°C, 90%) (equation 7).

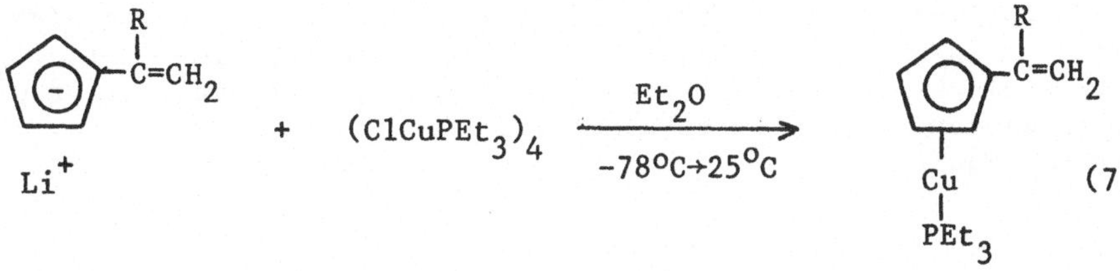

| 19 R = CH_3 | 30 R = CH_3 Y = 66% |
| 20 R = H | 31 R = H   Y = 90% |

The reaction of 19 and 20 with vanadium carbonyl derivatives
has not yet been studied. This would be a possible route to (η^5-
vinylcyclopentadienyl)tetracarbonylvanadium, 32. However, a number
of fulvenes have been reported to react directly with vanadium

hexacarbonyl to give good yields of the corresponding (η^5-alkenyl-cyclopentadienyl)tetracarbonylvanadium compounds.[50] Thus, by reacting 6-methylfulvene with vanadium hexacarbonyl, a 68% yield of $\underline{32}$ was obtained (equation 8).

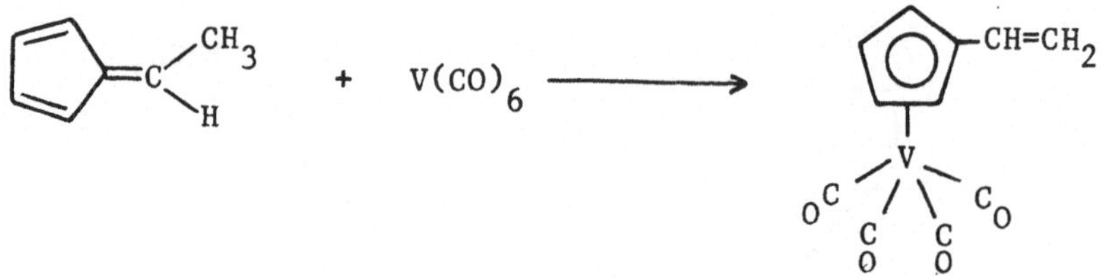

$$\text{(8)}$$

$$\underline{32}$$

The above reaction sequences demonstrate that the vinyl-substituted cyclopentadienyllithium reagents $\underline{19}$ and $\underline{20}$ can serve as valuable intermediates for a wide variety of new organometallic monomers and polymers. Polymerization studies of these monomers under free-radical or cationic initiation conditions are being studied and some results are discussed below. The availability of these new monomers is leading to a better understanding of how the organometallic moiety affects vinyl polymerization.

POLYMERIZATIONS

Homopolymerizations of $\underline{22}$ and $\underline{24}$ were carried out using AIBN initiation in benzene. While the kinetics of homopolymerization of these monomers have not yet been investigated, copolymerization of $\underline{22}$ with styrene (AIBN, in benzene) was studied at 50-75°C (equation 9).

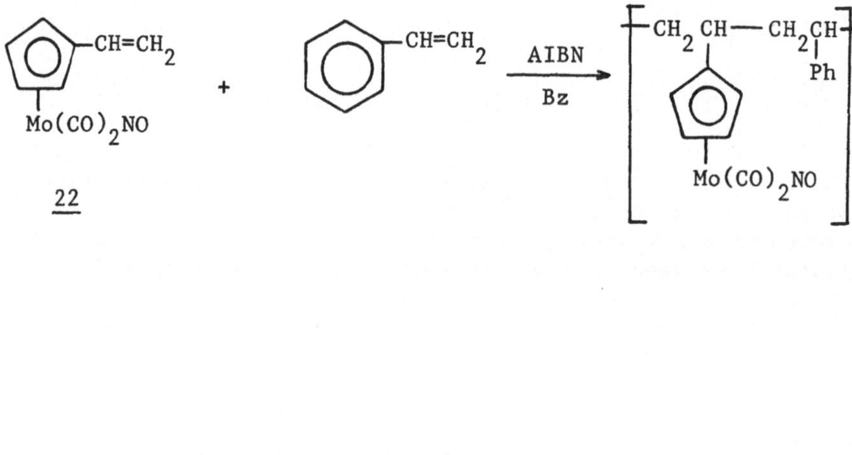

$$\text{(9)}$$

$$\underline{22}$$

Copolymers containing from 15-70 mol percent of 22 have been isolated with molecular weights of ca. 3×10^4. Bulk polymerizations lead to higher molecular weights. The Alfrey-Price value of "e" for 22 was obtained from styrene copolymerizations summarized in Table 1.

Table 1. Composition - Conversion Data for Copolymerization of
CpMo(CO)$_2$NO (M$_1$) With Styrene (M$_2$)

Run	M$_1$ in Feed (mol %)	% Conv.	M$_1$ in Copolym. (mol %)
1a	29.4	20.0	22.8
1b		21.7	24.6
1c		28.9	25.7
2a	50.9	10.7	50.0
2b		15.5	43.9
2c		34.4	49.1
3a	80.1	7.3	60.7
3b		13.1	66.5
3c		13.3	62.0
3d		19.8	70.8
3e		20.8	67.8

This composition-conversion data was fitted to the best values of r_1 and r_2 (r_1 = 0.31, r_2 = 0.83, where styrene = M$_2$) using the non-linear least squares method of Tidwell and Mortimer[51] as adapted to computer programs by Pittman and Rounsefell.[52] Remarkably, the value of "e" = -1.97, demonstrating that 22, like monomers 4 and 18, is an exceptionally electron-rich vinyl monomer. Moreover, this value of 22 is essentially identical to the "e" value of its chromium analog, 5 (e = -1.98), showing that the identity of the metal did not affect the copolymerization behavior when Mo is changed for Cr. This conclusion is reinforced by comparing the reactivity ratios for 5/styrene copolymerizations[14-15] (r_1 = 0.30, r_2 = 0.82) with those listed above for 22/styrene copolymerizations. The "Q" values for 5 and 22 are both 3.1, which is consistent with substantial delocalization in these α-cyclopentadienyl-metal radicals.

Cobalt monomer 26 was not successfully polymerized because competitive decomposition of the organometallic moiety occurred (equation 10). In contrast, its iridium analog, 29, appeared significantly more stable to polymerization conditions. It was

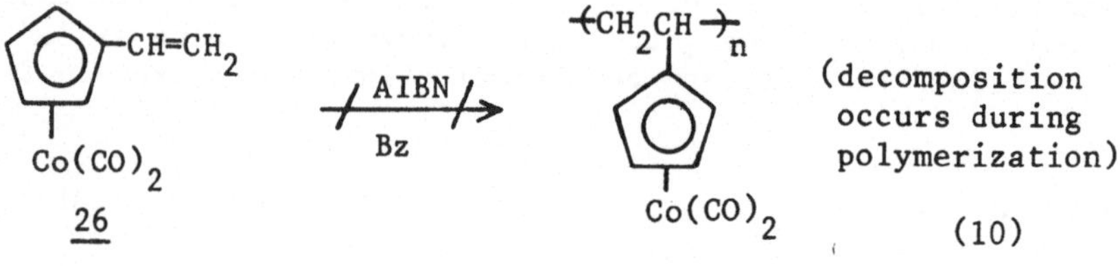

$$(10)$$

homopolymerized in benzene (AIBN) to give a polymer with over 50% by weight Ir. These polymers serve as catalysts for the hydrogen- ation of 1,5-cyclooctadiene at 100°C if they are first reduced by NAHBEt$_3$ (equation 11).

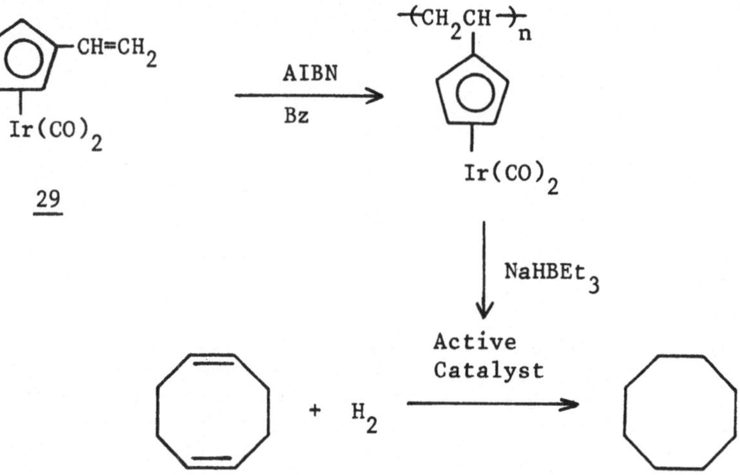

$$(11)$$

Since 29 was stable to free radical polymerization conditions, copolymerizations with styrene (M$_2$) were carried out to ascertain the reactivity ratios and the "e" value for comparison to the other (η^5-vinylcyclopentadienyl)-metal monomers 1, 4, 5, 18, and 22. Table 2 summarizes the composition-conversion data obtained from 29/styrene copolymerizations. From this data, the values of r$_1$ = 0.29 and r$_2$ = 0.68 were obtained. These are very similar to the r$_1$ and r$_2$ values obtained in copolymerizations of 5 and 22 with styrene. Again, variation of the metal and other ligands attached to the metal appear to have little effect on the copolymerization reactivity.

Table 2. Composition – Conversion Data for Copolymerization of $CpIr(CO)_2$ (M_1) With Styrene (M_2)

Run	M_1 in Feed (mol %)	% Conv.	M_1 in Copolym. (mol %)
1a	25.0	11.6	30.0
1b		20.3	22.4
1c		22.8	26.5
1d		27.7	22.2
1e		29.5	23.3
2a	42.0	9.65	40.3
2b		18.4	46.0
2c		19.0	39.2
2d		21.5	35.1
2e		35.9	41.8
3a	75.0	3.40	60.7
3b		7.69	57.5
3c		9.84	59.7
3d		27.0	65.0

The "e" value for iridium monomer 29, calculated from this data was -2.08, while "Q" = 4.1. As with all the other η^5-vinylcyclopentadienyl monomers studied, 29 has a very electron-rich vinyl group.

(η^5-Vinylcyclopentadienyl)tricarbonylmethyltungsten, 18, is the only organometallic monomer containing an alkyl group directly bonded to the metal by a sigma-bond whose polymerization has ever been studied.[40,53] Thus, it was examined in some detail. Homopolymerizations of 18 were sluggish under radical initiation. For example, solution homopolymerizations in benzene using AIBN initiation gave only 10-15% yields after 150-160 h at 18/AIBN ratios of 33. However, good yields of copolymers were obtained with acrylonitrile, methyl methacrylate, styrene, and N-vinyl-2-pyrrolidone at $18/M_2$ ratios of 30/70 (see Scheme 3). This group of comonomers spans a wide range of electronic properties from electron-poor to electron-rich vinyl groups. Rather low intrinsic viscosities were obtained, suggesting extensive chain-transfer may have occurred. Thus, more detailed homopolymerization studies were carried out in benzene. Anionic initiation attempts (Na^+ Naphth$^{\pm}$, THF, -78°C;

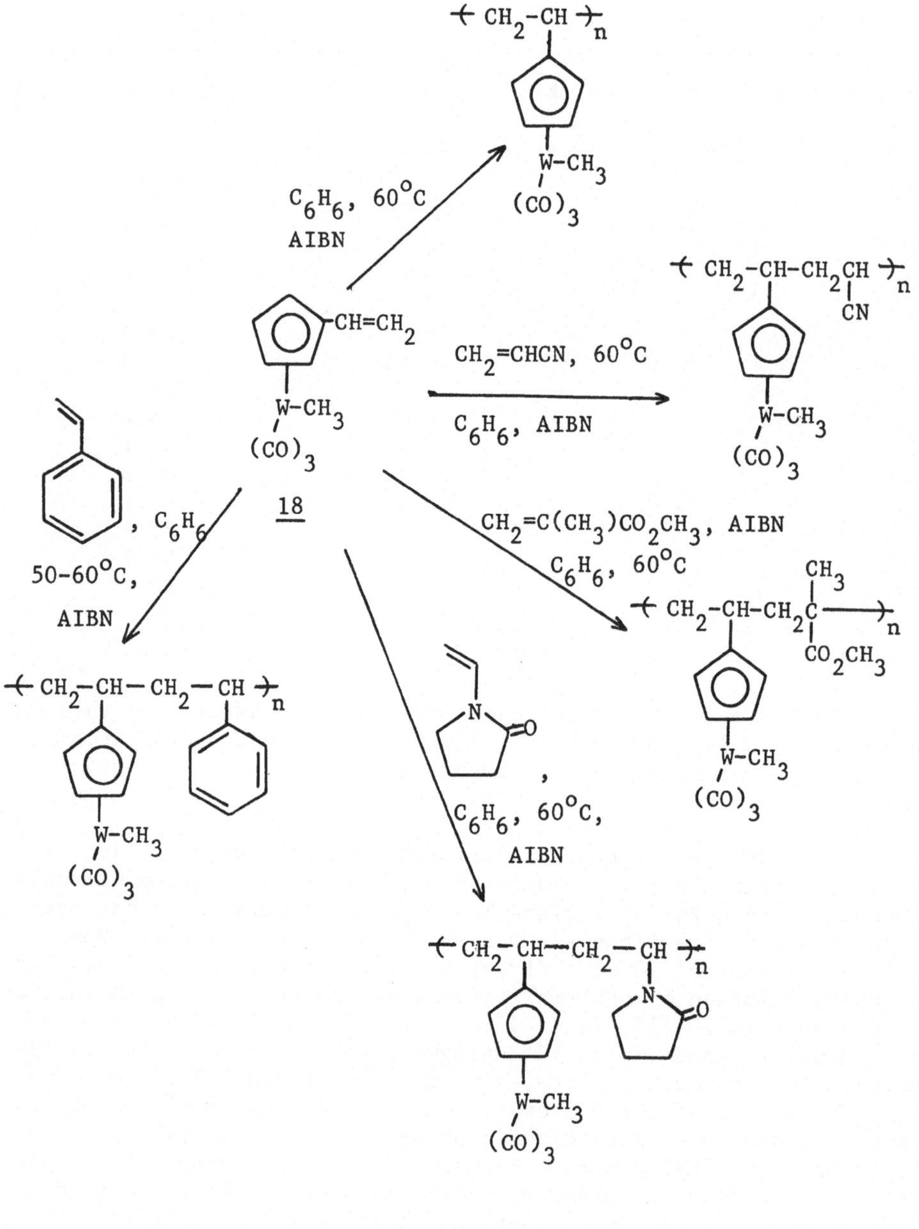

Scheme 3

BuLi, Bz, -78°C $\rightarrow$ $+25^{\circ}$C; LiAlH$_4$, THF, -78°C) failed, as did cationic initiation attempts with BF$_3$·OEt$_2$ in THF at 0°C $\rightarrow$ $+25^{\circ}$C.

Two potential routes exist for the homopolymerization of 18. First, the standard vinyl addition mechanism could occur. However, an internal hydrogen atom abstraction from the W-CH$_3$ group followed by termination or further polymerization by L$_n$W-CH$_2$· addition to another vinyl group can be envisioned (see Scheme 4). Thus, two limiting polymer structures, 33 and 36, are possible. The NMR spectrum of the homopolymer is only consistent with 33. The polymer exhibits a singlet at δ0.44 ppm for the W-CH$_3$ group and a multiplet centered at δ5.4 ppm for the cyclopentadienyl hydrogens. The respective area ratio is 3:4, in agreement with structure 33. Chemical shifts of the W-CH$_3$ group are the same for the polymer and monomer. Polymer structure 36 would be stable if it had formed, since η^5-C$_5$H$_5$W(CO)$_3$CH$_2$CH$_3$ can be sublimed without β-hydride elimination. The homopolymers were true polymers, with minimum $\overline{M}_n$ values of 15,000.

The rate law for the homopolymerizations of 18 in benzene at 60°C^{54} was V$_p$ = 1.13 x 10^{-2}[AIBN]$^{2.3}$[18]$^{0.8}$. The overall activation energy, E$_o$, of polymerization was 95 kcal/mol. In classic vinyl addition polymerizations the rate law is V$_p$ = k$_{obsd}$[M]^1[I]$^{0.5}$, and a condition necessary for this law is that f, the initiation efficiency, is independent of [M]. Since the order of monomer concentration with 18 is approximately 1, it appears that f is independent of [18].

The high order in initiator concentration and the sluggishness of homopolymerizations in both benzene and ethyl acetate could be due to chain-transfer and chain-termination by hydrogen abstraction from (a) the W-CH$_3$ group; (b) the cyclopentadienyl ring; or (c) the methine groups of the polymer backbone (i.e., benzylic-like). All of these possibilities were ruled out by studying styrene polymerizations in the presence of η^5-C$_5$H$_5$W(CO)$_3$CH$_3$, homopolymer 33, or homopolymers of 4. The molecular weights of the polystyrenes, obtained in the presence of varying amounts of these species, did not vary appreciably. If chain transfer by hydrogen abstraction from sites (a), (b) or (c) had occurred, the molecular weights of the polystyrenes would have been depressed (equation 12).

The reactivity ratio and "Q,e" values for 18 were obtained in styrene (M$_2$) copolymerizations using mol percents of 18 in the feed between 3.2 and 90.9. The values of r$_1$ = 0.16 (0.13-0.18) and r$_2$ = 1.55 (1.41-1.71) exhibited a rather large 95% joint confidence limit, but it is again quite clear that the vinyl group is very electron-rich (e = -1.98). We therefore see a remarkable similarity in the value of "e" to the values obtained with every other η^5-cyclopentadienyl monomer studied to date. Also, in agreement with other such monomers, a large value of "Q" (1.66) was found.

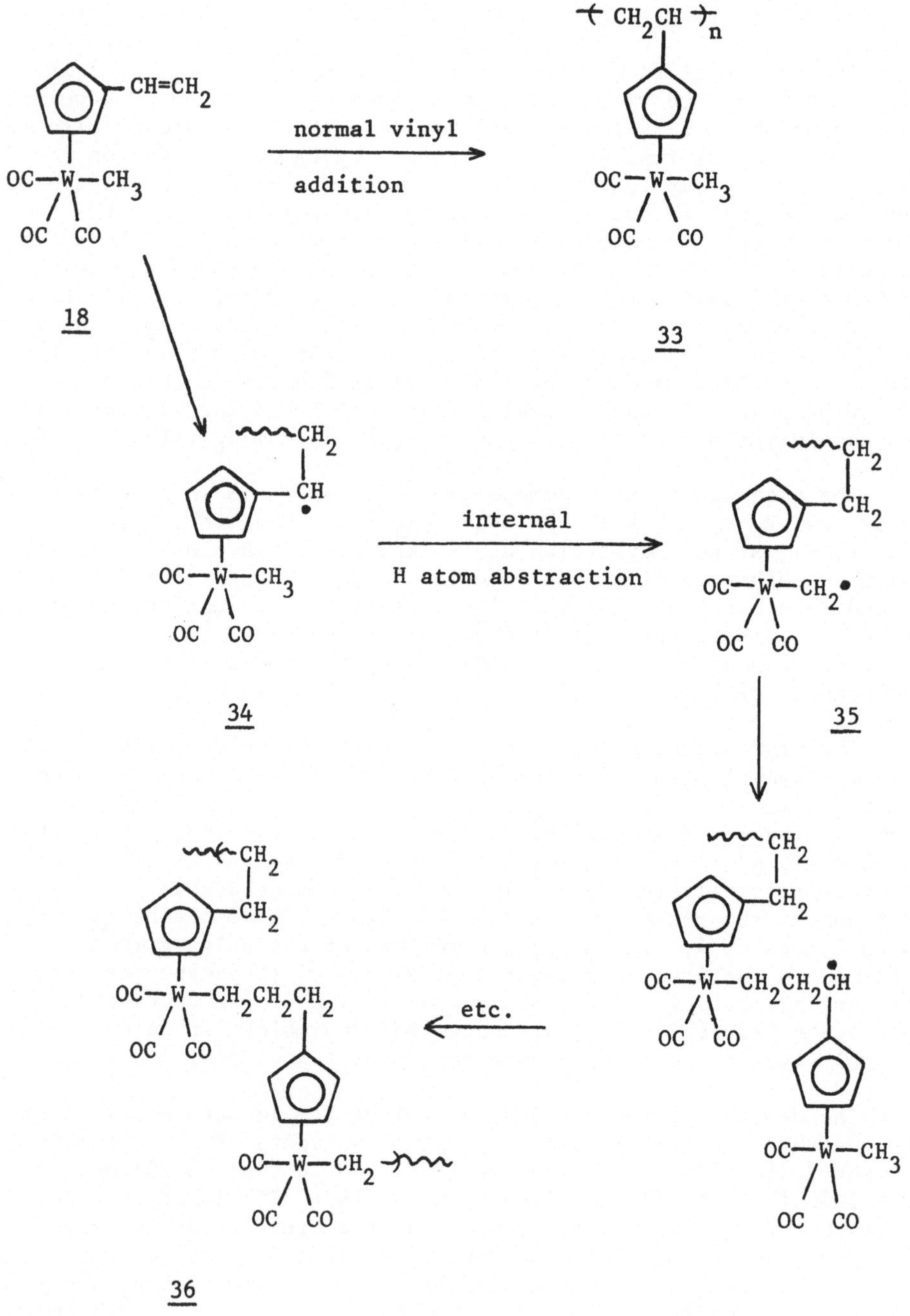
CH=CH2
OC—W—CH3
OC CO
normal vinyl
addition
CH2CH
n
18
33
CH2
CH
internal
H atom abstraction
CH2•
34
35
OC—W—CH2CH2CH
etc.
OC—W—CH2CH2CH2
OC—W—CH2
36

Scheme 4

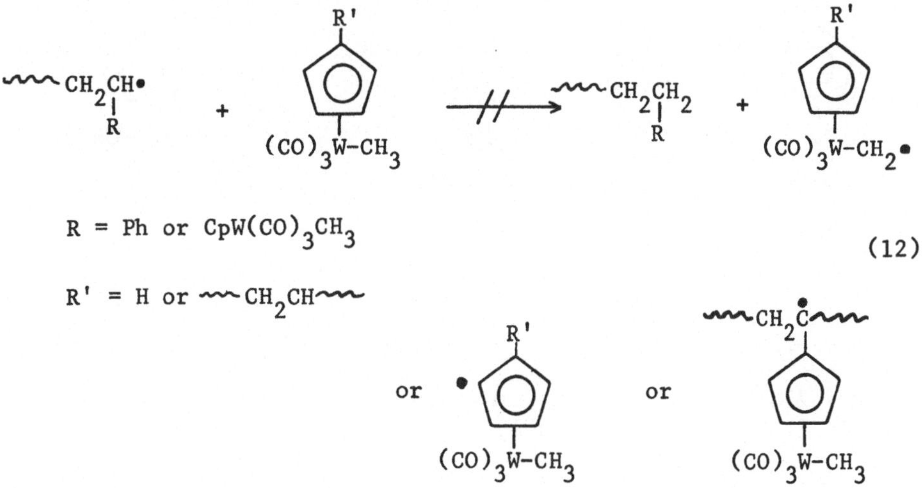

$$\text{~CH}_2\overset{|}{\underset{R}{C}}\text{H}\bullet \quad + \quad \underset{(CO)_3W-CH_3}{\overset{R'}{\bigcirc}} \quad \xrightarrow{\;\;/\!/\;\;} \quad \text{~CH}_2\overset{|}{\underset{R}{C}}\text{H}_2 \quad + \quad \underset{(CO)_3W-CH_2\bullet}{\overset{R'}{\bigcirc}}$$

R = Ph or CpW(CO)$_3$CH$_3$

R' = H or ~CH$_2$CH~

(12)

$$\text{or} \quad \bullet\underset{(CO)_3W-CH_3}{\overset{R'}{\bigcirc}} \quad \text{or} \quad \underset{(CO)_3W-CH_3}{\overset{\text{~CH}_2\overset{\bullet}{C}\text{~}}{\bigcirc}}$$

A picture of the polymerization reactivity of η^5-vinylcyclopentadienyl-metal monomers now emerges. The electronic interaction of the vinyl group with the cyclopentadienyl ring is the dominating factor, and to a first approximation, this interaction is independent of the nature of the metal and the other ligands attached to the metal. This conclusion is strikingly portrayed in Scheme 5. This interaction causes the vinyl group to be extremely electron-rich in vinyl polymerization. Furthermore, resonance stabilization of the radical center into the cyclopentadienyl ring is considerable, as is clearly demonstrated by comparing the "Q" values of the metallic monomers in Scheme 5 (1.7 to 4.1) to that of propene (0.002) and styrene (1.0). An organic monomer which appears to resemble the organometallic monomers most closely is 2-vinylthiophene, 37, which has a value of "Q" = 2.86.

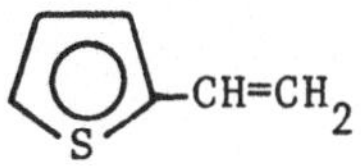

e = −0.80

Q = 2.86

37   38 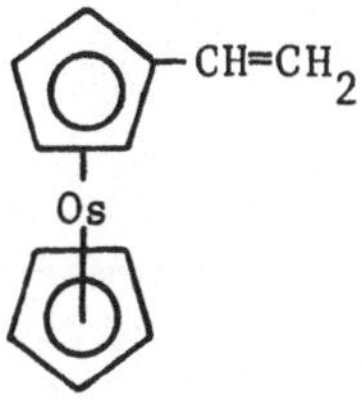   39

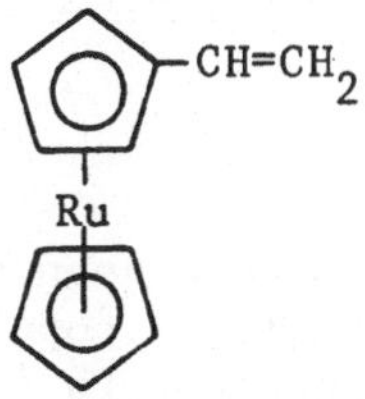

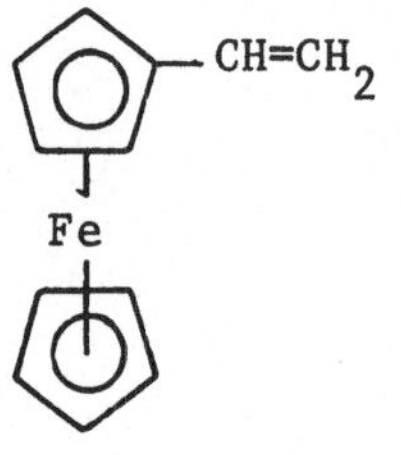

$\underline{1}$

e = -2.1

Q = 2.3

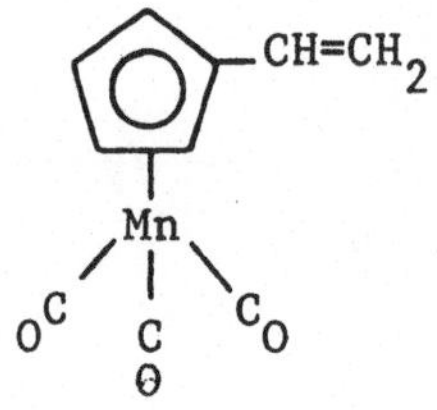

$\underline{4}$

e = -1.99

Q = 2.9

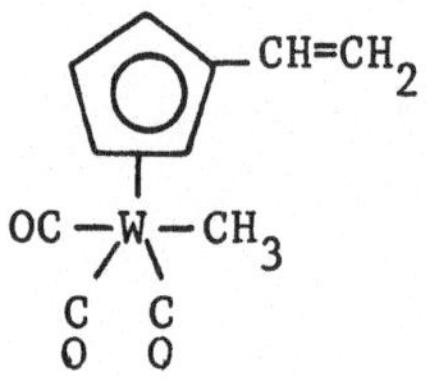

$\underline{18}$

e = -1.98

Q = 1.7

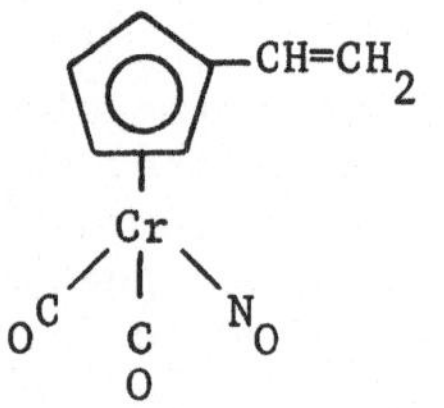

$\underline{5}$

e = -1.98

Q = 3.1

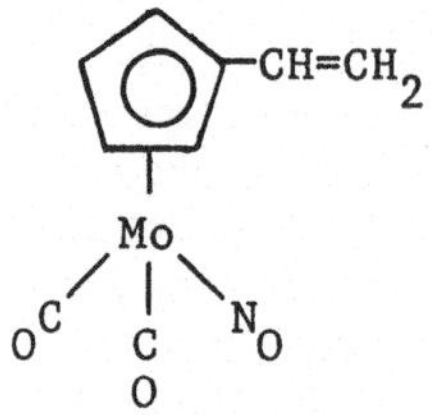

$\underline{22}$

e = -1.97

Q = 3.1

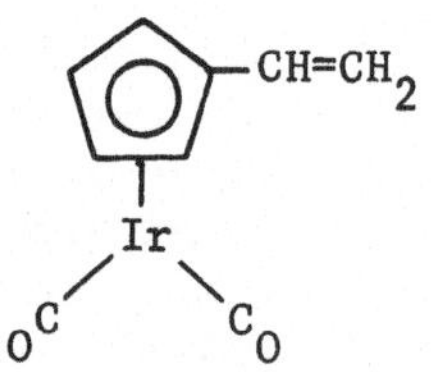

$\underline{29}$

e = -2.0

Q = 4.1

Scheme 5

Polymers containing $\underline{18}$ have been used successfully as metathesis catalysts. For example, a copolymer of $\underline{18}$ with styrene
(31 mol percent $\underline{18}$) was treated with isobutylaluminum chloride and
oxygen in hexane and trans-3-heptane. This gave an active metathesis system at 20°C which produced trans-4-octene, trans-3-
hexene, and other products (one identified as 4-ethyl-2-octene).
Thus, this polymer species appears to behave similar to polystyrene-
bound η^5-cyclopentadienyltricarbonylbenzyltungsten as a metathesis

catalyst when treated with i-BuAlCl$_2$.[55] The major difference
between our polymer and the benzyltungsten polymer is that the
former has a sigma-bonded methyl group while the latter has a
sigma-bonded benzyl group (to tungsten). The latter was a more
active metathesis catalyst, giving a 23% conversion of 3-heptene in
one hour compared to 3% for the polymer derived from 18 and styrene
under identical conditions.

Vinylruthenocene, 38, and vinylosmocene, 39, were received from
Professor J. Sheats (Rider College) and styrene solution copolymerizations (AIBN) were studied. At feed ratios of 5/95 38/styrene,
polymers with 3 mol % 38 and molecular weights of 1-2 x 10^5 were
obtained. However, at 40/60 feed ratios, 25-30 mol % of 38 was
incorporated into polymers of 8-20 x 10^4 molecular weight.
Reactivity ratio studies were carried out and the reactivity of 38
was compared to that of vinylferrocene, 1.

CONCLUSIONS

Several new and general synthetic routes to vinyl organometallic monomers have been developed in our laboratories. Many
of these monomers undergo radical-initiated homo- and copolymerizations to produce a variety of new metal-containing polymers.
The polymerization behavior of these monomers has now been defined
on a semi-quantitative basis. Preliminary studies indicate that
these new materials have useful properties and applications.

ACKNOWLEDGEMENTS

Acknowledgement is made to the Donors of The Petroleum Research
Fund, administered by the American Chemical Society (to M.D.R.),
to a National Science Foundation grant to the Materials Research
Laboratory, University of Massachusetts, and to the Army Research
Office (to C.U.P.) for support of this research. The experimental
assistance of Pam Stapleton, undergraduate research assistant at
the University of Alabama, is also gratefully acknowledged.

REFERENCES

1. C. E. Carraher, Jr., J. E. Sheats, and C. U. Pittman, Jr.,
    "Organometallic Polymers," Academic Press, New York (1978).

2. C. U. Pittman, Jr., Organomet. React. & Synth., 6, 1 (1977).

3. C. U. Pittman, Jr., in "Polymer Supported Reactions in Organic
    Synthesis," P. Hodge and D. C. Sherrington, eds., Wiley,
    New York (1980).

4. Y. Sasaki, L. L. Walker, E. L. Hurst, and C. U. Pittman, Jr.,
    J. Polym. Sci. Chem. Ed., 11, 1213 (1973).

5. M. H. George and G. F. Hayes, J. Polym. Sci. Chem. Ed., 13,
    1049 (1975).

6. C. U. Pittman, Jr., and G. V. Marlin, J. Polym. Sci. Chem. Ed.,
    11, 2753 (1973).

7. N. Takaishi, H. Imai, C. A. Bertelo, and J. K. Stille, J. Am.
    Chem. Soc., 100, 264 (1978).

8. M. H. George and G. F. Hayes, J. Polym. Sci. Chem. Ed., 14,
    475 (1976).

9. C. Aso, T. Kunitake, and T. Nakashima, Macromol. Chem., 124,
    232 (1969).

10. C. R. Simionescu, Macromol. Chem., 163, 59 (1973).

11. C. U. Pittman, Jr., and T. D. Rounsefell, Macromolecules,
    9, 936 (1976) and references therein.

12. C. U. Pittman, Jr., and T. D. Rounsefell, Macromolecules,
    9, 937 (1976).

13. C. U. Pittman, Jr., and T. D. Rounsefell, Macromolecules,
    11, 1022 (1978).

14. E. A. Mintz, M. D. Rausch, B. H. Edwards, J. E. Sheats, T. D.
    Rounsefell, and C. U. Pittman, Jr., J. Organometal. Chem.,
    137, 199 (1977).

15. C. U. Pittman, Jr., T. D. Rounsefell, E. A. Lewis, J. E.
    Sheats, B. H. Edwards, M. D. Rausch, and E. A. Mintz,
    Macromolecules, 11, 560 (1978).

16. C. U. Pittman, Jr., and B. Surynarayanan, J. Amer. Chem. Soc., 96, 7916 (1974).

17. C. U. Pittman, Jr., B. Surynarayanan, and Y. Sasaki in "Inorganic Compounds With Unusual Properties," Adv. in Chem. Ser. No. 150, R. B. King, ed., American Chemical Society, Washington (1976).

18. D. O. Cowan, J. Park, C. U. Pittman, Jr., Y. Sasaki, T. K. Matherjee, and N. A. Diamond, J. Amer. Chem. Soc., 94, 5110 (1972).

19. C. U. Pittman, Jr., and Y. Sasaki, Chem. Lett. Japan, 383 (1975).

20. A. B. Bocarsly, E. G. Walton, and M. S. Wrighton, J. Amer. Chem. Soc., 102, 3390 (1980).

21. C. U. Pittman, Jr., O. E. Ayers, and S. P. McManus, J. Macromol. Sci. Chem., A7(8), 1563 (1973).

22. C. U. Pittman, Jr., P. Grube, and R. M. Hanes, J. Paint Technol., 46, (597), 35 (1974).

23. For a review, see D. W. Macomber, W. P. Hart, and M. D. Rausch, "Functionally Substituted Cyclopentadienyl Metal Compounds," Adv. Organomet. Chem., 21, 1 (1982).

24. M. D. Rausch, E. O. Fischer, and H. Grubert, J. Amer. Chem. Soc., 82, 76 (1960).

25. F. A. Cotton and J. R. Leto, Chem. Ind. (London), 1368 (1958).

26. E. O. Fischer and K. Plesske, Chem. Ber., 91, 2719 (1958).

27. E. O. Fischer and W. Fellman, J. Organometal. Chem., 1, 191 (1963).

28. A. N. Nemseyanov, K. N. Anisimov, N. E. Kolobova, and L. I. Baryshnikova, Dokl. Akad. Nauk SSSR, 154, 646 (1964).

29. A. N. Nesmeyanov, N. E. Kolobova, K. N. Anisimov, and L. I. Baryshnikova, Izv. Akad. Nauk SSSR, Ser. Khim., 1135 (1964).

30. E. O. Fischer and K. Plesske, Chem. Ber., 93, 1006 (1960).

31. R. Riemschneider, O. Goehring, and K. Kruger, Monatsh., 91, 305 (1960).

32. R. Ercoli and F. Calderazzo, Chim. e Ind. (Milano), 42, 52 (1960)

33. E. O. Fischer and K. Plesske, Chem. Ber., 94, 93 (1961).

34. W. P. Hart, Ph.D. Dissertation, University of Massachusetts, Amherst (1981).

35. M. D. Rausch and R. A. Genetti, J. Org. Chem., 35, 3888 (1970).

36. K. Hafner, G. Schultz, and K. Wagner, Justus Liebigs Annalen Chem., 678, 39 (1964).

37. T. Okuyama, Y. Kenouchi, and T. Fueno, J. Amer. Chem. Soc., 100, 6162 (1978).

38. W. P. Hart, D. W. Macomber, and M. D. Rausch, J. Amer. Chem. Soc., 102, 1196 (1980).

39. J. M. Osgerby and P. L. Pauson, J. Chem. Soc., 4604 (1961).

40. D. W. Macomber, M. D. Rausch, T. V. Jayaraman, R. D. Priester, and C. U. Pittman, Jr., J. Organometal. Chem., 205, 353 (1981).

41. D. W. Macomber, Ph.D. Dissertation, University of Massachusetts, Amherst (1982).

42. G. R. Knox and P. L. Pauson, J. Chem. Soc., 4610 (1961).

43. J. Hine and D. B. Knight, J. Org. Chem., 35, 3946 (1970).

44. D. W. Macomber, W. P. Hart, M. D. Rausch, R. D. Priester, and C. U. Pittman, Jr., J. Amer. Chem. Soc., 104, 884 (1982).

45. N. E. Schore and B. E. LaBelle, J. Org. Chem., 46, 2306 (1981).

46. D. W. Bonds, Jr., C. H. Brubaker, Jr., E. S. Chandrasekaran, C. Gibbons, R. H. Grubbs, and L. C. Kroll, J. Amer. Chem. Soc., 97, 2128 (1975).

47. G. Bubitosa, M. Boldt, and H. H. Brintzinger, J. Amer. Chem. Soc., 99, 5174 (1977).

48. B. H. Chang, R. H. Grubbs, and C. H. Brubaker, J. Organometal. Chem., 172, 81 (1979).

49. P. Perkins and K. P. C. Vollhardt, J. Amer. Chem. Soc., 101, 3985 (1979).

50. N. Hoffman and E. Weiss, J. Organometal. Chem., 131, 273 (1977).

51. P. W. Tidwell and G. A. Mortimer, J. Macromol. Sci., Rev. Macromol. Chem., 5, 135 (1970).

52. C. U. Pittman, Jr., and T. D. Rounsefell, Comput. Chem. Instrum., 6 (1977).

53. C. U. Pittman, Jr., T. V. Jayaraman, R. D. Priester, Jr., S. Spencer, M. D. Rausch and D. W. Macomber, Macromolecules, 14, 237 (1981).

54. C. U. Pittman, Jr., R. D. Priester, Jr., and T. V. Jayaraman, J. Polym. Sci. Chem. Ed., 19, 3351 (1981).

55. S. Warwel and P. Buschmeyer, Angew. Chem. Int. Ed. Eng., 17, 131 (1978).

SYNTHESIS OF ORGANOMETALLIC POLYMERS FOR INERTIAL FUSION

APPLICATIONS

John E. Sheats*, Fred Hessel, Louis Tsarouhas,
Kenneth G. Podejko and Thomas Porter

Chemistry Department
Rider College
Lawrenceville, NJ

L.R. Kool[1a] and R.L. Nolen, Jr.[1b]

KMS Fusion, Inc., P.O. Box 1567
Ann Arbor, MI

INTRODUCTION

Nuclear fusion, the energy process operating in the sun, offers promise of production of almost unlimited energy without the toxic and radioactive wastes associated with nuclear fission. Harnessing nuclear fusion, however, has proven to be a challenging task that may not be completed for another thirty years. Because of the strong repulsive forces to be overcome in order for nuclei to fuse, the process will take place only at temperatures above 50,000,000°. No known materials can contain matter at this temperature. Thus the fusion reaction must be confined without its touching the walls of its container. Two approaches have been taken - Magnetic Confinement, which is currently being investigated at the Forrestal Laboratories in Princeton, and Inertial Confinement[2] which is being investigated at KMS Fusion, Inc. in Ann Arbor, Michigan, at the University of Rochester, at the National Laboratories at Los Alamos, New Mexico, and Livermore, California and elsewhere.

In the Inertial Confinement Process, commonly called Laser Fusion, a small hollow sphere containing deuterium and tritium at high pressure is placed at the focal point of a high-intensity laser beam. The energy - eight trillion watts per square centimeter (100 billion light bulbs on your fingernail) causes the spherical shell to vaporize outward and a shock wave to procede inward at one-tenth the speed of light, compressing the hydrogen to 1/1000th of its original volume and raising temperatures up to one-hundred-million degrees. Nuclear fusion takes place, producing helium and a neutron which can later be captured by a lithium shield,[3] releasing usable energy.

$$_1^2H + _1^3H \rightarrow _2^4He + _0^1n$$

deuterium + tritium = helium + neutron

$$_3^6Li + _0^1n \rightarrow _3^7Li + energy$$

As research on Inertial Confinement Fusion has progressed, target design has become more complex. The first targets were simple hollow spheres of silica or plastic.[5] As the energy of the laser beam was increased, one of the major problems encountered was the tendency of the laser beam to be reflected from the surface rather than to be absorbed. A double shell design was next employed with an outer layer composed of material which would absorb the energy of the laser beam and ablate away, generating a compressive force, and an inner layer which would contain the fuel. Later designs have contained multiple layers of materials with different responses toward the developing shock wave of high energy plasma. Between the layers were cushions of a low density foam which maintained the proper spacing. The spacing of the layers is critical since it controls the timing of the different events in the implosion - fusion sequence.

All shells must be uniform in thickness within 1% and the inner and outer surfaces of each spherical shell must be concentric within 1%. No warts or dents greater than 150 A° in height or depth can be tolerated.[7] A more detailed discussion of target design and the methods of target fabrication is beyond the scope of this article but further information can be found in References 4-20.

One complication encountered with the present target design is that when the outer shell begins to disintegrate, extremely high-energy (suprathermal) electrons and other types of radiation head toward the center and begin to heat the deuterium - tritium mixture before compression is complete.[21,22] This pre-heat is analogous to the knock in an automobile engine, which is caused by a premature explosion of the gasoline-air mixture while compression is still taking place. In both cases a great reduction in efficiency results.

The most recent target designs have therefore included preheat-shield layers, that is, layers between the ablator and fuel which can capture suprathermal electrons, thus reducing the extent of pre-heat.[4] Prospective pre-heat shields must be of low average atomic number $\bar{Z}$ with one to four atom % of atoms of atomic number 70-85

uniformly distributed on the molecular level. Clusters of the high
- Z atoms would cause inhomogeniety in the plasma as the ablation
progresses and would also leave gaps which could allow the electrons
to penetrate to the center.

Several approaches have been taken to preparing high-Z shields
including glow-discharge polymerization of organic monomers in the
presence of iodine or iodinated organic molecules,[17] glow discharge
polymerization in the presence of metal vapors[16] and glow discharge
polymerization in the presence of volatile organometallic compounds
such as tetramethyllead.[5] A different and highly promising approach
described in this paper is to prepare polymers from organometallic
vinylic monomers containing the metal with the desired atomic number
covalently bound to the monomer. The composition, molecular weight
and physical and mechanical properties of the polymer can be varied
as desired by including varying amounts of comonomers.

Vinylic polymerization of monomers containing first row transition elements has been studied extensively[29-32] but little has been
done with the second and third row transition elements because of
their greater cost. This paper reports the synthesis of monomers
containing lead, osmium, ruthenium and tungsten, attempts to polymerize them and a method for assessing the potential machinability
of the polymers.

EXPERIMENTAL

Monomer Synthesis

Triphenyl-p-styryllead, TSL. This monomer was synthesized by
the method of Noltes et al.[23] in 77% yield. The crude product was
recrystallized from ethanol, m.p. 107°. The IR and NMR spectra
agreed well with the literature values.[23] Elemental Analysis;
$C_{26}H_{22}$ Pb. Calcd.: C 57.65; H 4.07; Pb 38.25.

Found: C 56.71; H 4.10; Pb 36.65.

Vinyl Ruthenocene, VR. Ruthenocene, 5.00g was converted to
acetylruthenocene, mp 110-111°, in 89% yield by the procedure of
Hill and Richards.[25] The advantage of this procedure over conventional acylation with acetyl chloride and aluminum chloride is that
very little unreacted ruthenocene or 1,1'- diacetylruthenocene are
found in the product. Acetylruthenocene was reduced with sodium
borohydride in ethanol to hydroxyethylruthenocene which was converted
to vinyl ruthenocene, mp 51-52° in 40-50% overall yield by vacuum
sublimation from alumina at 150°.[26-27]

Vinylosmocene, VO. Osmocene, 4.4g (0.014 mole) Strem Chemical
Co., 250ml of acetic anhydride and 40ml 85% phosphoric acid were
stirred under N_2 for 4 hr at 95° and worked up as described for

acetyl ruthenocene. The crude residue was chromatographed on a
15-cm column of alumina. The first fraction eluted with benzene was
osmocene, 0.80g, (18%). A second fraction eluted with methylene
chloride gave 3.6g (72%) of acetylosmocene as a pale yellow powder
which was recrystallized from methanol, mp. 126°[40]. The IR and NMR
spectra closely resemble those of the ruthenocene derivative. A
third fraction eluted with ethanol contained traces of 1,1'-diacetyl
osmocene and polymeric tar.

Acetyl osmocene was reduced to hydroxyethylosmocene in 97%
yield by the procedure given for the ruthenocene derivative and
sublimed from alumina at 175° and 1mm to give a 55% yield of vinyl
osmocene, mp. 59°.[27] The IR and NMR spectra of the hydroxyethyl
osmocene and vinylosmocene resemble closely the corresponding ferrocene and ruthenocene derivatives and agree with previously published results.[25-26]

η^{5-} Cyclopentadienyl-η^{1}-methyl styryltricarbonyl tungsten
Sodium sand, 0.6g (0.026 mole) and 20ml freshly distilled tetrahydrofuran were stirred at 0° under N_2 and 3-4ml freshly cracked
cyclopentadiene was added dropwise. Gas evolution commenced and
the sodium dissolved completely after 3 hr. Tungsten carbonyl,
9.2g (0.026 mole) in 55ml of tetrahydrofuran was added. The solution
was refluxed 24 hr. Chloromethylstyrene 4.0g (0.026 mole) was added
dropwise at room temperature and the solution refluxed for 3 hr.
The solvent was evaporated in vacuo and the residue taken up in 100ml
benzene. The benzene solution was concentrated and chromatographed
on neutral alumina. The product eluted as a compact yellow band.
Dark colored impurities remained on the column. After the benzene
solution was reduced to a volumn of 25-50ml the product crystallized
as yellow needles - 5.24g (45%) mp 86°. An attempt to sublime the
material in vacuo caused polymerization. IR, cm^{-1} (KBr) 3110(w),
2000(s), 1855-1945(s), 1620(w), 1590(m), 1570(m), 1490(m), 1000(m),
910(m), 830(s), 730(m), 680(m), 460(s), 430(m). NMR, δppm. 2.9
(2H,s, CH_2-W); 4.4 (5H, s, Cp ring); 4.9 -6.7 (3H, m, vinyl); 6.9 -
7.4 (4H,m, phenyl).

Polymerizations - Unless otherwise noted, all polymerizations were
run in glass tubes under slight positive nitrogen pressure. After
introduction of solid monomer and initiator to the tube, it was
evacuated and back-filled with argon. Solvent and comonomer were
then introduced through a rubber septum and the tube was immersed
in a thermostatically controlled oil bath.

Dilute solution homopolymerization of TSL. TSL, 3g, triply crystallize
0.3g AIBN (Dupont VAZO 64) and 42 ml of freshly distilled benzene
were heated at 55°C for 72 hours. The polymer was isolated and
purified by multiple precipitation from methanol. Two grams of
polymer was isolated (66%; M_n = 3502 (VPO); T_g = 98°C (DSC). Torsional
braid analysis studies (TBA) were performed by J.K. Gillham,

Department of Chemical Engineering, Princeton University. The
polymer showed T_g = 140° and T_1 153°. Elemental Analysis:
$C_{26}H_{22}Pb$ Calcd: C 57.65; H 4.07; Pb 38.25   Found: C 58.15; H 4.44;
Pb 35.51.

Bulk thermal homopolymerization of TSL. TSL, 0.5g triply crystallized
was subjected to high vacuum (10^{-5}torr) and warmed slightly to remove solvent and air. After back-filling with dry argon, the tube
was heated in an oil bath at 124° for 16 hours. The monomer melted
to a liquid and polymerization occurred immediately upon melting.
The polymer softened slightly when heated to 200°C. This polymer
was too hard and brittle to machine.

Copolymerization of TSL with isopropylstyrene. TSL, 0.5 g,
0.02 g isopropylstyrene (Monomer-Polymer and Dajac Laboratories)
0.02 g of AIBN and 1 ml of benzene were heated at 70°C for 28 hr.
The polymer was purified by multiple precipitation in methanol.
T_g = 141°C (DSC); Softening point = 160-165°C. Films cast from
chloroform appear cloudy, phase-separated. This polymer did not
machine well because of brittleness.

Copolymerization of TSL with acrylonitrile. TSL, 0.35 g, 0.02 g
AIBN, 0.1 g acrylonitrile and one ml benzene were heated at 70°C
for 25.5 hours. The polymer was purified by precipitation in methanol
T_g = 159°C; softens at 165°C. A C≡N stretch was apparent at 2240
cm^{-1} in the IR spectrum. Elemental analysis: C 60.99; H 4.87; N
7.96; Pb 26.35. Acrylonitrile in polymer: 30% by weight. This
polymer was too brittle to be machinable, and yellowed somewhat on
heating.

Copolymerization of TSL with octadecyl methacrylate. TSL, 0.3 g,
0.02 g AIBN, 0.1 g octadecyl methacrylate and one ml benzene were
heated at 70°C for 25.5 hours. The polymer was purified by precipitation in methanol. The purified material softens at 46°C. T_g
is below room temperature. The IR spectrum had a strong absorption
at 1726 cm^{-1} (C=O stretch). Elemental analysis: C 64.43; H 7.08;
Pb 24.89; O 3.56. This copolymer was 34.7% by weight of octadecyl
methacrylate, based on oxygen content. It formed smooth, strong
films from the melt.

Copolymerization of TSL with n-lauryl methacrylate and divinyl
benzene. TSL, 3.0g, 0.6 g n-lauryl methacrylate, 0.09 ml divinyl
benzene, 0.12 g AIBN, and 18 ml benzene were heated at 70°C for 70
hours. The polymer was purified by repeated precipitation from
methanol. It softens at 90°C. No T_g was apparent on DSC. Elemental
analysis: C 62.46; H 5.82; Pb 28.53; O 3.05, which corresponds to
25.3% lauryl methacrylate M_n = 9613, M_w = 17286, Dispersity = 1.80.

Copolymerization with octadecylmethacrylate, large scale. TSL, 10g,
3.3g octadecylmethacrylate, 0.67g AIBN and 30 ml benzene were

heated at 70°C for 28 hours. The polymer, when precipitated from
methanol, softens at 105°C. T_g =114°C.

Copolymerization of vinyl osmocene with octadecyl methacrylate.
Vinyl osmocene, 3g, 0.6 g octadecyl methacrylate, 0.12 AIBN and 12
ml benzene were heated at 70°C for 73 hours. Yield: 1.44 g (40%).
The polymer softens at 100°C. No T_g is apparent on DSC.

Copolymerization of CMT and octadecylmethacrylate. CMT, 0.44 g,
0.15 g octadecylmethacrylate, 0.04 g AIBN and 5 ml benzene were
heated at 70°C. The polymerization was re-initiated with 0.02 g
AIBN every two hours during the day, with no re-initiation during
the evening. The polymerization was run for 96 hr. The polymer
was purified by precipitation from methanol. The softening point
of the polymer was 96°C. No T_g was apparent on DSC. Elemental
analysis: C 65.74; H 8.78; N11.42; W 10.53; O 13.25. From the
percentages of N and W in the sample we can calculate the presence
of two AIBN end groups, 3.9 octadecyl methacrylate units and one
CMT ($M_n \sim 2000$).

RESULTS AND DISCUSSION

     The goal of this research has been to produce a variety of
organometallic polymers with low average atomic number $\bar{Z}$ with 1-4
atomic % of an element with atomic number number 70-85. A wide
range of organometallic polymers have been reported previously,[27-30]
but relatively few polymers have been prepared with atomic numbers
in this range. Many of the ones that have been reported contain
large amounts of O and N or elements with intermediate atomic number
such as P, S or Cl. This paper, although only a preliminary report,
shows that we have made significant progress in producing machinable
polymers containing lead, tungsten and osmium.

Monomer Synthesis. The monomers described in this paper, with one
exception had all been reported previously. The synthesis of triphenyl-p-styryl lead, TSL, from p-chlorostyrene via the Grignard
reagent procedes smoothly (23,32,34) in 87% yield. (Scheme 1).

                              Scheme 1

      $ClC_6H_4CH=CH_2$ $\xrightarrow[\text{THF}]{\text{Mg}}$ $ClMgC_6H_4CH=CH_2$

      $(C_6H_5)_3PbCl$
      $\xrightarrow{\hspace{2cm}}$ $(C_6H_5)_3$ Pb $C_6H_4CH=CH_2$

The synthesis of vinylruthenocene VR and vinylosmocene VO requires acetylation, reduction and dehydration by pyrolysis from alumina (Scheme 2). The procedure of Hill and Richards[25] was used rather than more conventional acetylation with acetyl chloride and aluminum chloride because it gave near quantitative yield of acetylruthenocene with no formation of 1,1' - diacetylruthenocene. In the case of osmocene acetylation is only 70-80% complete. Increasing the reaction time produces extensive decomposition. Thus chromatography is required to remove the tar formed and to separate osmocene and acetylosmocene. In general higher temperatures and longer reaction times are required for osmocene as compared to ferrocene and ruthenocene, in agreement with previous observations.[26,27]

Scheme 2

$$(C_5H_5)_2M \quad \xrightarrow[H_3PO_4]{(CH_3CO)_2O} \quad C_5H_5MC_5H_4COCH_3$$

$$\xrightarrow[\text{ethanol}]{NaBH_4} \quad C_5H_5MC_5H_4 \; CH(OH)CH_3 \quad \xrightarrow[150-180°]{Al_2O_3}$$

$$C_5H_5 \; M \; C_5H_4CH=CH_2 \qquad VR \; M = Ru$$

$$VO \; M = Os$$

η^6 - Cyclopentadienyl - η^1-methylstyryl tricarbonyltungsten, CMT, was synthesized as shown in Scheme 3.

Scheme 3

$$NaC_5H_5 + W(CO)_6 \xrightarrow[THF]{} C_5H_5W(CO_3)^-$$

$$\xrightarrow{ClCH_2C_6H_4CH=CH_2} C_5H_5W(CO)_3 \; CH_2C_6H_4 \cdot CH=CH_2$$

CMT

Tungsten - containing monomers

$CH_2=CH - C_5H_4 \; W(CO)_3CH_3$, $CH_2=CH-C_5H_4 \; W(CO)_2NO$ and $CH_2=C(CH_3)$

$C_5H_4 \; W(CO)_2NO$ have been synthesized previously by Rausch et al.[37-39] by a multistep procedure in low overall yields and preliminary studies of polymerization performed. This procedure can be done

in one pot and gives better yields. The main difficulty is that
the bond between the tungsten and the CH_2 group is weakened by the
phenyl group, hence cleavage may take place during radical-initiated
polymerization. Not enough studies have been performed to determine
whether cleavage of this type actually occurs.

Organolead polymers. TSL was polymerized and copolymerized under a
variety of conditions described in the Experimental Section and
summarized in Table I. Homopolymerization in dilute solution
yielded brittle, low molecular weight polymer (M_w=3500). A similar
result had been obtained by Fujita[35] for the analogous organopalladium
monomer (M_n <11000 , DP<17).

Bulk thermal polymerization yielded polymer which was only
partially soluble and softened at 210°C. Concentrated solution
polymerization produced polymer with T_g of 190° (DSC) which was
soluble in chloroform and less brittle. Since it was impossible
to machine homopolymer samples, we decided to copolymerize the
lead-containing monomer with isopropylstyrene, acrylonitrile, lauryl
methacrylate and octadecyl methacrylate. Acrylonitrile was chosen
because it generally produces copolymers with styrene with impact
strength superior to polystyrene.[36] The other monomers were selected
because they have bulky side chains which would act as internal
plasticizers. The methacrylate monomers contain oxygen, which was
avoided in the selection of the high-Z-containing monomers, so the
quantities added were kept to a minimum.

In almost every case the copolymers produced had lower softening
points and glass transition temperature than those of the homopolymer.

We are unaware of any clear correlation between measured physical
properties, such as glass transition temperature or impact strength,
and the practical machinability of a polymer. Polymeric hemishells
have been produced by single-point diamond machining of partially
cross-linked polystyrene. We obtained a sample of this polymer from
Rockwell International, Rocky Flats, Colo. where the micromachining
was to be performed, to serve as a basis for comparison with the
various copolymers we produced. We devised a simple method for
qualitatively assessing machinability which involves the following:
The purified polymers were melted and pressed between two glass
slides. After cooling, the slides were separated carefully and the
resultant thin films were scratched lightly with a sharp stylus.
The microscopic "tool marks" thus produced were examined for smooth-
ness and evenness as compared with the machinable sample supplied
by Rockwell.

Figure 1 is a photomicrograph of a film of poly(triphenyl-p-
styryllead) homopolymer produced as described above. Note the
brittleness of this polymer which is manifest as cracks throughout
the film surface. This can be compared with Fig. 2, which is a

TABLE I

POLYMERS AND COPOLYMERS OF TRIPHENYL-P-STYRYLLEAD, TSL

Comonomer	Wt ratio TSL: comonomer	Yield	Wt. % TSL	Atom % Pb	T_g (DSC) °C	T_s, °C
none[a]		67%	100	2.0	98	
none[b]		---	100	2.0	140(TBA) >200	
isopropylstyrene[a]	25:1	---	---	---	141	160–165
acrylonitrile[a]	3.5:1	---	70	1.3	159	165
octadecyl-methacrylate[a]	3:1	45	65	0.9	<25	46
Lauryl-methacrylate and divinylbenzene[a]	5:1:0.15	---	75	1.2	---	90
n-butyl methacrylate[a]	5:1	---	---	---	114	105

a benzene is solvent   b no solvent

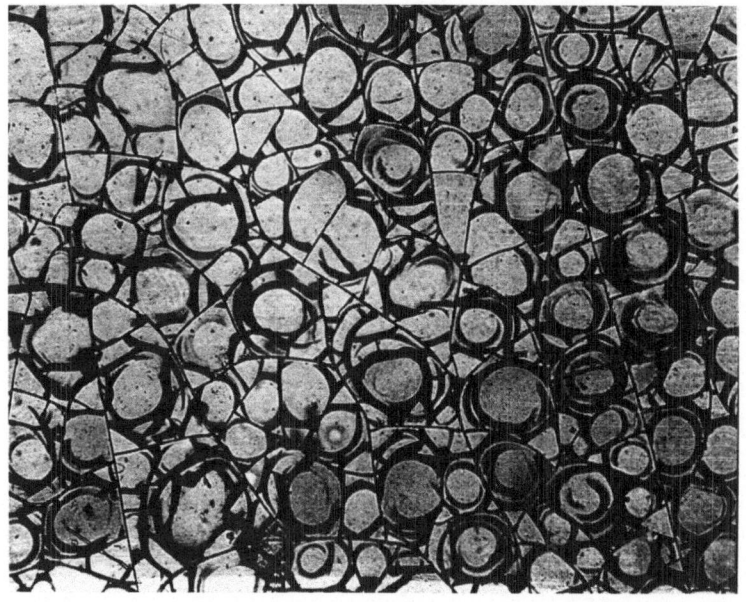

Figure 1: Hot-pressed film of lead homopolymer 100X.

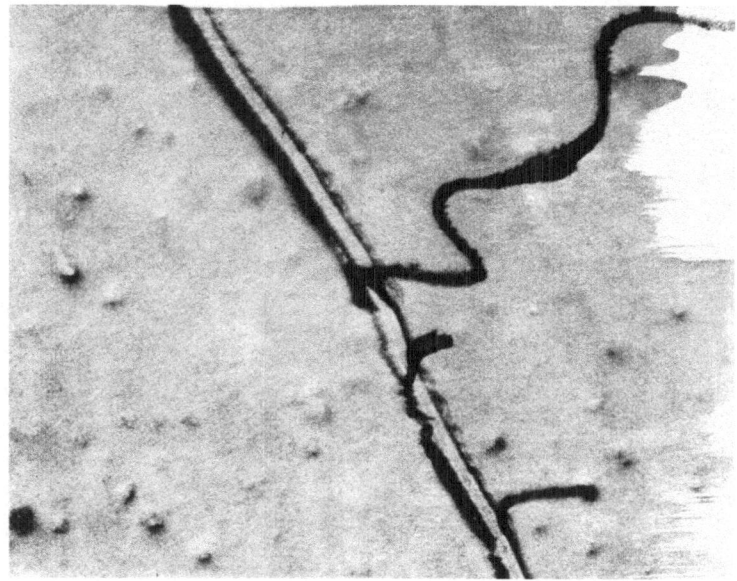

Figure 2: Tool mark in tungsten polymer 100X.

photomicrograph of a tool mark in a machinable copolymer. Note the smooth, even removal of polymer, which is characteristic of machinable polymers.

Samples of polymers which appeared to be most machinable were supplied to Rockwell. Fig. 3 is a SEM photomicrograph of a hemishell produced from a copolymer of triphenyl-p-styryl lead and octadecyl methacrylate, which was the most machinable of the lead-containing copolymers studied. Reactivity ratios were measured for these two monomers using IR spectra of copolymers of varying compositions polymerized to low conversions. The values were calculated to be: R_{Pb} = 1.05 and R_{ODM} = 0.396. This indicates that the lead-containing monomer has higher activity to copolymerization than octadecylmethacrylate, with the monomer becoming more readily attached to its own radical than to an ODM radical.

Hemishells were also produced from this polymer by vacuum molding. This was accomplished by placing a free-standing polymer film on a silicon hemispherical cavity mold.[9] The mold was then evacuated to remove any trapped air. Next, the mold was heated above the softening point of the polymer and atmospheric pressure was admitted, causing the film to be pushed into the mold. Finally, the interconnecting matrix was removed by grinding with calcium carbonate, which is softer than silicon, and the hemishells were removed from the mold with ultrasonic agitation and/or immersion in liquid nitrogen. Fig. 4 is a photomicrograph of a hemishell made by this method.

Organoruthenium polymers

The results of an earlier study of polymerization of vinylruthenocene are summarized in Table II. The full experimental details have been published elsewhere.[24] No attempt was made at the time these studies were performed to assess machinability. The homopolymer of VR is highly brittle, with T_g too high to measure. Copolymerization produces materials that form clear films which coat metal surfaces readily and do not crack easily.

Organoosmium and organotungsten polymers. Since these monomers were not available in large quantities, extensive copolymerization studies were not performed. Copolymerization with octadecyl methacrylate yielded polymers which were similar to those obtained with the lead monomer and appear to be acceptably machinable. In the case of the CMT copolymer, it was difficult to obtain polymer of high molecular weight in initial polymerizations, presumable due to chain transfer to tungsten during polymerization. Higher molecular weights were obtained by running polymerization for very long times with multiple reinitiation. End-group analysis shows two AIBN end groups, 3.9 ODM units and one CMT for Mn $\sim$ 2000, DP=5 (0.33 atom % W).

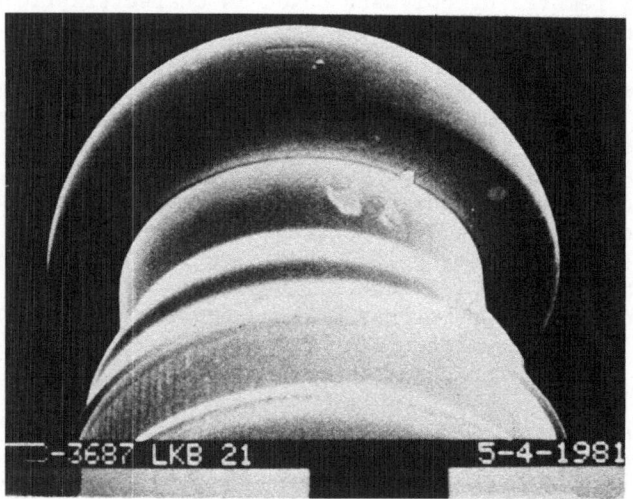

Figure 3: SEM photomicrograph of machined hemishell.

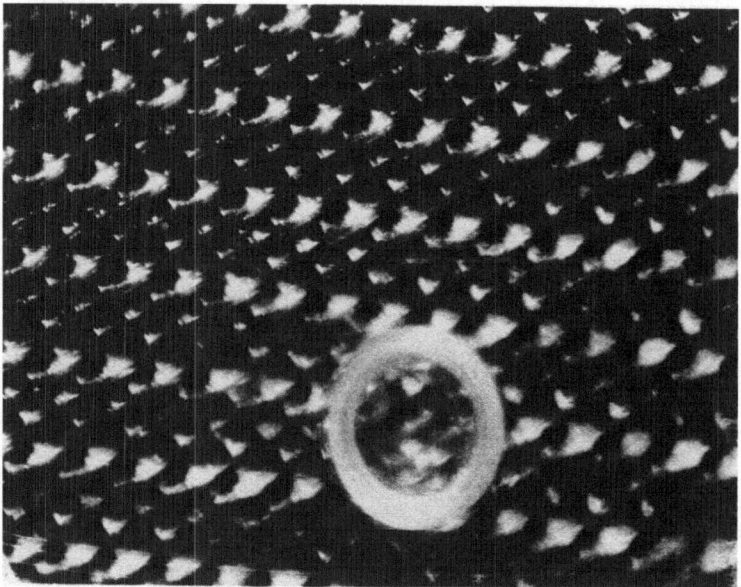

Figure 4: 300 micron diameter vacuum molded hemishell.

TABLE II

Polymers and Copolymers of Vinylruthenocene, VR

Comonomer	Wt. Ratio VR: Comonomer	Yield %	M_n	M_w	Wt. % VR	Atom % Ru	η_i dlg^{-1}	T_g °C
none		95	5,860	18,670	100	4.0		>250
"		93	19,660	118,660	100	4.0	0.15	>250
Methyl acrylate	1:1	76	26,210	139,670	45.2	1.4	0.103	91
	2:1	85	8,650	114,000	60.1	2.0	0.095	111
Styrene	1:1	45	6,520	26,830	34.0	1.1	0.096	99–117
	2:1	57	5,980	29,870	54.5	1.9	0.129	105–131
Vinylpyrrolidinone	2:1	--	3,130	9,650	55.9	1.8	----	203–217

REFERENCES

1a Present address Chemistry Dept. University of Massachusetts, Amherst, MA   01002

1b Present address Radian Co., P.O. Box 9948, 8501 MoPac Blvd., Austin, TX   78766

2. R. J. Nuckolls, L. Wood, A. Thiessen and G. Zimmerman, Nature, 239, 139 (1972).

3. W. R. Meier and W. B. Thomson, Proc. Top. Meet. Technol Controlled Nucl. Fusion Vol. 3, 297-307 (1978).   Chem Abs. 92, 118069 (1980).

4. R. L. Nolen, Jr., L. B. Kool and P. J. Evans, J. Vac. Sci. Technol., 20, 1121 (1982)

5. R. Liepins, M. Campbell, J. S. Clements, J. Hammond and R. J. Fries, Ibid. 18, 1218-26 (1981).

6. A. T. Young, D. K. Moreno and R. G. Marsters, Ibid., 20, 1094-7 (1982).

7. D. M. Stupin, K. R. Moore, G. D. Thomas and R. L. Whitman, Ibid. 20, 1071-4 (1982).

8. L. B. Kool, R. L. Nolen and K. W. Sherwood, Ibid. 18(3) 1233-7 (1981).

9. K. D. Wise, M. G. Robinson and W. J. Hillegas, Ibid. 18(3) 1179-82 (1981).

10. H. W. Deckman, Ibid., 18(3) 1171-4 (1981).

11. D. W. Carroll and W. J. McCreary, Ibid., 20(4) 1087-90 (1982).

12. T. Norimatsu, A. Furusawa, M. Yoshida, Y Izawa and C. Yamanaka, Ibid. 18(3) 1288-9 (1981).

13. I. S. Goldstein, F. Kalk and J. Trovato, Ibid., 18(3) 1175-78 (1981).

14. J. M. Kendall, M. C. Lee and T. G. Wang, Ibid., 20(4) 1091-3 (1982).

15. B. A. Brinker and J. R. Miller, Ibid. 20(4) 1079-81 (1982).

16. K. W. Bieg, Ibid., 18(3) 1231-2 (1982).

17. R. L. Crawley, L. B. Kool and R. L. Nolen, Ibid., 18(3), 1255-7 (1982).

18. R. L. Nolen, L. B. Kool and J. E. Sheats, Proceedings of Tenth International Conference on Organometallic Chemistry, Toronto, Canada, August 9-14, 1981 p. 129.

19. L. B. Kool and R. L. Nolen, Ibid. p 127

20. Proceedings two earlier symposia on inertial confinement fusion have been published.
Digest of Topical Meeting on Inertial Confinement Fusion, Feb. 26-28, 1980, San Diego, CA, American Optical Society, Washington, D.C., 1980.

Digest of Technical Meeting on Inertial Confinement Fusion, Feb. 7-9, 1978. American Optical Society, Washington, D.C. 1978.

21. E. K. Storm, Phys. Rev. Lett., 40, 1570-3 (1978)

22. R. E. Kidder, Nucl. Fusion, 21(2), 145-51 (1981).

23. J. G. Noltes, H. A. Budding and G. J. M. Van der Kerk, Recuil Trav. Chem Pays Bas, 79, 408, 1976 (1960).

24. J. E. Sheats and T. C. Willis, Org. Coatings and Plastics Chem., 41(2) 33-7 (1979). Full paper, Journal of Polymer Science, in press.

25. E. A. Hill and J. H. Richards, J. Amer. Chem. Soc., 83, 3840-6 (1961).

26. M. D. Rausch and A. Siegel, J. Organometal. Chem., 11, 317 (1968)

27. G. K. Buell, W. E. McEwen and J. Kleinberg, J. Amer. Chem. Soc., 84, 40 (1962).

28. J. K. Gillham, AICHE Journal, 20, 1066 (1974).

29. J. E. Sheats in Kirk-Othmer Encyclopedia of Chemical Technology" John Wiley & Sons, New York, 1981, Vol. 15 pp 184-219.

30. J. E. Sheats, J. Macromol Sci. - Chem. A15(6) 1173-99 (1981).

31. C. E. Carraher, Jr., J. E. Sheats and C. U. Pittman, Jr., Eds. Organometallic Polymers, Academic Press, N.Y. 1978.

32. C.E. Carraher, Jr., J.E. Sheats and C.U. Pittman, Jr., Advances in Organometallic Polymer Science, J. Macromol. Sci. Special Symposium Issue A16, 1981; Hardback Marcel Dekker, New York, 1982

33. J. R. Leebrick and H. E. Ramsden, J. Org. Chem., 23, 935 (1958).

34. M. M. Koton and L. F. Dukubina, Vysokomol. Soyed. (Russian Journal of Polymer Science) 6(10), 1791-4 (1964).

35. N. Fujita and K. Sonagshira, J. Polym. Sci, Polym. Chem. Ed., 12, 2845-56 (1974).

36. W. J. Roff and J. R. Scott, Handbook of Common Polymers, CRC Press, 1971, p. 51-2.

37. C. U. Pittman, Jr., T. V. Jayaraman, R. D. Priester, Jr., S. Spencer and M. D. Rausch and D. W. Macomber, Macromolecules 14, 237 (1981).

38. D. W. Macomber, M. D. Rausch, T. V. Jayaraman, R. D. Priester and C. U. Pittman, Jr., J. Organometal. Chem. 205, 353 (1981).

39. D. W. Macomber, W. P. Hart, M. D. Rausch, R. D. Priester and C. U. Pittman, Jr., J. Amer. Chem. Soc., 104, 884 (1982).

40. M. D. Rausch, E. O. Fischer and H. Grubert, J. Amer. Chem. Soc., 82, 76 (1960).

SYNTHESIS AND NMR CHARACTERIZATION OF COPOLYMERS OF α-FLUOROSTYRENE

WITH METHYL ACRYLATE

Ramendra N. Majumdar and H. James Harwood

Institute of Polymer Science
The University of Akron
Akron, Ohio 44325

INTRODUCTION

Polymers of α-halostyrenes are unstable because they contain labile halogen atoms which are readily lost as hydrogen halide[1]. Although a few studies on the synthesis and applications of homopolymers and copolymers of α-chlorostyrene[2-10], α-bromostyrene[11,12] and α-fluorostyrene[13-16] have been reported, no systematic studies on the copolymerization behavior of these monomers or on the microstructures of their polymers are known. Most of these studies have dealt with homopolymerization of α-halostyrenes in emulsion or in suspension or with copolymerizations (mostly graft) involving very small molar percentages of α-halostyrene.

We have recently found that pure α-fluorostyrene can be polymerized by the conventional free radical method and that poly-(α-fluorostyrene) solutions can be stabilized by weak or highly hindered bases[17]. This finding enabled us to investigate the microstructure of poly(α-fluorostyrene) by ^{13}C-NMR spectroscopy. This study has been extended to copolymers of α-fluorostyrene with methyl acrylate because of our interest in the cotacticity of statistical and alternating copolymers of styrene and substituted styrenes with acrylates and methacrylates. The present paper covers the 300 MHz [1]H-, 20 MHz ^{13}C- and 282 MHz ^{19}F-NMR studies of poly(α-fluorostyrene-co-methyl acrylate)s.

EXPERIMENTAL

α-Fluorostyrene was prepared by bromofluorination of styrene using N-bromoacetamide and anhydrous hydrofluoric acid in anhydrous

ether followed by dehydrobromination with KOH[18]. It was distilled
immediately prior to use, b.p. 43-44°C (13 mbar) [Literature[18],
b.p. 45-46°C (14 mbar)]. The ^{1}H-NMR spectrum of the monomer contains aromatic (7.1-7.7 ppm) and olefinic (4.4-5.3 ppm) proton
resonances having relative intensities of 5:2. Chemical shifts and
coupling constants for the olefinic protons were determined from
the 60- and 300 MHz ^{1}H-NMR and 282 MHz ^{19}F-NMR spectra of the monomer by use of the LAOCN3 program[19]. Designating the olefinic protons cis and trans to the fluorine atom in the monomer as A and B,
respectively, the following values were obtained: δ_A=4.74 ppm;
δ_B=4.92 ppm; J_{AB}=3.2 Hz; J_{A-F}=16.9 Hz; and J_{B-F}=49 Hz. Figures 1
and 2 show that these parameters lead to simulated spectra that
agree with observed spectra.

Methyl acrylate was freshly distilled over calcium hydride
immediately prior to use, b.p. 80°C.

Copolymerizations were conducted in vials sealed under vacuum
at 60°C in bulk using AIBN as initiator. Polymerization mixtures

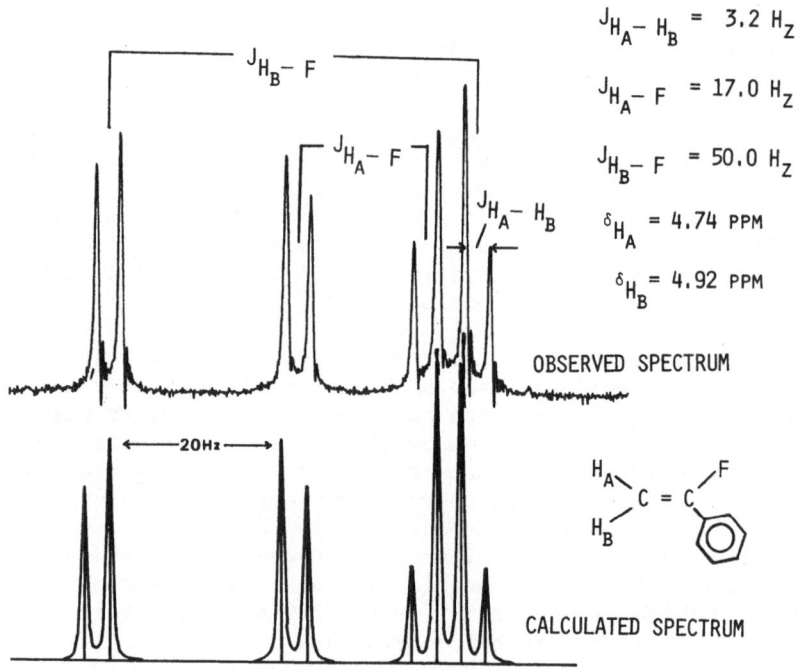

Fig. 1.   60 MHz olefin proton resonance of α-fluorostyrene. The
calculated spectrum is based on the following parameters:
δ_A=4.74 ppm; δ_B=4.92 ppm; J_{AB}=3.2 Hz; J_{AF}=16.9 Hz;
J_{BF}=49 Hz.

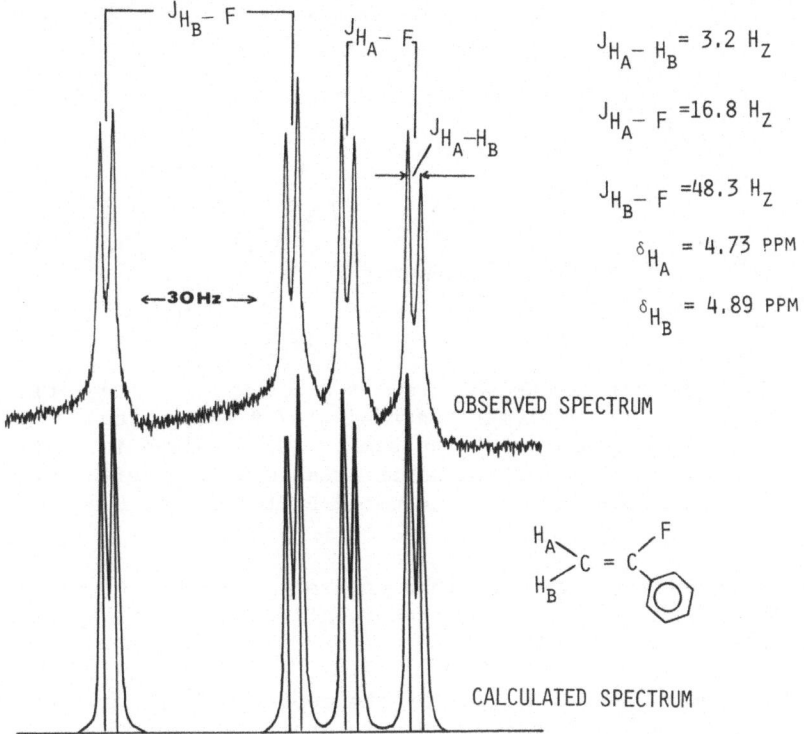

$J_{H_A - H_B} = 3.2\ H_Z$

$J_{H_A - F} = 16.8\ H_Z$

$J_{H_B - F} = 48.3\ H_Z$

$\delta_{H_A} = 4.73\ PPM$

$\delta_{H_B} = 4.89\ PPM$

OBSERVED SPECTRUM

CALCULATED SPECTRUM

Fig. 2.   300 MHz olefin proton resonance of α-fluorostyrene.
The calculated spectrum is based on the parameters
given in Figure 1.

were poured into methanol to precipitate the copolymers. These
were washed thoroughly with methanol and were dried under vacuum.
Copolymer compositions were determined from the relative ratios of
aliphatic and aromatic proton resonance areas observed in 60 MHz
^{1}H-NMR spectra of the copolymers in $CDCl_3$ solution.

300 MHz ^{1}H-NMR spectra of the copolymers in $CDCl_3$ solution
were recorded at ambient temperature and used in studies on their
methoxy proton resonances. Three methoxy proton resonance areas
were defined as indicated in Figure 3 and were measured by cutting
and weighing. Resonances observed at δ=3.33-3.86 ppm, δ=2.91-3.33
ppm and δ=2.65-2.91 ppm will be designated A, B and C in subsequent
discussion.

20 MHz ^{13}C-NMR spectra of the polymers in $CDCl_3$ solution were
recorded at ambient temperature using a Varian CFT-20 NMR Spectrometer. Pulse widths of 14-19 μsec., aquisition times of 0.5 sec.

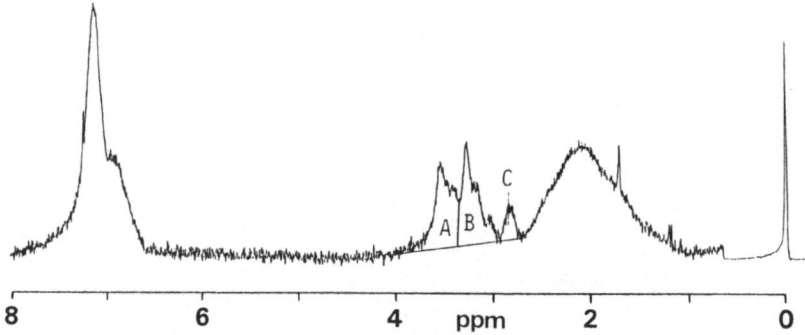

Fig. 3.   300 MHz ^{1}H-NMR spectrum of a methyl acrylate-α-fluoro-
           styrene copolymer containing 47.9 mole percent
           α-fluorostyrene. The sample is dissolved in CDCl$_3$.
           Methoxy proton resonance areas used for coisotacticity
           determination are indicated by letters A, B and C.

and pulse intervals greater than one second were generally employed.
Chemical shifts (δ_c) are reported relative to the TMS signal
(δ_c=0.0).

      282 MHz ^{19}F-NMR spectra of the copolymers in chloroform so-
lution were recorded at room temperature. p-Difluorobenzene was
used as the internal standard and its chemical shift in chloroform
solution relative to CFCl$_3$ was found to be +120.2 ppm. [Following
the proposed convention[20], the chemical shift of CFCl$_3$ is taken to
be zero and resonances upfield from it are assigned positive
values. These are represented by the symbol δ_f in subsequent
discussion].

      Compositional features of the copolymers, such as monomer
contents, dyad distributions, and monomer centered triad fractions
were calculated from monomer feed compositions, conversions obtained in copolymerization experiments and reactivity ratios of
0.78 and 0.26 for αFS and MA, respectively, using methods and a
computer program we have described previously[21]. For ease of
discussion, αFS will be represented by S and MA will be represented
by M when dyads and triads are discussed subsequently. Monomer
centered triad fractions, which represent the fraction of monomer
units of a given type, eg. M, in a particular triad environment,
eg. SMS, will be designated as F_{xyz}, where xyz is the triad sequence and y is the central monomer unit under consideration.

      Approximate M-centered triad fractions, F_{MMM}, $F_{(MMS + SMM)}$
and F_{SMS}, can also be calculated from average monomer feed compo-
itions using the following equations, where P_{MS} is the probability
of a given M unit being followed by an S unit. The average ratio, $\overline{X}$,

of α-fluorostyrene to methyl acrylate in the monomer mixture can be calculated as described by Joshi[22].

$$F_{MMM} = (1 - P_{MS})^2$$

$$F_{MMS} = F_{SMM} = P_{MS} (1 - P_{MS})$$

$$F_{SMS} = P_{MS}^2$$

$$P_{MS} = 1/(1 + r_M/\bar{X})$$

Gel permeation chromotograms of some of the polymers were measured with a Waters Association Model 440 Gel Permeation Chromatograph equipped with an absorbance detector. Tetrahydrofuran (THF) was used as a solvent and a calibration based on polyisobutylene standards was used. The approximate number and weight averages are $\bar{M}_n$=136,000 and $\bar{M}_w$=296,000 for the copolymer having 19.6 mole % α-fluorostyrene; $\bar{M}_n$=60,200 and $\bar{M}_w$=98,000 for the copolymer having 48.1 mole % α-fluorostyrene; and $\bar{M}_n$=16,400 and $\bar{M}_w$=24,800 for poly-(α-fluorostyrene). A similar poly(α-fluorostyrene) sample that was analyzed earlier using a polystyrene calibration had $\bar{M}_n$=4,900 and $\bar{M}_w$=10,400[17].

RESULTS AND DISCUSSION

Copolymerization Studies

Table I lists monomer feed compositions, copolymer compositions and conversions obtained in the copolymerization experiments. The copolymerization diagram of the system (Fig. 4) shows a tendency towards alternation with an azeotropic point at 70 mole % MA. Reactivity ratios for the αFS-MA copolymerization system, determined by the Kelen-Tüdös method[23], were r_{MA}=0.26 and $r_{\alpha FS}$=0.78. The KT-plot is shown in Figure 5. Average monomer feed compositions[24] were used for this determination whenever the conversion was above 10 wt. percent. Almost identical values of the reactivity ratios were obtained when calculated by the Tidwell-Mortimer[25] method. The reactivity ratio product for this copolymerization system ($r_{MA} \cdot r_{\alpha FS} \approx 0.2$) indicates a tendency for alternation.

Based on Q and e values of 0.45 and +0.64 for methyl acrylate[26], Q and e values for α-fluorostyrene, were calculated to be 0.9 and -0.7, respectively. This indicates that α-fluorostyrene, although slightly less reactive than styrene, behaves as a relatively strong electron donor, comparable to styrene itself. This is interesting since vinyl fluoride is a strong electron acceptor. Table II compares the Q and e values of some relevant substituted ethylenes with those of α-fluorostyrene.

Table I. Data of α-Fluorostyrene-Methyl
Acrylate Copolymerization System

Mole % α-Fluorostyrene			Conversion
Monomer Feed	Copolymer		Wt.%
8.4	19.6[a]	20.1[b]	14.3
16.76	33.4	32.9	6.7
23.46	40.1	39.5	8.3
34.4	47.9	47.4	22.2
37.1	48.1	48.4	37.4
66.06	68.8	68.9	3.7
85.7	84.5	84.5	6.3

(a) Observed   (b) Calculated from monomer feed composition and conversion

^{1}H-NMR Studies

As can be seen in Figure 3, the methoxy proton resonance of the copolymers is observed as a complicated pattern that seems to be influenced by triad and pentad monomer sequence effects as well as by stereosequence effects. The methoxy proton resonance patterns were separated into three contributions as shown in this Figure. As was done previously in other studies on copolymers of styrenes with acrylate and methacrylate esters[27,28], the highest field component, C, was assigned to SMS triads having coisotactic MS and SM placements. The central component, B, was assigned to

Table II. Q and e Values of Some Substituted
Ethylenes[26]

Monomer	Q	e
Ethylene	0.016	+0.05
Vinyl Fluoride	0.008	+0.72
Styrene	1.00	−0.80
α-Fluorostyrene	0.90	−0.7
α-Methylstyrene	0.97	−0.81

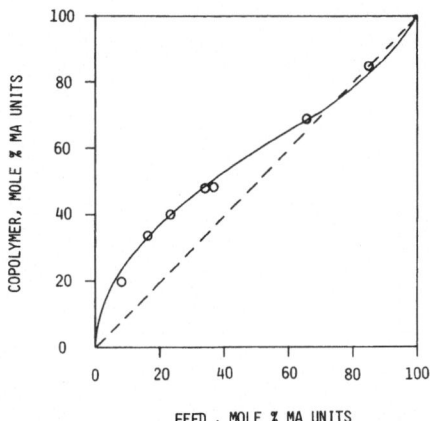

Fig. 4.   Copolymerization diagram for the α-fluorostyrene/
methyl acrylate system.

SMS, SMM or MMS triads having one coisotactic MS or SM placement.
Other methoxy proton resonances were assigned to component A.

    According to the assignments, the following equations relate
the fractions of methoxy proton resonance observed in the A-, B-
and C-resonance areas (F_A, F_B and F_C), methyl acrylate centered
triad fractions (F_{MMM}, F_{SMM}, F_{MMS} and F_{SMS}) and the probability
σ_{MS} that an MS or SM placement is coisotactic[28].

$$\frac{F_A - F_{MMM}}{F_{SMS}} = (1-\sigma_{MS}) \frac{(F_{SMM} + F_{MMS})}{F_{SMS}} + (1-\sigma_{MS}) \qquad (1)$$

$$\frac{F_B}{F_{SMS}} = \sigma_{MS} \frac{(F_{SMM} + F_{MMS})}{F_{SMS}} + 2\sigma_{MS}(1-\sigma_{MS}) \qquad (2)$$

$$\frac{F_C}{F_{SMS}} = \sigma_{MS}^2 \qquad (3)$$

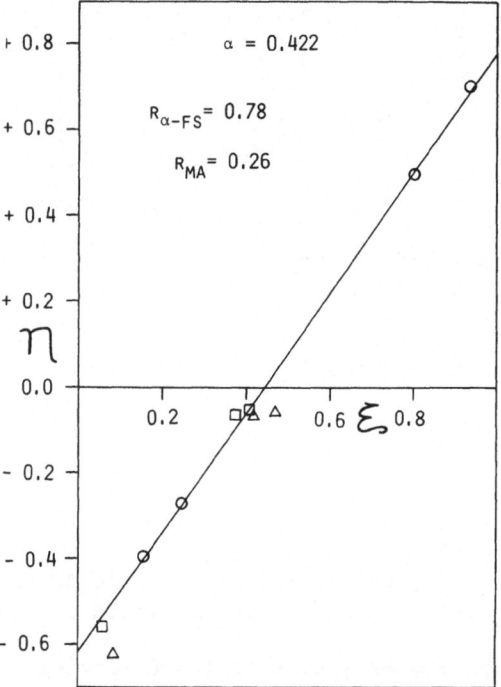

Fig. 5.   Kelen-Tüdös plot of data for the α-fluorostyrene/methyl
           acrylate system. (O) low conversion data, (▲) high
           conversion data, (☐) high conversion data calculated
           from average monomer feed.

Methyl acrylate centered triad fractions were used in combination with measured F_A, F_B and F_C values to test the applicability of Eqs. 1-3 to describe the methoxy proton resonance patterns. Table III lists calculated MA-centered triad fractions, F_A, F_B and F_C values used in this determination. Figure 6 shows plots of the left-hand parts (L.H.P.) of Eqs. 1-3 vs. $(F_{MMS}+F_{SMM})/F_{SMS}$. The plots are linear and their slopes and intercepts define a σ_{MS} value of 0.47. This suggests that MS and SM placements in the copolymers have nearly equal probabilities of being coisotactic or cosyndiotactic. Since poly(methyl acrylate)[29] and poly(α-fluorostyrene)[17] have been shown to have random stereosequence distributions when prepared by free radical initiated polymerizations, the copolymers may be considered to have atactic stereosequence structures.

It should be noted that Ito and Yamashita previously reported σ_{MS} values of $\sim$0.8 for copolymers of styrene with methyl acrylate and benzyl acrylate[27], based on studies of the 60 MHz ^{1}H-NMR spectra of the copolymers. However, the radically prepared homopolymers of styrene[30-32], methyl acrylate[29] and benzyl acrylate[33] have σ_{MS} values of approximately 0.5. Recent 300 MHz ^{1}H-NMR studies[33] on copolymers of styrene with methyl acrylate or benzyl acrylate indicate that σ_{MS} for such copolymers should also be $\sim$0.5. The finding of a σ_{MS} value of $\sim$0.5 for α-fluorostyrene-methyl acrylate copolymers is thus expected on the basis of the σ_{MS} values of the corresponding radically prepared homopolymers.

Table III. MA-Centered Triad Fractions and
Relative Methoxy Proton Resonance
Areas of the Copolymers

Mole % αFS in Copolymer	Calculated MA-Centered Triad Fractions			Relative Methoxy Proton Resonance Areas		
	F_{MMM}	$F_{(MMS+SMM)}$	F_{SMS}	F_A	F_B	F_C
19.6	0.583	0.181	0.056	0.819	0.168	0.013
33.4	0.329	0.245	0.182	0.603	0.335	0.052
40.1	0.219	0.249	0.282	0.555	0.336	0.067
47.9	0.121	0.227	0.424	0.464	0.449	0.088
48.1	0.112	0.222	0.444	0.478	0.436	0.086
68.8	0.014	0.104	0.778	0.284	0.528	0.187
85.4	0.002	0.039	0.919	0.289	0.514	0.196

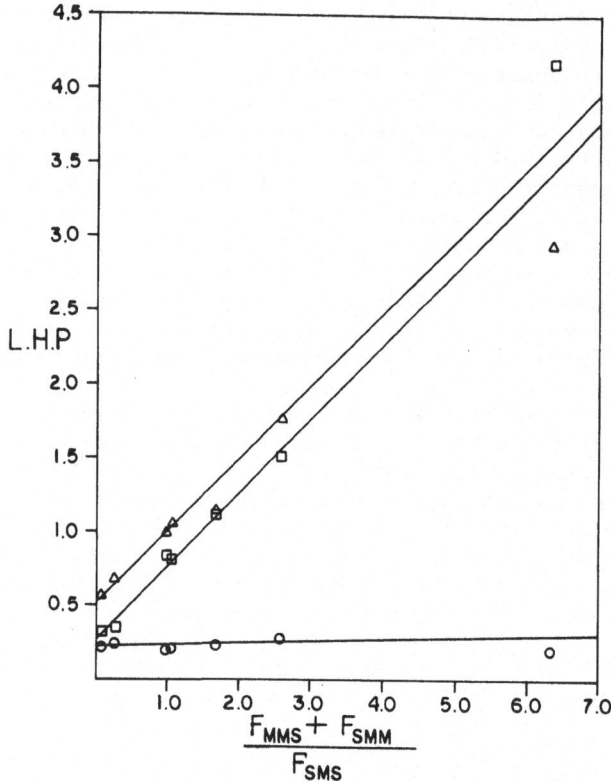

Fig. 6.   HR Plot of methoxy proton resonance area data for
           α-fluorostyrene-methyl acrylate copolymers. ▲, □ and
           O define plots of left-hand parts (LHP) of Eqs. 1, 2
           and 3, respectively, versus $F_{(MMS + SMM)}/F_{SMS}$.

^{13}C-NMR Spectra

Figures 7A, 7B and 8A show 20 MHz ^{13}C-NMR spectra of several
methyl acrylate-α-fluorostyrene copolymers in CDCl$_3$ solution at
room temperature. Many of the resonances evident in these spectra
are sources of microstructure information.

The carbonyl carbon resonance of the copolymers is observed
as a group of closely spaced signals occurring at $\delta_c \simeq 175.3$ ppm.
The carbonyl carbon resonance patterns are not well enough resolved

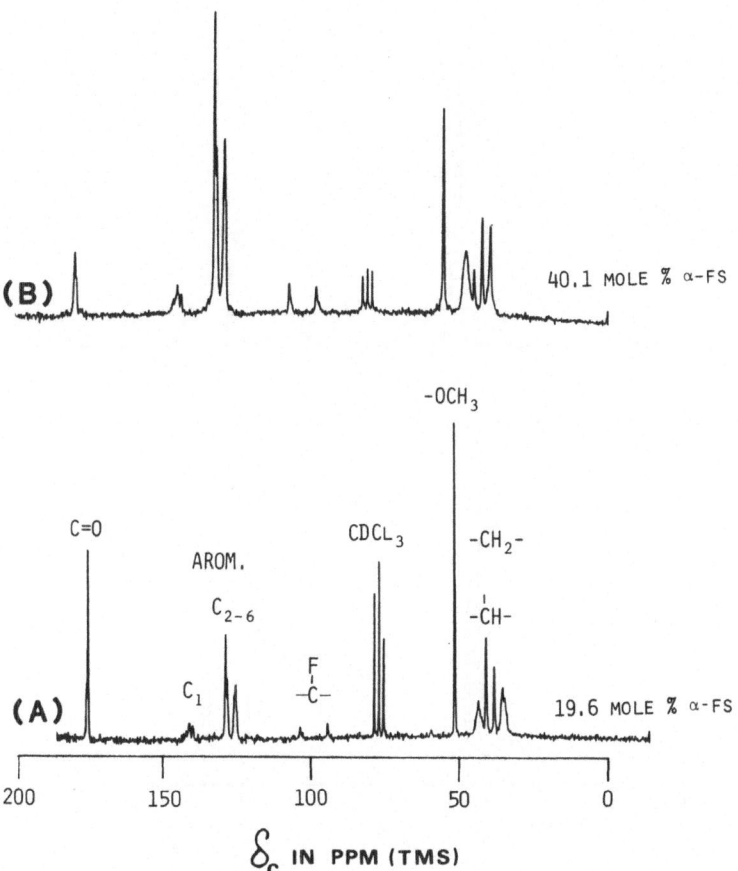

Fig. 7.   20 MHz ^{13}C-NMR spectra of α-fluorostyrene-methyl acry-
           late copolymers containing 19.6 and 40.1 mole percent
           α-fluorostyrene. The copolymers are dissolved in CDCl$_3$

to enable a reliable interpretation to be made in terms of sequence
effects or sequence-stereosequence effects, although the patterns
can be roughly divided into three regions having relative areas
that are approximately the same as calculated MA-centered triad
fractions. On the basis of this correlation, triad resonances are
probably being observed, with SMS, (SMM+MMS) and MMM appearing in
order of increasing field. The width of the MMM signal is about

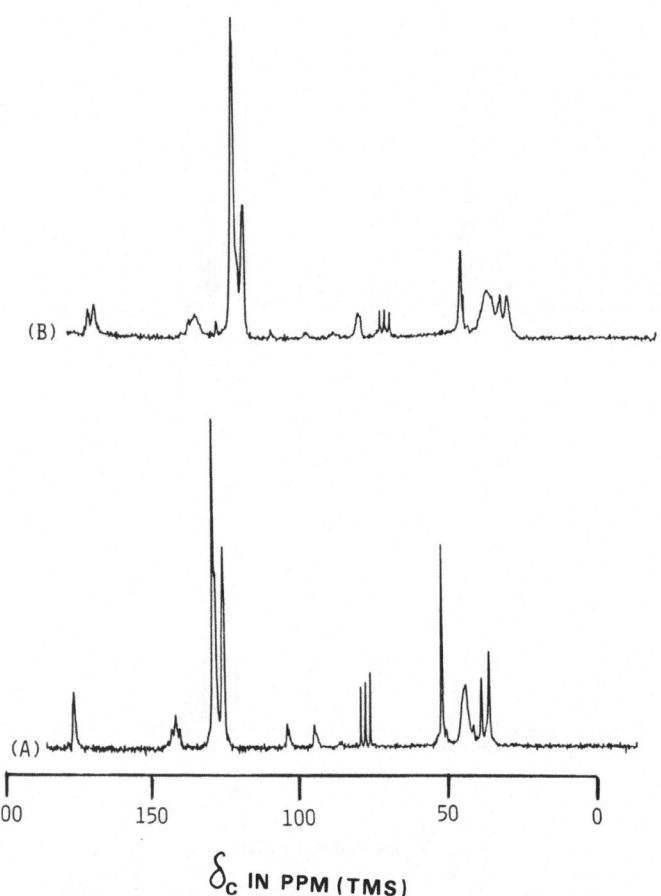

Fig. 8.   20 MHz ^{13}C-NMR spectra of an α-fluorostyrene-methyl
           acrylate copolymer containing 48.1 mole percent
           α-fluorostyrene. Spectrum A is of a freshly prepared
           solution in CDCl$_3$ Spectrum B is of the solution
           after standing for 12 days at room temperature.

the same as that of the carbonyl signal in the spectrum of polyMA,
indicating that signals due to MMM stereosequences are contained
within this region. The (SMM + MMS) resonance seems to consist of
two signals, suggesting a sensitivity to stereosequence or pentad
effects. The SMS signals are too weak to provide information on

this point. It is interesting to note that samples that have decomposed in solution (i.e., lactonized, <u>vide infra</u>) show two or
three well resolved carbonyl carbon resonances occurring over a
5 ppm range. This observation should be useful for studies on the
lactonization process.

The quaternary aromatic carbon resonance of the copolymers
(δ_c=140.8 ppm) is observed as a group of four resonances, the relative intensities of which are influenced by C-F couplings, by
sequence distribution effects and probably also by stereosequence
distribution effects. In the spectrum of poly(α-fluorostyrene),
this resonance region contains two strong signals separated by
28 Hz (1.4 ppm), which corresponds to J_{CCF} for this polymer. The
δ_c values for the centers of these patterns are 143.3 and 142.4 ppm.
One can thus assume that CF carbons centered in SSS triads in the
copolymers will also be observed at these δ_c values. In the spectrum of the copolymer containing 19 mole percent αFS, most (∿90%)
of the αFS units are centered in MSM triads. In the spectrum of
such a copolymer (Figure 7A) the CF resonance region contains two
strong signals having δ_c values of 140.5 and 139.2 ppm. The separation of these signals (25 Hz) is about the same as the J_{CCF}
value observed for polyαFS. These signals are thus attributed to
MSM triads. Figure 7A also contains a small signal at δ_c=141.4 ppm
and it should be noted that the signal at δ_c=140.5 is slightly more
intense than that of δ_c=139.2 ppm. It thus seems reasonable to believe that CF carbons in (MSS + SSM) triads are responsible for the
signal at δ_c=141.4 ppm and for some of the resonance observed in
the δ_c=140.5 ppm region. Based on this assignment, J_{CCF} for MSS +
SSM triad is ∿25 Hz, in agreement with that estimated for SSS and
MSM triads. The quaternary aromatic carbon resonance pattern observed for polyαFS[17] is complex, due to stereosequence effects, but
it is dominated by two strong signals occurring at δ_c=143.3 and
142.4 ppm. To a first approximation these may be taken as the
chemical shifts of quaternary aromatic carbons centered in SSS
triads in the copolymers. If overlapping of the various resonances
discussed above is taken into account, the quaternary carbon resonance of the copolymers may be considered to occur in four general
areas, labled A, B, C and D, in order of increasing field, as is
shown in Figure 9. Based on the above discussion, these resonances
are assigned as indicated in Table IV, which also shows that their
relative areas agree reasonably well with those calculated from
triad fractions. A more complete interpretation of this region
should include consideration of stereosequence effects, but this is
not possible at the present time.

The resonance of other aromatic carbons occurs between
δ_c = 124-130 ppm but these patterns are almost the same in all the
spectra and are not considered useful sources of microstructure
information.

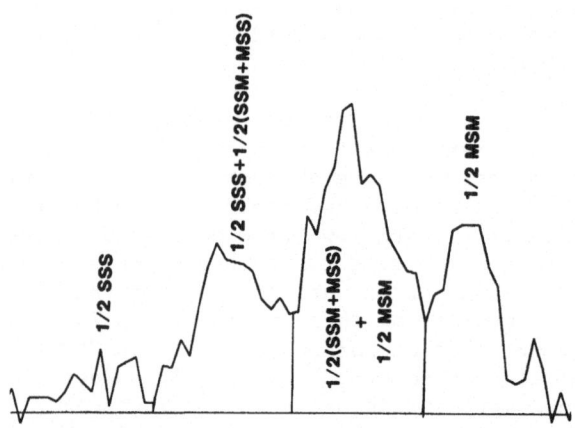

Fig. 9.   Quaternary aromatic carbon resonance of an α-fluoro-
           styrene-methyl acrylate copolymer containing 48 mole
           percent α-fluorostyrene. Assignments for the various
           regions are given in Table IV.

The resonance of CF carbons of αFS units in the copolymers is
observed as two identical patterns centered at δ_c=∿94 and ∿103 ppm,
due to C-F coupling. The J_{C-F} constant deduced from this splitting
(177 Hz), is very close to that (190 Hz) determined for the CF carbon resonances of polyαFS. Since the CF carbon resonances of polyαFS occur at δ_c=92.1 and δ_c=101.6 ppm and are not split appreciably
by stereosequence effects, it seems that the CF carbon resonance
patterns of the copolymers may be useful for sequence distribution
measurements: CF-carbon resonances of SSS and MSM triads should be
observed at (δ_c=92.4 and 101.9 ppm) and (δ_c=94.3 and 103.2 ppm),
respectively; those associated with (MSS + SSM) triads should be
observed in the vicinity of δ_c=93 and 102.5 ppm. As can be seen in
Figure 10, this seems to be the case for the copolymer containing
48 mole percent αFS; the observed CF-carbon resonance pattern seems
to be consistent with αFS triad fractions calculated for this
sample, which are F_{MSM}=0.52, $F_{(SSM+MSS)}$=0.40, and F_{SSS}=0.08.

The resonance of methoxy carbons is observed as a sharp singlet that coincides with the resonance of CH$_2$ carbons centered in SS
dyads. Thus, the chemical shift of the methoxy carbons (δ_c=52.2 ppm)

Table IV. Assignments and Relative Areas of Quaternary
Aromatic Carbon Resonances of α-Fluorostyrene-
Methyl Acrylate Copolymers

Resonance Region (ppm)	Assignment	Relative Area for Copolymers With					
		19% αFS		40% αFS		48% αFS	
		Obsd.	Calc.	Obsd.	Calc.	Obsd.	Calc.
A. (142.4)	½ SSS	–	0.002	0.04	0.020	0.06	0.038
B. (141.1)	½ SSS + ½(SSM + MSS)	0.06	0.058	0.14	0.174	0.25	0.238
C. (140.5)	½(SSM + MSS) + ½ MSM	0.58	0.449	0.53	0.482	0.46	0.462
D. (139.2)	½ MSM	0.36	0.443	0.30	0.330	0.23	0.262

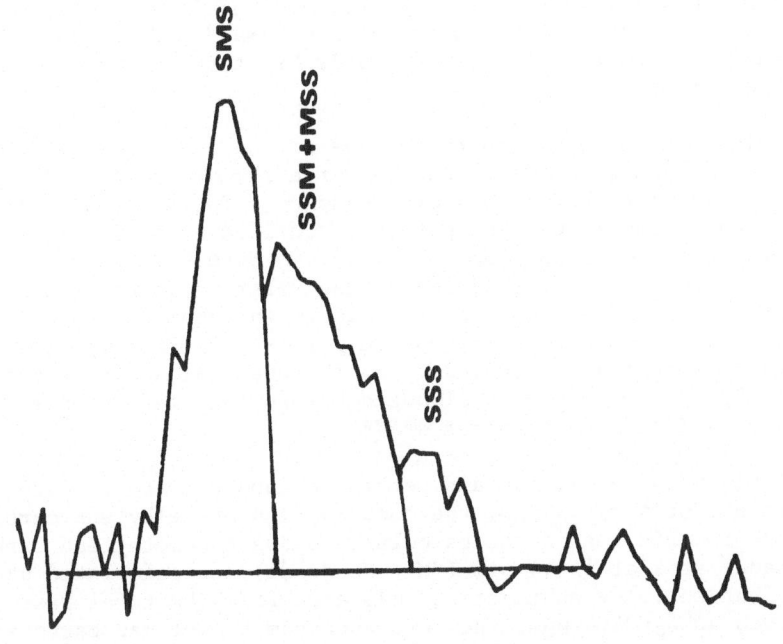

Fig. 10. High field CF-carbon resonance pattern of an α-fluorostyrene-methyl acrylate copolymer containing 48.1 mole
percent α-fluorostyrene, showing possible contributions
due to α-fluorostyrene-centered triads.

is overlapped with that of the broad methylene carbon resonance of polyαFS (δ_c=55.7-46.2 ppm)[17].

The resonances of methylene carbons and of the methine carbons of MA units (δ_c=45-32 ppm are partially overlapped but are sufficiently well defined that assignments may be made based on dyad distributions and on MA-centered triad fractions calculated for the copolymers and on assignments made previously for polyMA[34] and polyαFS[17]. The three sharp signals observed between δ_c=46-32 ppm in Figures 7A, 7B and 8A are attributed to methine carbons because the resonances of such carbons are narrower than those of methylene carbons in the spectra of polystyrene and poly(alkyl acrylates)[34]. With increasing content of αFS in the copolymers, the approximate ratio of the broad peaks to sharp peaks in the δ_c=46-32 ppm region decreases. This further proves that the sharp peaks belong to methine carbons and that the broad peaks are due to methylene carbons.

On the basis of the relative intensities of the methine carbon resonances of the copolymers and the chemical shift of the methine carbons in polyMA (δ_c=41 ppm), the methine carbons centered in MMM, (SMM + MMS) and SMS triads in the copolymers are assigned chemical shifts of δ_c=41.1, 38.6 and 35.4 ppm, respectively. As in the ^{13}C-NMR spectra of polyMA[34] and atactic polystyrene[35], these signals are not split by stereosequence effects in 20 MHz spectra.

The methylene carbon resonances of the copolymers are observed as broad patterns that occur in three general areas. The broad patterns are attributed to stereosequence effects and also to carbon-fluorine coupling. The patterns centered at δ=52, 44, and 35 ppm are assigned to SS, (SM + MS), and MM dyads, respectively. These assignments are justified by the relative intensities of the resonances in the various spectra and by the chemical shifts observed for the methylene carbons in polyαFS (δ_c=51 ppm)[17], and polyMA (δ_c=35 ppm)[34]. In addition, the chemical shift assigned to (MS + SM) dyads, δ_c=44, is almost equal to the average of the chemical shifts for the other dyads.

Since the methylene and methine carbon resonances were severly overlapped at δ_c=35.4 ppm, the contribution of methylene carbon resonance to this region was calculated using the methylene carbon resonance area at δ_c=44 ppm (SM + MS dyads) and calculated dyad distributions. By subtracting this area from the total resonance observed at δ_c=35.4 ppm, the resonance due to methine carbons in this region was determined. This was used, along with the other methine carbon resonance areas to determine the proportions of methine carbon resonance occurring at δ_c=41.1, 38.8 and 35.4 ppm in the various spectra. These were then compared with MA-centered triad fractions that were calculated for the copolymers. Good agreement was obtained between these relative resonance areas and calculated triad fractions, as is shown in Table V.

It is thus seen that the carbonyl-, quaternary aromatic-, CF-, methine- and methylene-carbon resonances of αFS-MA copolymers can provide information about their structures. In particular, the methine carbon resonances seem to be the most useful for sequence distribution measurement.

^{19}F-NMR Studies

282 MHz ^{19}F-NMR spectra of poly-αFS and of several αFS-MA copolymers in CDCl$_3$ solution at room temperature are shown in Figure 11. A large number of signals are evident and these have not been assigned completely. The spectra can be divided broadly into regions assignable to αFS-centered triads. Resonance observed from δ_f=113.8-118 ppm is assigned to αFS units centered in MSM triads. Thus, Table VI shows that the proportion of resonance observed in this region agrees well with F_{MSM} values calculated for the copolymers. Since the ^{19}F-resonance of poly-αFS is observed at δ_f=106-111 ppm, αFS units centered in SSS triads can be expected to be observed in this region. The fine structure observed in this pattern has been attributed to triad stereosequences[17]. The copolymer containing 40 mole percent αFS is calculated to have only 3.5 percent of its units present in SSS triads. The resonance observed at δ_f=106-113.8 ppm in the ^{19}F-spectrum of this copolymer is therefore largely due to αFS units centered in (SSM + MSS) triads. The ^{19}F-resonance of such triads is thus substantially overlapped with that of SSS triads and no effort was made to measure the relative concentrations of these individual triads from the ^{19}F-spectra.

Table V. Comparison of Methine Carbon Resonance Areas to Triad Fractions Calculated for the Copolymers.

Mole % αFS in Copolymer	Relative Methine Carbon Resonance Areas			MA-Centered Triad Fractions*		
	δ_c=35.5	δ_c=38.6	δ_c=41.1	F_{SMS}	$F_{(SMM+MMS)}$	F_{MMM}
19.6	0.00	0.39	0.61	0.06	0.36	0.58
40.1	0.27	0.49	0.24	0.28	0.50	0.22
48.1	0.44	0.45	0.11	0.43	0.45	0.12

*Calculated from monomer feed compositions, conversions, and reactivity ratios of 0.78 and 0.26 for αFS and MA, respectively, using the program we have described previously[21].

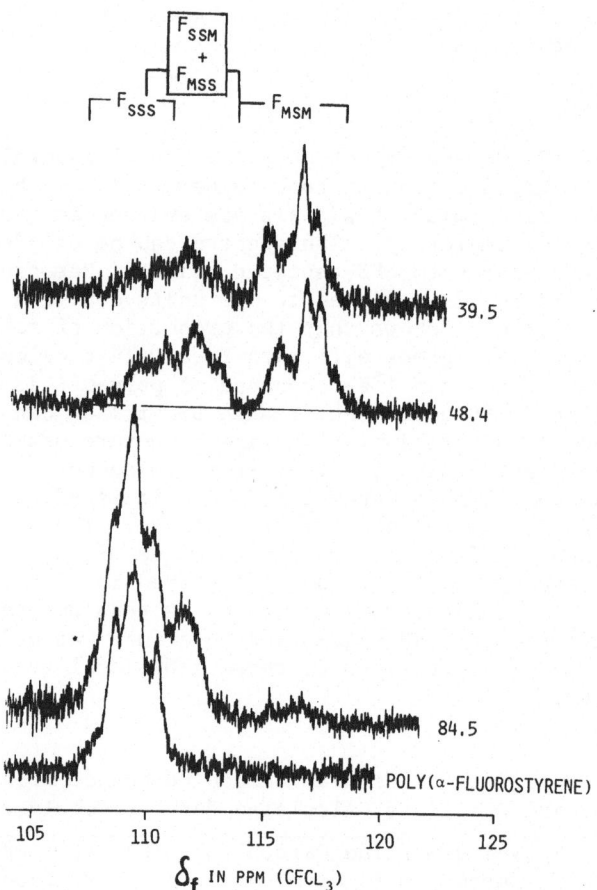

Fig. 11.   282 MHz ^{19}F-NMR spectra of poly(α-fluorostyrene) and of
several α-fluorostyrene-methyl acrylate copolymers in
CDCl$_3$ solution at room temperature. The α-fluorostyrene
contents of the copolymers are indicated next to each
spectrum.

Table VI. Comparison of ^{19}F-Resonance Observed at
δ_f=113.8–118.0 ppm with F_{MSM}

Mole % αFS in Copolymer	Proportion of Resonance Observed at δ_f = 113.8–118.0 ppm	F_{MSM}
100	0	0
84.5	0.03	0.031
48.4	0.53	0.527
39.5	0.69	0.651

Copolymer Stability

Although the copolymer with 19 mole percent αFS was very stable in solution and did not seem to change on storage in CDCl$_3$ solution for six months, copolymers with higher αFS contents were not stable in solution and they become increasingly unstable as their αFS contents increased.

Sample decomposition was indicated by the loss of CF carbon resonances from the ^{13}C-NMR spectra of the copolymers and by the appearance of new resonances at δ_c=177.2 and 85.4 ppm. Figure 8 compares the ^{13}C-NMR spectrum of a freshly prepared solution of poly(MA-co-αFS) containing 48.1 mole percent αFS with that of the solution after it had aged for 12 days at room temperature. The loss of the CF carbon resonances and the occurrence of new carbon resonances on aging is clearly evident. PolyαFS is so unstable in solution that no CF carbon resonances were evident in its spectra after only one day aging.

This phenomenon prevented ^{13}C-NMR measurements from being made on copolymers with greater than 48 mole percent αFS. It is possible that this limitation may be obviated by the use of hindered tertiary amine stabilizers such as 2,6-di-tert-butylpyridine, since such stabilizers work well for polyαFS solutions[17].

PolyαFS seems to decompose by HF elimination and this mode of decomposition may also be expected to operate in the copolymers with high αFS contents. However, the infrared spectrum of a decomposed copolymer (Figure 12) contained a strong carbonyl absorption at 1765 cm^{-1} (γ-lactone) that was not present in the spectrum of the original copolymer. In addition, the appearance of a new carbonyl carbon resonance at δ_c=177.2 ppm and the appearance of a new resonance at δ_c=85.4 ppm in the ^{13}C-NMR spectra of aged copolymers suggest that the degradation process involves γ-lactone formation,

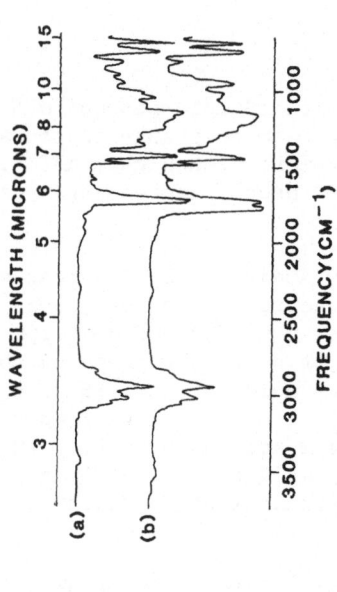

Fig. 12.   Infrared spectra of (A) a methyl acrylate-α-fluorostyrene copolymer containing 40.1 mole percent α-fluorostyrene and (B)
the same copolymer after spontaneous degradation in CHCl₃
solution.

<u>viz.</u>

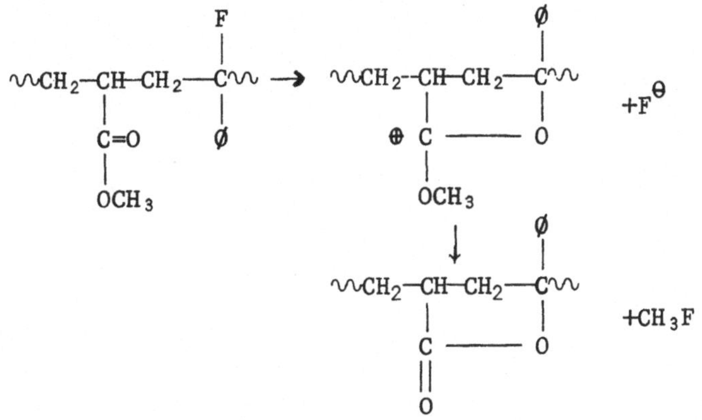

Similar processes are known for copolymers of acrylate and methacry-late esters with vinyl chloride and vinyl bromide[36-40]. The carbon resonances observed at δ_c=177.2 and 85.4 ppm are close to those reported for the carbonyl and methine carbon resonances of γ-substituted γ-butyrolactones[41].

It is interesting that the copolymers containing 19 mole per-cent αFS are less prone to undergo this reaction than those with higher αFS contents. This may indicate that the dehydrofluorination reaction, which is expected to occur when long sequences of αFS units are present in the copolymers, yields by-products that pro-mote the lactonization process. It could be that HF catalyzes the lactonization process. Johnston[42] has shown that Lewis acids can do this. More work must be done before a good understanding of the degradation process is obtained.

Glass Transition Temperature

Glass transition temperatures of several of the polymers were measured using a DuPont Thermal Analyzer. They show a linear de-pendence on composition as shown in Figure 13.

ACKNOWLEDGEMENT

This work was supported by a grant from The National Science Foundation (DMR-80-10709). The authors are grateful to Messr's Everett R. Santee, Jr. and Moris K. Niknam for their assistance in recording some of the spectra and to Mrs. Louise Hutchison for neat and careful typing.

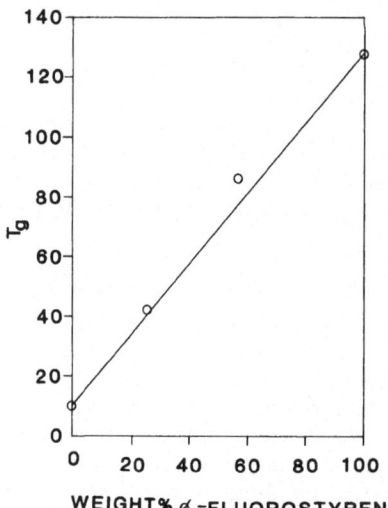

Fig. 13. Glass transition temperatures (T_g) of poly-α-fluoro-
          styrene, poly(methyl acrylate) and methyl acrylate-α-
          fluorostyrene copolymers as a function of α-fluoro-
          styrene content (weight percent).

REFERENCES

1. C.F. Raley and R.J. Dolinski, Reactive halogenated monomers
    in: "Functional Monomers", R.H. Yocum and E.B. Nyquist,
    Eds., Marcel Dekker, Inc., New York (1973).
2. W.C. Mast and C.H. Fisher, Vulcanization of chlorine containing
    acrylic elastomers, Ind. Eng. Chem., 40: 107 (1948).
3. T.-J. Suen and A.M. Schiller, Sulfonated (hydroxymethyl)
    acrylamide polymers and copolymers, Brit. Patent 738,047
    (Oct. 5, 1955); Chem. Abstr. 50: 8252 (1956); U. S. Patent
    2,761,856 (Sept. 4, 1956); Chem. Abstr. 50: 15130 (1956).
4. J.P. Regeaud, α-Chlorostyrene and α-cyanostyrene and their
    phenomena, Chim. Mod. 4: 91 (1959); Chem. Abstr. 53: 18902
    (1959).
5. W.J. Blank and G.J. Pietsch, Priming electrodeposited acrylic
    resin coatings to increase their adhesion to epoxy coatings,
    U. S. Patent 3,619,399 (Nov. 9, 1971); Chem. Abstr. 76:
    73863 (1972).
6. M.M. Mednikov, V.I. Budkin, V.N. Semenov, V.A. Krol, and
    N.F. Khakhalina, Low-molecular-weight styrene copolymers,
    U.S.S.R. Patent 382,107 (Feb. 2, 1972); Chem. Abstr. 77:
    20645 (1972).
7. M.M. Mednikov, V.A. Krol, V.I. Budkin, V.N. Semenov, and
    N.F. Khakhalina, Low-molecular-weight styrene copolymers,
    U.S.S.R. Patent 328,105 (Feb. 2, 1972); Chem. Abstr. 77:
    20647 (1972).
8. C.L. Meredith and G.A. Von Bodungen, High-impact-strength EPDM
    graft copolymers, U.S. Patent 3,657,395 (April 18, 1972);
    Chem. Abstr. 77: 35551 (1972).
9. C.L. Meredith and G.A. Von Bodungen, EPDM rubber-modified
    plastic compositions, U.S. Patent 3,671,608 (June 20, 1972);
    Chem. Abstr. 77: 89474 (1972).
10. E. Agouri, R. Laputte, and J. Rideau, Low-density polyethylene
    masses for manufacturing films, Ger. Offen. 2,731,040
    (Jan. 19, 1978); Chem. Abstr. 88: 153510 (1978).
11. T. Goto, E. Sakaoka, T. Hiraoka, and S. Rokuwatari, Transparent,
    impact-resistant rubbers, Japan Patent 7031,677 (Oct. 13,
    1970); Chem. Abstr. 74: 14020 (1971).
12. A.D. Pomogailo, A.P. Lisitskaya, A.I. Kuzaev, and F.S.
    D'yachkovskii, Polymerization of ethylene in the presence
    of macromolecular carbonium salts of complex catalyst com-
    ponents, Kompleks. Metalloorgan. Katalizatory Polimeriz.
    Olefinov. Chernogolovka 1980: 66; Chem. Abstr. 85: 43768
    (1981).
13. K. Matsuda, J.A. Sedlak, J.S. Noland, and G.C. Gleckler,
    α-Fluorostyrene: preparation, properties, and polymerization,
    J. Org. Chem., 27: 4015 (1962).
14. J.S. Noland, Homopolymers of α-fluorostyrene, U.S. Patent
    3,207,733 (Sept. 21, 1965); Chem. Abstr. 63: 18292 (1965).

15. T. Sata, S. Murakami, and Y. Murata, Diaphragm for electrolysis cell, Ger. Offen. 2,504,622 (Aug. 7, 1975); Chem. Abstr. 83: 199502 (1975).

16. Tokuyama Soda Co., Ltd., Cation-exchanging membrane for brine electrolysis, Jpn. Kokai Tokkyo Koho 81 38,490 (Apr. 13, 1981); Chem. Abstr. 95: 123027 (1981).

17. R.N. Majumdar, M.K. Niknam, H.A. Nguyen, and H.J. Harwood, Stabilization and NMR spectra of poly(α-fluorostyrene), Makromol. Chem., Rapid Commun. 3: 421 (1982).

18. L. Eckes and M. Hanack, Herstellung von Vinylfluoriden, Synthesis 1978: 217.

19. A.A. Bothner-By and S.M. Castellano, LAOCN3, in: "Computer programs for chemistry", Vol. 1, D.F. DeTar, Ed.; W.A. Benjamin, Inc., New York (1968).

20. G. Filipovich and G.V.D. Tiers, Fluorine N.S.R. spectroscopy. I. Reliable shielding values, Ø, by use of CCl_3F as solvent and internal reference, J. Phys. Chem. 63: 761 (1959).

21. H.J. Harwood, A FORTRAN II program for conducting sequence distribution calculations, J. Polym. Sci., Part C, No. 25: 37 (1968).

22. R.M. Joshi, A brief survey of methods of calculating monomer reactivity ratios, J. Macromol. Sci., Chem. 7: 1231 (1973)

23. T. Kelen and F. Tüdös, Analysis of the linear methods for determining copolymerization reactivity ratios. I. A new improved linear graphic method, J. Macromol. Sci., Chem. 9: 1. (1975).

24. S.S. Rao, S. Ponratnam, S.L. Kapur, and P.K. Iyer, Kelen-Tüdös method applied to the analysis of high conversion copolymerization data, J. Polym. Sci., Polym. Lett. Ed. 14: 513 (1976).

25. P.W. Tidwell and G.A. Mortimer, An improved method of calculating copolymerization reactivity ratios, J. Polym. Sci., Part A, 3: 369 (1965).

26. R.Z. Greenley, An expanded listing of revised Q and e values, J. Macromol. Sci., Chem. 14: 427 (1980).

27. K. Ito, S. Iwase, K. Umehara, and Y. Yamashita, Copolymer microstructure by high resolution NMR studies, J. Macromol. Sci., Chem. 1: 891 (1967).

28. H.J. Harwood and W.M. Ritchey, Methoxy and α-methyl proton resonance in styrene-methyl methacrylate copolymers, J. Polym. Sci., Part B, 3: 419 (1965).

29. T. Suzuki, E.R. Santee, Jr., H.J.Harwood, O. Vogl, and T. Tanaka, Measurement of tetrad configurations in poly-(methyl acrylate) by 300 MHz PMR spectroscopy, J, Polym., Sci., Polym. Lett. Ed., 12: 635 (1974).

30. L. Shepherd, T.K. Chen, and H.J. Harwood, Epimerization of isotactic polystyrene, Polym. Bull. 1: 445 (1979).

31. D.L. Trumbo, T.K. Chen, and H.J. Harwood, Observation of triad stereosequence in a polystyrene derivative, Macromolecules 14: 1138 (1981).

32. T. Kawamura, T. Uryu, and K. Matsuzaki, Reinvestigation of the stereoregularity of polystyrene by 100 MHz ^{13}C-NMR spectroscopy, Makromol. Chem., Rapid Commun. 3: 661 (1982).

33. M.K. Niknam and H.J. Harwood, Unpublished results.

34. J.C. Randall, "Polymer Sequence Determination. Carbon-13 NMR Method", Academic Press, New York (1977).

35. T.K. Chen, T.A. Gerken, and H.J. Harwood, Methylene carbon resonance spectra of epimerized isotactic polystyrene, Polym. Bull., 2: 37 (1980).

36. N.L. Zutty and F.J. Welch, Synthesis of vinyl polymers containing α-substituted γ-butyrolactone groups in their backbone, J. Polym. Sci., Part A 1: 2289 (1963).

37. N.W. Johnston and H.J. Harwood, Intersequence cyclization reaction in methyl methacrylate - vinyl halide copolymers and terpolymers, J. Polym. Sci., Part C, No. 22: 591 (1969).

38. N.W. Johnston and H.J. Harwood, Intersequence cyclization in brominated methyl methacrylate - butadiene copolymers, Macromolecules 3: 20 (1970).

39. N.W. Johnston and H.J. Harwood, Intersequence cyclization in methyl methacrylate - vinyl chloride-styrene terpolymers, Macromolecules 2: 221 (1969).

40. F. Shepherd and H.J. Harwood, Equimolar alternating vinyl chloride-methyl methacrylate copolymers: synthesis and proof of structure, J. Polym. Sci., Polym. Lett. Ed. 9: 419 (1971).

41. H. Pyysalo, J. Enqvist, E. Honkanen, and A. Pippuri, Identification of volatile lactones by NMR. I. Synthesis of γ- and δ-lactones and assignments in the carbon-13 NMR spectra of 4-hydroxyhexanoic and 5-hydroxyheptanoic acid lactones, Finn. Chem. Lett. 1975: 129; Chem. Abstr. 84: 69087 (1976).

42. N.W. Johnston, Catalyzed cyclization of vinyl chloride-acrylate, -fumarate, and -methacrylate copolymers, Polym. Prepr., Am. Chem. Soc., Div. Polym. Chem. 13: 1065 (1972); Chem. Abstr. 81: 64515 (1974).

QUINODIMETHANE POLYMERS

J.E. Mulvaney

Chemistry Department
University of Arizona
Tucson, AZ 85721

INTRODUCTION

p-Quinodimethane[1] also known as p-xylylene is the parent compound of a family which has received renewed consideration recently.

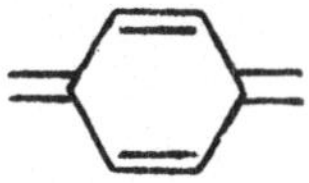

1

Many years ago Thiele and Balhorn[1] attempted to prepare 1 according to eq. 1 but instead obtained an insoluble white powder which they described as poly(p-xylylene) and left the matter at that.

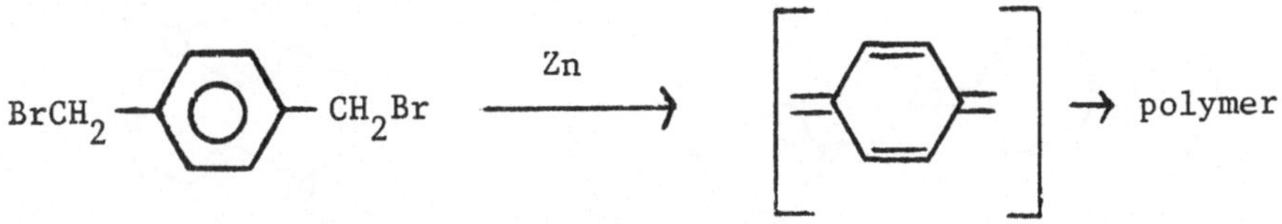

Eq. 1

However Thiele[1] and later Staudinger[2] were able to isolate a relatively stable tetraphenyl derivative of 1, namely 2

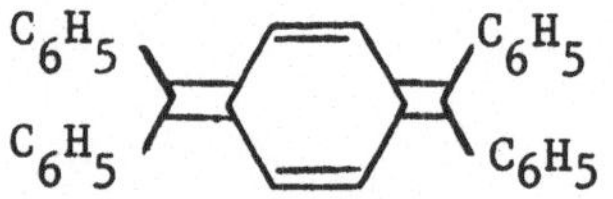

This research was motivated by Gomberg's discovery of the stable triphenylmethyl radical which led to the question does 2 exist in in the quinoid form or as a biradical 3. Subsequently a number of other tetraphenylquinodimethanes and related compounds were isolated some of which are in Fig. 1

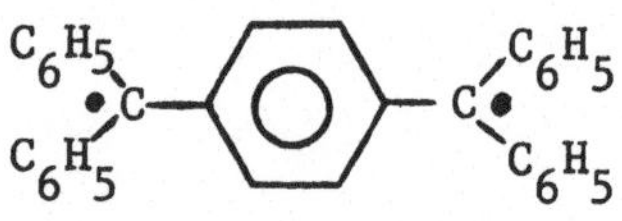

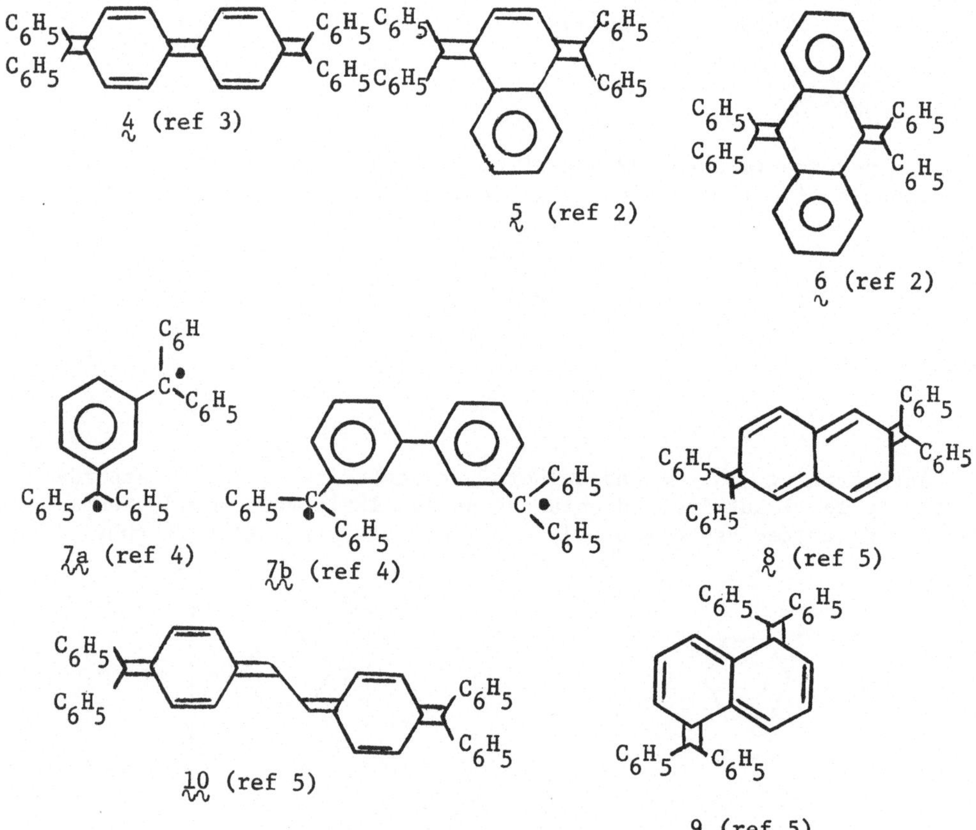

Fig. 1.   Early examples of quinodimethanes.

Early investigations indicated that 2 consisted of less than 0.2% biradical 3, but that compound 4, the Chichibabin hydrocarbon, was 5-10% biradical based upon p to o-hydrogen conversion[7a] or esr[7b]. However a more recent esr study[6] has led to the conclusion that the paramagnetic species in 4 is in fact the dimer. On the other hand because of their reactivity and the impossibility of drawing class-ical valence bond structures, with no unpaired electrons, the Schlenk hydrocarbons (7a,b) were considered to be diradicals.

The compounds in Fig. 1 are colored and highly reactive. Al-though they apparently do not homopolymerize, little or nothing has been done with regard to copolymerizations.

HOMOPOLYMERIZATIONS

p-Quinodimethane was first isolated and properly character-ized by Szwarc[8] who obtained it by the pyrolysis of p-xylene. Al-though the monomer is stable in the vapor phase, in the condensed phase even at low temperatures it polymerizes spontaneously to poly-p-quinodimethane 11 as well as the cyclic dimer (p-cyclophane) and trimer.[8] (Eq. 2).

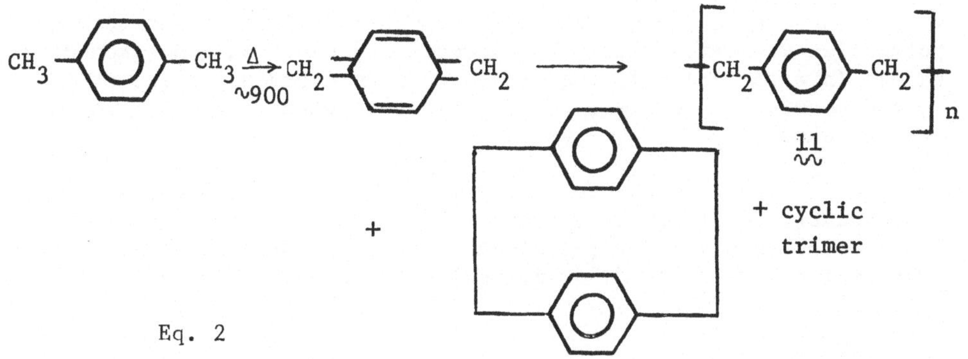

Eq. 2

Poly-p-quinodimethane prepared by the pyrolysis of p-xylene is crystalline and soluble only at very high temperatures (~300°C) in substances which may cause it to degrade[8]. Furthermore the polymer

is loosely cross-linked[8, 39]. A large number of other p-dimethyl
aromatic compounds have also been found to undergo pyrolysis to
highly reactive substituted p-quinodimethanes which also polymerize to produce materials which are generally insoluble or
soluble only with much difficulty[8].

More recently the question of biradical vs. quinoid structure
was examined with regard to p-quinodimethane and the related compounds 12 and 13. In all three cases both low temperature solution spectra (ir,uv,nmr) and solid state spectra (77°K) indicate

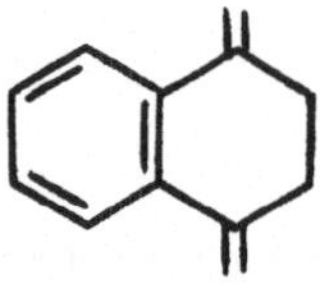

12 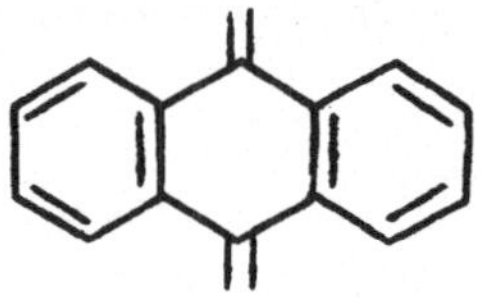   13

that the quinoid structure is predominant and that biradicals, if
formed, are very short-lived. For example, no esr signals were
obtained and the nmr signals were not broadened[9]. In addition the
photoelectron spectrum of p-quinodimethane[10] and a ring substituted dimethyl p-quinodimethane[11] were consistent with the
quinoid structure.

Because of its high reactivity, p-quinodimethane does not copolymerize with either electron acceptor or electron donor monomers
except in a few cases in the presence of an enormous excess of the
second monomer[8]. However, copolymerization with O_2 or SO_2 can be
accomplished[8].

A much better method for preparing poly-p-quinodimethane was
subsequently developed by Gorham[39] who found that low-pressure
(~0.1 torr) pyrolysis of p-cyclophane at 600°C produces linear poly-
p-quinodimethane free of low molecular weight by-products.

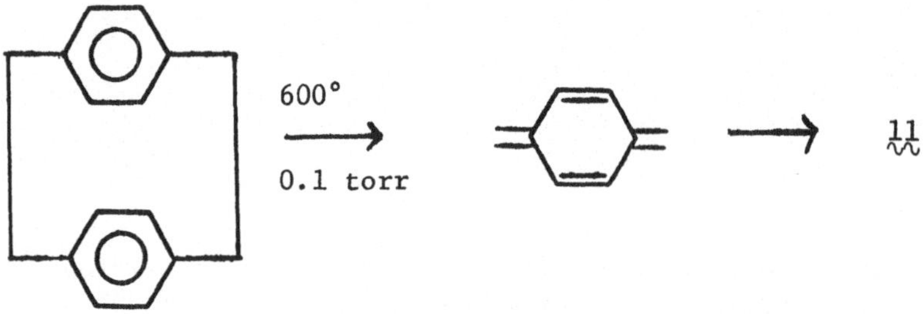

The p-quinodimethane monomer is stable in the vapor phase but polymerizes on a surface maintained below 30°C. Gorham has extended
this method to the pyrolysis of many substituted p-cyclophanes to
produce substituted poly-p-quinodimethanes

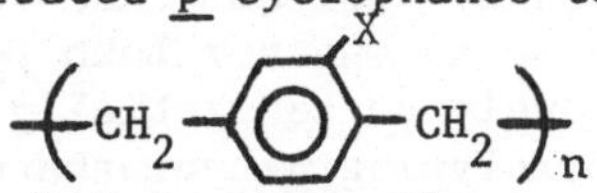

$$\left(CH_2 - \text{⬡} - CH_2 \right)_n^{X}$$

in which X has been Cl, Br, CH$_3$, C$_2$H$_5$ and n-butyl among others.
Poly-p-quinodimethane prepared by the Gorham method is soluble in
chlorinated biphenyls or benzyl benzoate at temperatures above
200°C. The substituted polymers are considerably more soluble.
For example polymethyl-p-quinodimethane is soluble in hot toluene
and poly-n-butyl-p-quinodimethane is soluble in chloroform at room
temperature[39]. This process and the morphology of the polymers
have been reviewed[41,42].

    By copyrolysis of suitably substituted p-cyclophanes one can
prepare p-quinodimethane copolymers.[41]

    Commercially, Union Carbide makes p-cyclophane by the 900°C
pyrolysis of toluene in the presence of steam [41]. The best
laboratory preparation for p-cyclophane is probably the pyrolysis
of   as described by Weinberg and

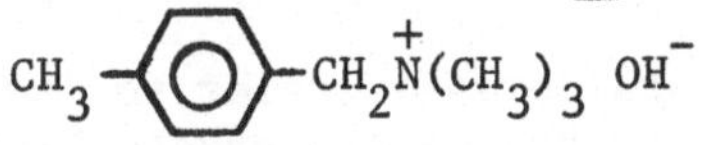

$$CH_3 - \text{⬡} - CH_2 \overset{+}{N}(CH_3)_3 \ \overset{-}{OH}$$

Fawcett[40]. P-Cyclophane monomers are sold by Union Carbide under
the trade name Parylene, and they are particularly useful for
coating irregularly shaped particles by the Gorham vapor deposition
method.

    A question arises concerning the p-cyclophane pyrolysis as to
whether p-quinodimethane is actually formed. For example only one
bond in the p-cyclophane may break to produce

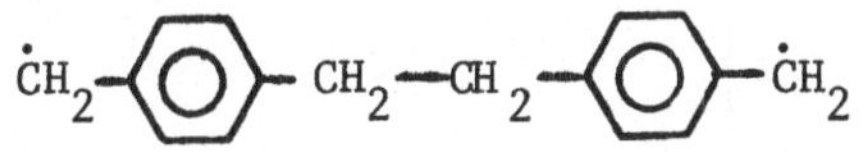

$$\dot{C}H_2 - \text{⬡} - CH_2 - CH_2 - \text{⬡} - \dot{C}H_2$$

which then polymerizes by polyrecombination, Gorham[39] has answered
this question by polymerizing 13a, which produces a mixture of two
homopolymers as shown below.

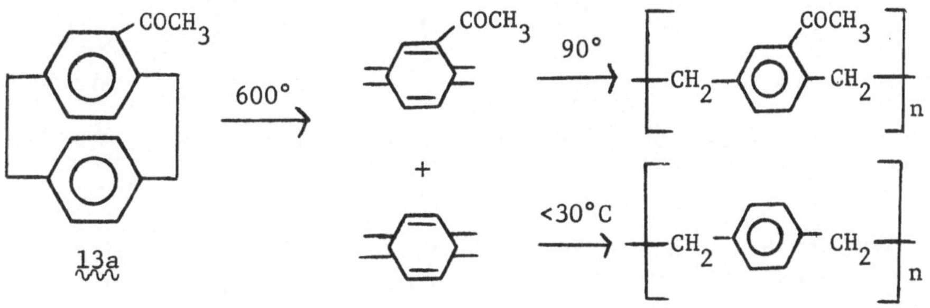

By allowing the pyrolysis vapors to pass through tubing maintained
at 90°C and subsequently through a 20°C tube two homopolymers were
obtained. Gorham acknowledges that this does not answer the question as to whether both bonds in the p-cyclophane are broken simultaneously or sequentially. In fact, it should be pointed out that
by pyrolyzing p-cyclophanes at 200°-250°C Cram[31] has established
unequivocal evidence for the breakage of only one bond (see below).
It should also be pointed out that the concerted cleavage of p-
cyclophane the p-quinodimethane is a $\pi^6_s + \pi^6_s$ process which is

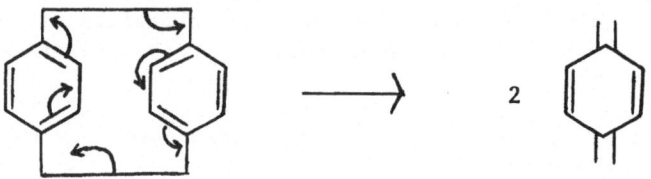

orbital symmetry forbidden.[43] Thus both experimentally and theoretically there is compelling reason to believe that bond cleavage
in cyclophanes proceeds in a stepwise manner.

Chlorinated quinodimethanes are considerably easier to isolate
than p-quinodimethane itself. They are obtained as crystalline
solids and their reactivity decreases with increasing chlorine
content. Fig. 2

Figure 2. Chlorinated p-Quinodimethanes

Compounds 14, 15 and 16 can be stored at low temperatures but
upon warning they spontaneously homopolymerize to the insoluble polymers 14A, 15A and 16A.

In sharp contrast the perchlorinated monomer 17 can be isolated in excellent yield, does not polymerize under any conditions, and it does not react with oxygen or a variety of vigorous chemical reagents[14],[15]. The failure of 17 to polymerize is due largely if not completely to the tremendous steric crowding in the hypothetical polymer, not only of the adjacent aliphatic chlorines but the crowding of the o- chlorines also. One can not assemble the polymer with space filling scale atomic models. Similarly the tetraphenyl perchlorinated derivative 18 is totally unreactive[16].

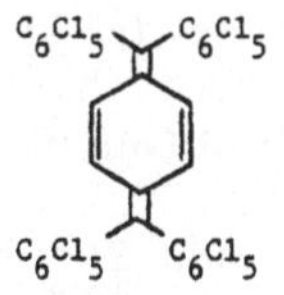

18

7,7,8,8-Tetracyanoquinodimethane (TCNQ) 19 is easily isolated and does not homopolymerize.[17]

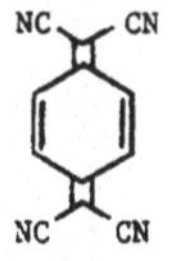

19

Replacement of cyano groups by carboalkoxy as in 20 or 21 gives compounds which are isolable but nevertheless undergo thermal or radical polymerization to the expected polymers.

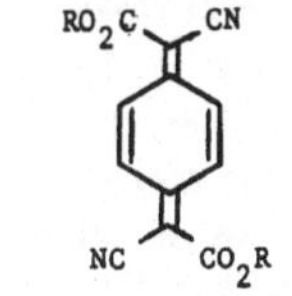

20 a, R=CH$_3$

b, R=CH$_2$CH$_3$

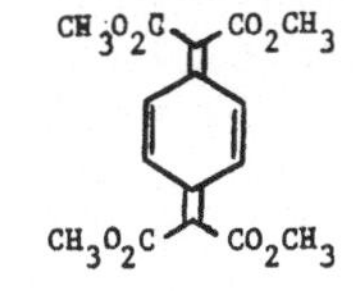

21

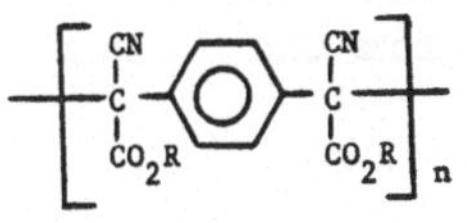

22 a, R=CH$_3$

b, R=CH$_2$CH$_3$

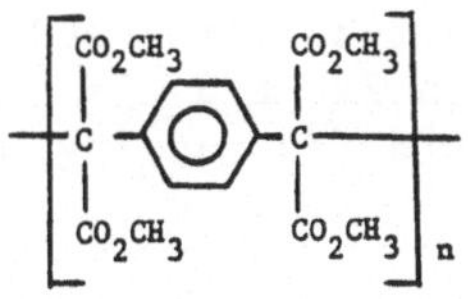

23

Compound 21 was originally synthesized by Acker and Hertler[17] who briefly described its properties. Subsequently 21 was investigated at greater length by Iwatsuki and Itoh[18] and by Hall and

Bentley[19]. Although 21 can be stored at 3° for extended periods,
it polymerizes at room temperature in diffuse light or upon heating.
Azobisisobutyronitrile or n-butyllithium will also initiate polymerization. In common with other quinodimethane polymers, 23 is
insoluble in organic solvents. However 23 does dissolve in
sulfuric, chlorosulfonic and trifluoromethanesulfonic acids evident-
ly by forming an oxonium ion poly salt. This was shown by the viscosity concentration plot in which the inherent viscosity increases
with dilution[19]. One might anticipate on the basis of a steric
effect and resonance stabilization of the resulting radical that the
bond between the sp^3 carbon atoms in 23 would undergo homolysis
relatively easily. In accord with this anticipation heating 23 at
150° and 0.1 torr caused sublimation of monomer 21 in 10-15% yield[19].

    The cyanocarboalkoxy compounds 20a,b were prepared by the oxida-
tion of 24 with Br_2[20] or N-chlorosuccinimide[21].

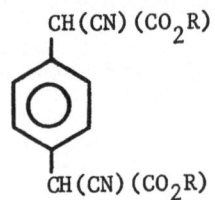

$$CH(CN)(CO_2R)$$

$$CH(CN)(CO_2R)$$

24 a   (R=CH$_3$)

   b   (R=C$_2$H$_5$)

    Compounds 20,a,b polymerize spontaneously[20],[21] but the polymeri-
zations are accelerated by n-butyllithium, tertiary amines or AIBN,
but not by BF_3[20], yielding polymers 22 a,b. Polymer 22a (R=CH$_3$) is
insoluble in organic solvents, again like other quinodimethanes, but
polymer 22b (R=C$_2$H$_5$) is soluble in organic solvents such as chloroform and acetonitrile. This is apparently the only quinodimethane
homopolymer (without a substituent on the benzene ring) which is
soluble in organic solvents (polymer 23 is sulfuric acid soluble, see
above). The solubility may be due to the fact that polymer 22b is
the first quinodimethane homopolymer containing adjacent chiral centers which may exist in either the meso or the R,S form and thus make
the polymer less crystalline and more soluble. The difference in
solubility between 22b and 22a may result from the larger ethyl group
preventing well ordered packing of chains.

REACTIONS OF QUINODIMETHANES WITH VINYL MONOMERS

    As mentioned previously p-quinodimethane is so highly reactive
that attempted copolymerizations with vinyl monomers yield only
homopolymers unless a very large excess of the vinyl monomer is
used[8].

However, p-quinodimethanes with electron withdrawing groups react with vinyl monomers, in most cases with no added initiator to produce either alternating copolymers or homo vinyl polymers (eq. 3).

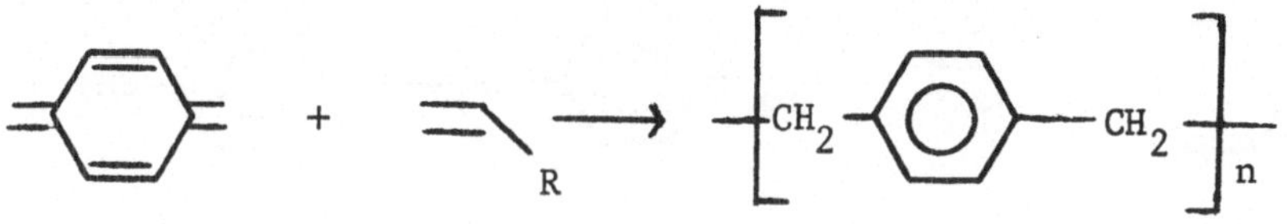

Eq. 3

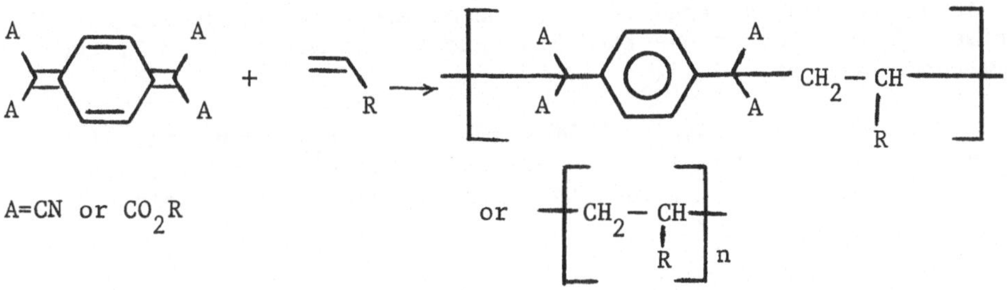

A=CN or CO$_2$R   or

These reactions usually proceed in solution between room temperature and 60°-70°C giving good yields in a few hours or days of low to moderate molecular weight polymer. Results are presented in TABLE I.

The only cases in which homo vinyl polymer are obtained are those in which vinyl ethers or N-vinyl carbazole are the comonomers, Cases VII, VIII and X, Table I.

MECHANISM

Because vinyl ethers are homopolymerized only by cationic initiators[28] it has been concluded that the propagating species in Cases VII, VIII and X, Table I is a carbocation which reacts exclusively with vinyl ether monomer to produce homopolymer (see Table 1 for refs.). The same may be said for N-vinyl carbazole as a comonomer. The quinodimethanes all bearing electron withdrawing groups are unreactive toward a carbocation. In contrast the alternating nature of the copolymers in all other cases clearly indicates that the propagating species is a radical.

With regard to initiation in the spontaneous polymerizations let us consider p-quinodimethane itself. Errede and Szwarc considered dimerization to form the biradical 25 (eq. 4) to be unlikely as the initiation step because p-quinodimethane does not polymerize

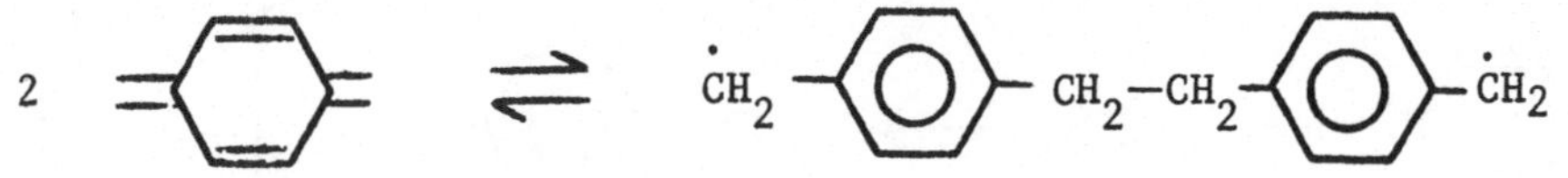

$$2 \; \text{[diagram]} \; \rightleftharpoons \; \dot{C}H_2 - \text{[ring]} - CH_2 - CH_2 - \text{[ring]} - \dot{C}H_2$$

Eq. 4

25

in the vapor phase. They suggested that initiation had to be termolecular or perhaps even higher in order to produce 26, the argument being that 25 readily dissociates to quinodimethane because by breaking the central bond four new bonds are created whereas with trimer 26 breaking of one bond leads to the formation of only two new bonds (eq. 5).

However more recent work by Cram has clearly shown that diradical 25 persists even at elevated temperatures[31]. Pyrolysis of cyclophane 28, Figure 3, results in the formation of 25 which is

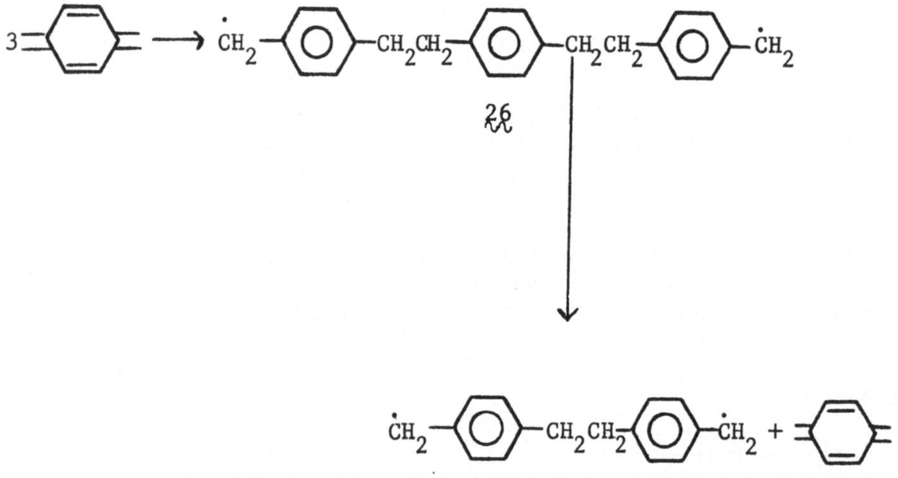

$$3 \; \text{[diagram]} \; \longrightarrow \; \dot{C}H_2 - \text{[ring]} - CH_2CH_2 - \text{[ring]} - CH_2CH_2 - \text{[ring]} - \dot{C}H_2$$

26

$$\dot{C}H_2 - \text{[ring]} - CH_2CH_2 - \text{[ring]} - \dot{C}H_2 + \text{[diagram]}$$

Eq. 5

trapped by p-diisopropyl-benzene or dimethyl fumarate. Furthermore the optically active compound 28a racemizes via a diradical without cleavage of the second CH_2-CH_2 bond (Fig. 3).

Table 1
Polymers from Quinodimethanes and Vinyl Monomers

Abbreviations:

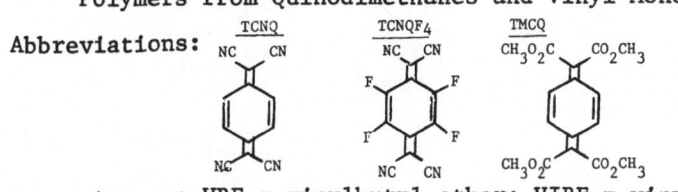

sty = styrene; VBE = vinylbutyl ether; VIBE = vinyl isobutyl ether; CEVE = B-cloroethyl vinyl ether; PVE = phenyl vinyl ether; VOAc = vinyl acetate; MMA = methyl methacrylate; MA = methyl acrylate; AN = acrylonitrile; VCbz = N-vinyl carbazole.

Case	Monomer 1	Monomer 2	Solvent	Product	Ref.
I		many other vinyl monomers	several	Homopolymer of quinodimethane except with large excess of vinyl monomer	8
II		sty	?	copolymer	13
III		sty	toluene	alternating copolymer	30
IV		sty	C_6H_6	alternating copolymer	13
V	TMCQ	VCb or pCH$_3$Osty or sty or CEVE or PVE or VIBE or TCNQ	several	alternating copolymer	18, 19
VI	TCNQ	sty or CEVE or PVE or VOAc	CH$_3$ CN	alternating copolymer	22
VII	TCNQ	VBE or VIBE or VCbz	CH$_3$ CN	homopoly VBE or VIBE or VCbz	23, 24, 28, 27
VIII	TCNQ	CEVE		homopoly CEVE	25
IX	TCNQF$_4$	sty or CEVE or PVE or VOAc or MMA or MA	CH$_3$ CN	alternating copolymer	26
X	TCNQF$_4$	VBE or VIBE	CH$_3$ CN	homopoly VBE or VIBE	26

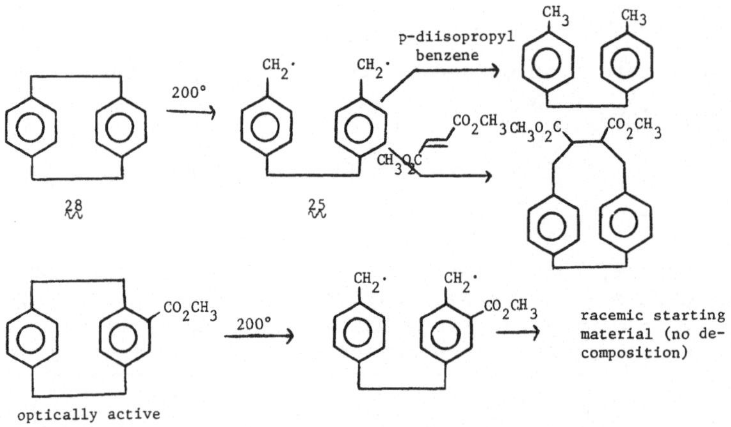

Figure 3 Pyrolysis of Cyclophanes (Ref 31)

In addition to the nature of the initiation process, several questions remain concerning the homopolymerization of p-quinodimethane. The suggested mode of propagation involving biradical coupling[8] remains to be established. Concerning termination it has been pointed out that biradical coupling is not a true termination, but apparently no one has considered radical disproportionation as shown in equation 6.

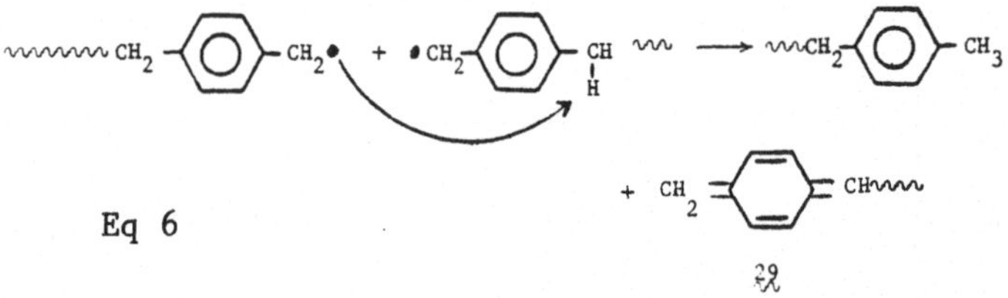

Eq 6

If this occurs at both ends of the chain a true termination can occur. Furthermore the presence of quinodimethane units as in 29 which could react with growing radicals would result in branching or even cross-linking which could be a factor in the insolubility of not only homopolymers but also the copolymer of hexachloroquinodimethane 15 and styrene when the styrene content is less than 50%[13]. It must be mentioned however, that not all growing chains terminate because poly-p-quinodimethane contains one unpaired electron for every thousand mer units.[39]

Returning to the initiation process H.K. Hall and his collaborators[32][35] have been investigating the spontaneous copolymerization of vinyl monomers containing donor and acceptor groups. They have presented strong evidence for the existence of a tetramethylene intermediate 30 (Eq. 7) which is a resonance hybrid of a zwitterion and a spin paired biradical, the terminal carbon atoms of which sense each other by through bond coupling[36].

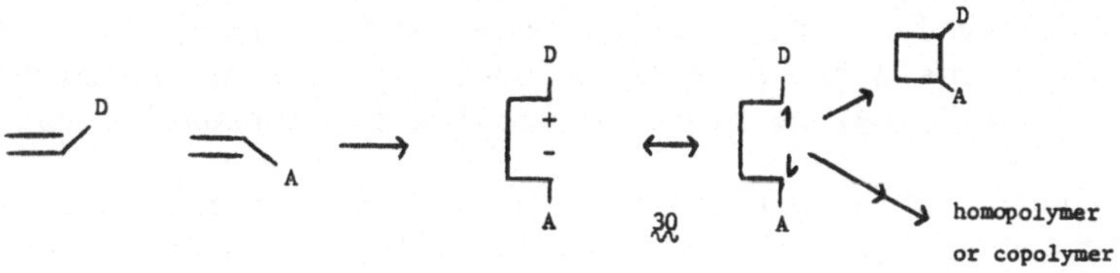

This intermediate 30 has been identified in many small molecules reactions such as cycloaddition[36]. The tetramethylene 30, depending
upon conditions, may cyclize or react with more monomer to form
either a copolymer or a homopolymer. It has been proposed that a
largely zwitterionic 30 produces cationic homopolymers whereas a
largely biradical tetramethylene produces an alternating
copolymer[32,35].

The dimerization of a quinodimethane to the biradical 25 is
analogous to the tetramethylene 30 and may represent the initiation
step in p-quinodimethane homopolymerization[32].

In the case of copolymerizations of a quinodimethane and a
vinyl monomer the reaction may proceed analogously. For example in
Table I the experiments resulting in alternating copolymer may be
explained by formation of a largely biradical analog of the tetramethylene (a bis phenylene tetramethylene). This scheme is presented in Fig. 4.

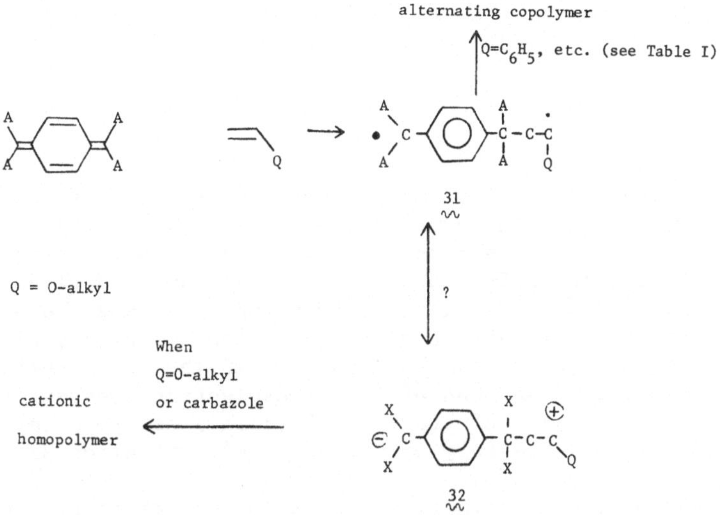

Figure 4.   Reaction of p-Quinodimethanes and Vinyl Monomers.

If the substituent Q in the vinyl monomer is O-alkyl or carbazole and the A group is CN, a zwitterionic intermediate 32 may be
formed which produces the cationic homopolymer (cases VII, VIII, X,
Table I). If the A group is CN or $CO_2 CH_3$ in the quinodimethane
then biradical 31 may be produced leading to alternating copolymer.

The concept of a zwitterion as a reactive intermediate has also
been proposed by Stille[24a] and Iwatsuki.[37] For example treatment of
TCNQ and 1-methoxycyclopentene[37] followed by work-up with aqueous acid
yielded oligomer 33, (Eq. 8).

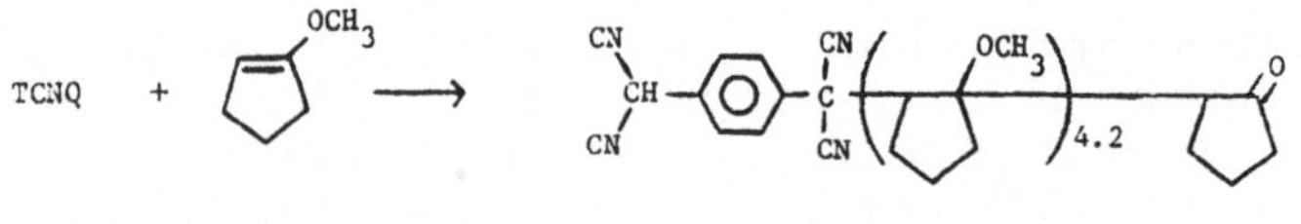

TCNQ   +   [structure with OCH₃]   ⟶   [structure]

33

Oligomer 33 was suggested to arise from a zwitterion analogous
to 32, namely 34.

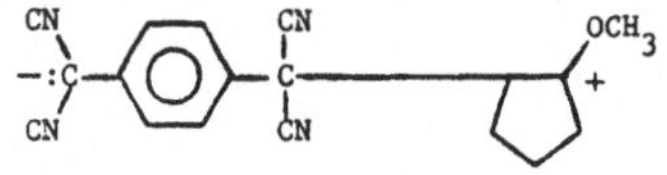

34

Vinyl β-chloroethyl ether presents an interesting case in which
either homopolymer or alternating copolymer may be obtained. In
contrast to vinyl akyl ethers Case VII, Table I which give homopolymer with TCNQ in $CH_3 CN$, the β-chloroethyl compound gives alternating copolymer Case VI, Table I presumably because the electron
withdrawing inductive effect of the chlorine leads to a biradical
intermediate 31 rather than 32. However if the copolymerization
is carried out in a solvent of higher dielectric constant (ethylene
carbonate) homopolymer of vinyl β-chloroethyl ether is formed,
Case VIII, Table I possibly by formation of zwitterion 32.

In many of the copolymerizations a charge transfer complex between the monomers has been detected, and it is certainly

possible if not probable that charge transfer complex formation pre-
cedes 31 or 32. Furthermore the charge transfer complex may dissociate via complete electron transfer to a cation-radical anion-
radical pair 35 prior to bond formation leading to 32. For example
in the case of TCNQ and t-butyl vinyl ether[24a] the radical anion
of TCNQ was detected by its electronic absorption spectrum.

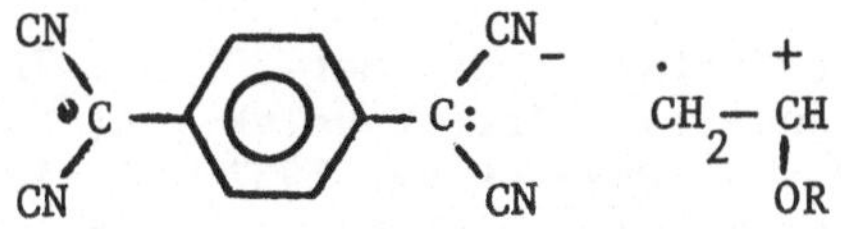

In a somewhat related area it should be pointed out that
DDQ[24a,38] 36 forms alternating copolymers with styrene but with
vinyl ethers homopolymerization of the vinyl ether occurs. When
a 1:1 mixture of DDQ and methyl vinyl ether were mixed in CH_3CN
followed by quenching in CH_3OH product 37 was obtained (eq. 9).
Good evidence was presented to show that 37 arises via charge transfer complex formation followed by complete electron transfer and
bond formation to produce 38, an intermediate analogous to 31 and
32.

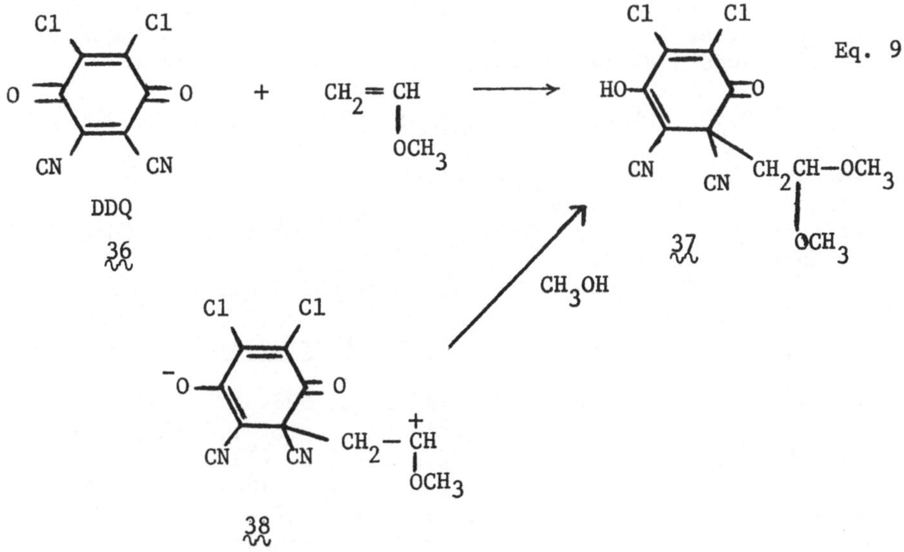

Eq. 9

REFERENCES

1.   Thiele, J.; Balhorn, H., Ber. 1904, 37, 1463.
2.   Staudinger, H., ibid, 1908, 41, 1355.
3.   Chichibabin, A.E., ibid., 1907, 40, 1810.
4.   Schlenk, W.; Brauns, M., ibid, 1915, 48, 716.
5.   Wittig, G.; Wiemer, W., Liebigs, J. Ann. 1930, 483, 144.
6.   van der Hart, W.J.; Oosterhoff, L.J., Mol. Physics 1970,
        18, 281.
7.   (a) Schwab, G.M.; Ngliardi, N., Ber., 1940, 73B, 95.
        (b) Hutchinson, S.A., Jr.; Kowalsky, A; Pastor, R.C.;
        Wheland, G.W., J. Chem. Phys. 1952, 20, 1485).
8.   The literature concerning p-xylylene through 1958 has been
        admirably reviewed. Errede, L.A.; Szwarc, M., Quart.
        Rev. Chem. Soc., 1958, 12, 301.
9.   Williams, D. J.; Pearson, J.M.; Levy, M., J. Am. Chem. Soc.,
        1970, 92, 1437. Pearson, J.M.; Six, H.A.; Williams,
        D.J.,; Levy, M., ibid., 1971, 93, 5034.
10. Koenig, T.; Wielesek, R.; Suell, W.; Balle, T., ibid., 1975,
        97, 3226.
11. Koenig, T.; Southworth, S.; ibid., 1977, 99, 2807.
12. Gilch, H.G., Angew. Chem. (Intern. Ed.), 1965, 4, 598.
        Gilch, H. G., J. Polym. Sci., Part A-1, 1966, 4, 438.
13. Iwatsuki, S.; Kamiya, H., Macromolecules, 1974, 7, 732.
14. Ballester, M.; Castaner, Anales Real Soc. Espan. Fis. Quim.,
        1960, 56B, 207.
15. Ballester, M.; Castaner, J.; Riera, J., J. Amer. Chem. Soc.,
        1966, 88, 957.
16. For a review of the chlorocarbon work of the Ballester group
        see, M. Ballester and S. Olivella in "Polychloroaromatic
        Compounds", H. Suschitzky, Ed., Plenum Press, 1974.

17. Acker, D.S.; Hertler, W.R., J. Amer. Chem. Soc., 1962, 84, 3370.

18. Iwatsuki, S.; Itoh, T., Macromolecules, 1980, 13, 983.
19. Hall, H. K., Jr.; Bentley, J. H., Polymer Bull., 1980, 3, 203.
20. Iwatsuki, S.; Itoh, T.; Nishihara, K.; Faruhashi, H., Chem.
        Lett., 1982, 517.
21. Hall, H. K., Jr.; Cramer, R. J.; Mulvaney, J. E., Polymer
        Bulletin, 1982, 7, 165.
22. Iwatsuki, S.; Itoh, T.; Horinchi, K., Macromolecules, 1978,
        11, 497.
23. Iwatsuki, S.; Itoh, T.; ibid., 1979, 12, 208.
24. (a) Tarvin, R.F.; Aoki, S.; Stille, J.K., ibid., 1972, 5 663.
        (b) Aoki, S.; Stille, J.K.; ibid., 1970, 3, 473.
25. Iwatsuke, S.; Itoh, T.; Sadaike, S., ibid., 1981, 14, 1608.
26. Iwatsuki, S.; Itoh, T., ibid., 1982, 15, 347.
27. Scott, H.; Miller, G.A.; Labes, M.M. Tetrahedron Lett. 1963,
        1073.

28. Eley, D.D.; "The Chemistry of Cationic Polymerization,"
        Plesch, P.H.; Ed., Pergamon Press, Elmsford, N.Y., 1963.
29. Errede, L. A.; Hoyt, J. M., J. Amer. Chem. Soc., 1960, 82, 436.
30. Iwatsuki, S.; Inoue, K., Macromolecules, 1977, 10, 58.
31. Reich, H.J.; Cram, D.J., J. Amer. Chem. Soc., 1969, 91, 3517.
32. For a review see Hall, H. K., Jr., Angew. Chem. (Intern.
        Ed.), in press 1983.
33. Hall, H.K., Jr.; Sentman, R. C., J. Org. Chem.,1982, 47, 4572.
34. Rasoul, H.A.A.; Hall, H.K., Jr., ibid., 1982, 47, 2080.
35. Hall, H. K., Jr.; Reineke, K.,; Ried, J. H.; Sentman, R. C.;
        Miller, D., J. Polym. Sci., Polym. Chem. Ed., 1982,
        20, 361.
36. For reviews concerning tetramethylenes in small molecule
        reactions see, for example: Huisgen, R., Acc. Chem.
        Res., 1977, 10, 199; Borden, W.T., in "Reactive
        Intermediates" (Wiley), 1981, 2, 175.
37. Kindo, A.; Iwatsuki, S., J. Org. Chem., 1982, 47, 1965.
38. Iwatsuki, S.; Itoh, T., J. Polym. Sci. Poly. Chem. Ed.,
        1980, 18, 2971.
39. W. F. Gorham, J. Polym. Sci., Part A-1, 1966, 4, 3027.
40. Winberg, H.E.; Fawcett, E.S., Org. Syn., 1962, 42, 83. But
        also see Ito, Y.; Miyata, M.; Nakatsuka, M.; Saegusa, T.,
        J. Org. Chem., 1981, 46, 1043.
41a. Gorham, W.F.; Adv. Chem. Ser., 1969, No. 91, 643.
41b. Gorham, W.F.; "Encyclopedia of Polymer Science and Technology",
        N. Bikales, Ed., Interscience, Vol. 15, p. 98 (1971).

42. Niegisch, W.D., ibid., Vol. 15, p. 113.
43. Woodward, R.B.; Hoffmann, R., "The Conservation of Orbital
        Symmetry", Verlag Chemie and Academic Press, 1970.

ALKYLLITHIUM-INITIATED POLYMERIZATION OF MYRCENE

NEW BLOCK COPOLYMERS OF STYRENE AND MYRCENE

Roderic P. Quirk and Tzu-Li Huang

Michigan Molecular Institute
1910 W. St. Andrews Road
Midland, Michigan 48640

INTRODUCTION

Alkyllithium-initiated, anionic polymerization of vinyl monomers
is a very useful synthetic method since the major variables
affecting polymer properties can generally be controlled, i.e.,
molecular weight, molecular weight distribution, copolymer composition, diene microstructure, molecular architecture, and chain-
end functionality.[1-3] This control is a direct consequencce of the
fact that in the absence of reactive impurities these polymerizations are termination and chain transfer free; therefore, the
products of these polymerizations are polymer chains with
carbanionic chain ends ("living polymers").[4] An illustrative
mechanistic description is shown in Scheme I for the sec-
butyllithium-initiated polymerization of styrene. The product of
this polymerization is poly(styryl)-

Scheme I

initiation

$$1/2(\underline{sec}\text{-}C_4H_9Li)_4 + 2CH_2{=}CHC_6H_5 \xrightarrow{k_i} (\underline{sec}\text{-}C_4H_9CH_2\underset{\underset{C_6H_5}{|}}{C}HLi)_2$$

propagation

$$\underline{sec}\text{-}C_4H_9CH_2\underset{\underset{C_6H_5}{|}}{C}HLi + n\ CH_2{=}CHC_6H_5 \xrightarrow{k_p} \underline{sec}\text{-}C_4H_9{-}(CH_2\underset{\underset{C_6H_5}{|}}{C}H)_n{-}CH_2\underset{\underset{C_6H_5}{|}}{C}HLi$$

PsLi, $\underline{1}$

lithium(PsLi, $\underline{1}$) which has an active anionic chain end. It should be
noted that both simple alkyllithium and polymeric organolithium compounds are associated and cross-associated in hydrocarbon solution.[1]
This association complicates kinetic analysis of initiation and
propagation rates.[5]

If the initiation rate is of the same order of magnitude as the
propagation rate, then all chains grow for approximately the same
time interval until all of the monomer has been consumed. Under
these circumstances the molecular weight distribution will approach
the Poisson distribution, and virtually monodisperse molecular
weight distributions ($\overline{M}_w/\overline{M}_n$ 1.0) can be obtained.[1,6] Furthermore,
polymers with predictable molecular weights are obtained under these
conditions also;[6] the predicted stoichiometric number average molecular weight at 100% conversion (M_n^{stoic}) can be calculated from the
relationship shown in eq. 1.

$$\overline{M}_n^{stoic} \quad = \quad \frac{gm\ of\ monomer}{moles\ of\ initiator} \qquad (1)$$

On of the most unique aspects of living anionic polymerizations
is the ability to synthesize block copolymers by sequential monomer
addition.[7] Thus, the product of the alkyllithium-initiated polymerization of styrene (Scheme I) can be reacted with a diene such as
butadiene or isoprene to produce a living diblock copolymer as shown
in eq. 2 for butadiene. It is important to note that the alkyllithium-initiated poly-

$$PsLi + CH_2=CHCH=CH_2 \rightarrow Ps\underset{m-1}{[CH_2CH=CHCH_2]}CH_2CH=CHCH_2Li \qquad (2)$$

Ps-PBd-Li, 2

merization of dienes produces high 1,4-microstructure in hydrocarbon media, unlike other alkali metal derivatives which produce considerable 1,2-(and 3,4- for isoprene) enchainement.[8] After all of the diene has been consumed (eq. 2), the living diblock copolymer (PsPBd-Li, 2) can be reacted with more styrene monomer to form triblock copolymer (eq. 3). These SBS triblock copolymers (3) exhibit unique properties because of

$$Ps-PBd-Li + nCH_2=CHC_6H_5 \rightarrow Ps-PBd-Ps-Li \xrightarrow{CH_3OH} Ps-PBd-Ps \qquad (3)$$

2   3

$$Ps-PBd-Ps = \underset{n}{[CH_2-CH]}\underset{m}{[CH_2CH=CHCH_2]}\underset{n}{[CH_2CH]}$$
$$\underset{C_6H_5}{\qquad} \qquad 3 \qquad \underset{C_6H_5}{\qquad}$$

the incompatability of the rigid polystyrene and elastomeric polydiene segments which results in microphase spearation.[9] Triblock copolymers such as poly(styrene-b-butadiene-b-styrene), 3, within certain composition limits exhibit properties characteristic of vulcanized rubber with only physical crosslinking which can be broken reversibly by heating, i.e. they are thermoplastic elastomers.[10] These polymers have found applications in adhesives, sealants, molded articles and as additives for imparting toughness in thermoplastic resins.[9]

One of the limitations of anionic polymerization is the types of monomers which can be polymerized without termination and chain transfer. The most well-behaved systems are the vinyl monomers based on styrene and butadiene which are available primarily from petroleum feedstocks. Several years ago we began a search for anionically polymerizable monomers which would be available from renewable natural resources. One of the monomers uncovered (rediscovered) in this search was myrcene (4), 7-methyl-3-methylene-1,6-octadiene,

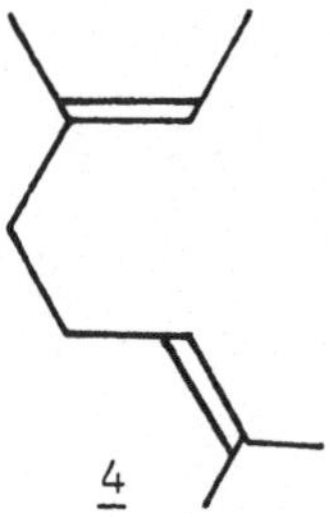

4

which is available from a renewable natural resource, pine
trees.[11,12] Myrcene is obtained from the pyrolysis of β-pinene(5),
eq. 4,[11,13] which is a major component of turpentine. Myrcene is
closely related

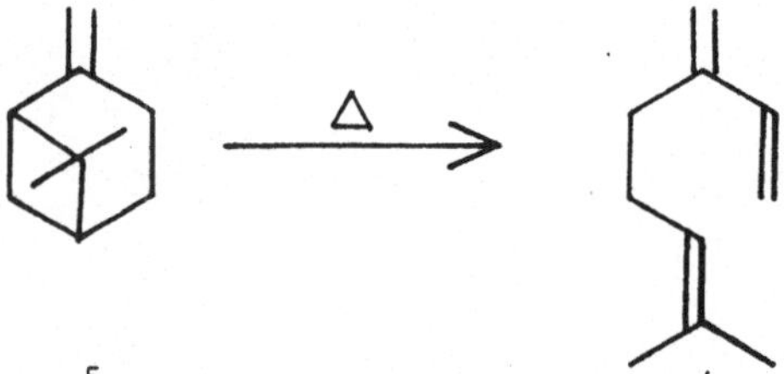

(4)

5   4

to isoprene (6) in structure and it would be expected that the diene

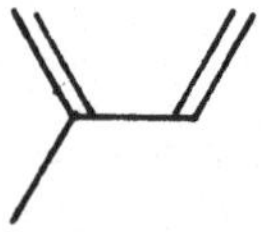

6

functionality would thus be susceptable to anionic polymerization.
A brief description of the alkyllithium-initiated polymerization of
myrcene was reported by Marvel and Hwa,[14] although they obtained
only low molecular weight poly(myrcene) and their characterization
studies were limited. A more recent study of the anionic polymerization and copolymerization of myrcene reported obtaining higher
molecular weight polymers; however, the molecular weight distributions were rather broad (1.50-1.64) and the living nature of these
polymerizations was not documented.[15]

Myrcene is an interesting diene (actually triene) monomer since it has an isolated double bond in its alkyl side chain. Thus, this monomer provides a unique opportunity to study the effects of internal π-bond interactions in living anionic polymerizations and also to examine the effects of this functionality on the physical properties of poly(myrcene) and various block copolymers as reported herein.

EXPERIMENTAL

Chemicals. Myrcene (SCM Chemical Company) was purified by stirring and degassing over freshly crushed CaH_2 on a high vacuum line followed by distillation onto sodium dispersion. After stirring for a few hours, final purification involved distillation from this dispersion directly into calibrated ampoules using a high boiling-point monomer distillation apparatus.[1] The ampoules were heat sealed and stored in a freezer at $-20°C$. Styrene and benzene were purified as described previously.[16] sec-Butyllithium (Lithium Corporation of America, 12.0 wt% in cyclohexane) was used as the initiator. Concentrations of active lithium were determined using the double titration method with 1,2-dibromethane.[17]

Polymerizations. Polymerizations were carried out at $30°C$ in all glass, sealed reactors using breakseals and standard high vacuum techniques.[1] Block copolymers were synthesized by sequential monomer additions. The polymerizations were terminated by addition of small amounts of dry, degassed methanol. Polymers were isolated by precipitation from methanol and dried in a vacuum oven at room temperature.

Polymer Characterization. Number average molecular weights ($\overline{M}_n$) of polymers were determined using an automatic membrane osmometer (Shell Development Design) in toluene solution at $35.5°C$. Membranes (Sartorious Membranfilter, SM 11539) were cautiously conditioned from ethanol into methyl ethyl ketone and finally toluene. Weight average molecular weights ($\overline{M}_w$) of poly(myrcene) were determined using a light scattering photometer FICA, 50 in tetrahydrofuran (THF) solution at 25°C. The refractive index increment (dn/dc) for the calculation of $\overline{M}_w$ from the light scattering data was determined at a wavelength of 546 nm using the method and apparatus similar to those described by Penther and Noller.[18] The value of dn/dc of poly-(myrcene) was found to be 0.1311 ml/g, which is about ten fold larger than that reported by Sivola.[15] The viscosities of all polymers were measured using an Ubbelohde viscometer in THF solution at 25°C. The intrinsic viscosity was obtained from the specific viscosity extrapolated to zero concentration.

Gel permeation chromatographic analyses (GPC) were obtained using either a Waters Model 200 GPC equipped with four, 4-foot styragel columns (7 x 10^5–5x10^6, 1.5–5 x 10^4, 3.0 x 10^3, and 8.0 x 10^2 Å) or a Perkin Elmer Model 601 HPLC with three μ–styragel columns (10^5, 10^4 10^3 Å) after calibration with standard poly(styrene) samples. Weight average molecular weights ($\overline{M}_w$) for the triblock copolymers were obtained from a universal calibration curve of ([ ]$\overline{M}_w$) versus solvent elution volume.[19]

^{1}HNMR spectra of all polymers were obtained in 5% CCl_4 solutions using a Varian EM–360L NMR spectrometer with TMS as internal reference. Infrared spectra were obtained from samples cast from solutions onto KBr plates using a Beckman IR–20A-X spectrometer.

Glass transition temperatures of all polymers were measured using either a DuPont 910 Differential Scanning Calorimeter (DSC) or a DuPont Dynamic Mechanical Analyzer (DMA). Samples were initially quenched from room temperature to approximately -150°C and then heated at a rate of 1°C/min. Samples for DMA measurements were prepared by casting films onto thin brass shims.[20]

Test specimens for tensile property measurements (Instron tester) were prepared by either casting from dilute solution or molding at 150°C, 12,000 psi for 5 minutes in an hydraulic press and annealing at a rate of 1°C/minute. The cast film samples were stored in a vacuum oven at 60°C for 3 days prior to testing.

RESULTS AND DISCUSSION

Poly(myrcene). The sec-butyllithium-initiated polymerization of myrcene in benzene solution at 30°C produces poly(myrcene) with predictable molecular weight and narrow molecular weight distribution ($\overline{M}_n$ = 58,000, $\overline{M}_w$ = 63,000; $\overline{M}_w/\overline{M}_n$ = 1.09). The yields of poly-(myrcene) and its block copolymers were generally ca. 90%, because the commercial myrcene (perfume and flavor grade) used in this study contains about 10% of impurities which are readily observable by gas phase chromatography.[21] Although our purification procedure removes impurities which could effect termination and transfer (vide infra), the principal impurities (β-pinene and limonene) appear to survive and are carried through the polymerization. The measured glass transition temperature (DSC) for poly(myrcene ($\overline{M}_n$ = 58,000) is -64°C, which is slightly higher than that reported recently for an analogous anionically prepared poly(isoprene) ($\overline{M}_n$ = 37,000) with a T_g of 69°C.[22] The large pendant side chain (4-methylpent-3-enyl group) in poly(myrcene) would be expected to increase the available free volume, relative to the methyl group in poly(isoprene), thus decreasing the T_g as observed in the poly(1-alkene) series.[23] The observed T_g does not reflect this type of effect, but the T_g is in the range for useful elastomers.

The infrared spectra of poly(myrcene) prepared in benzene show absorptions at 825cm^{-1} and 890cm^{-1} which can be attributed to 1,4- and 3,4-microstructures, respectively.[15] Since no absorptions appear at 910cm^{-1} and 1010cm^{-1}, 1,2-microstructures appear to be

absent using the peak assignments of Sivola.[15] Sivola[15] reported
that peaks at $910cm^{-1}$ and $1010cm^{-1}$ were observed in the infrared
spectra of poly(myrcene) polymers prepared in tetrahydrofuran.

^{1}H-NMR spectra (60MHz) of poly(myrcene) show vinyl proton
resonances at $\delta 5.1$ppm which correspond to the unsaturated side
chain protons for all microstructures $[-C\underline{H} = C(CH_3)_2]$ superimposed
on the C_3 proton in the 1,4-structure $\begin{bmatrix} C \\ >C=CH-C \\ C \end{bmatrix}$ (7)

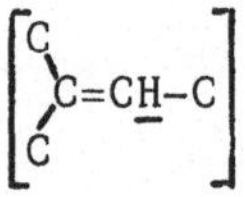

and at $\delta 4.8$ppm which corresponds to the methylene protons in the
3,4-structure $[>C = CH_2]$ (8).

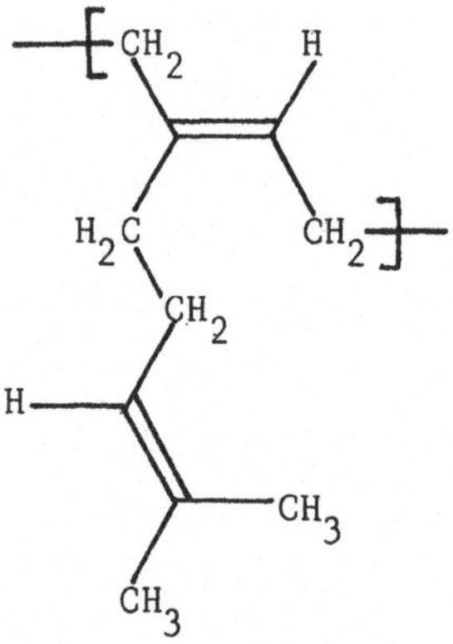

7

1,4-MICROSTRUCTURE

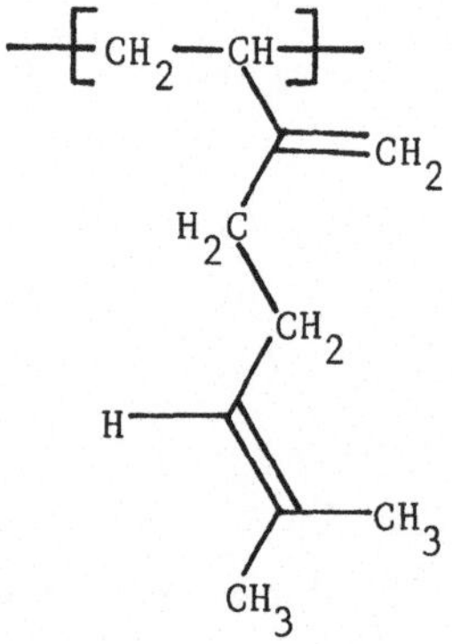

8

3,4-MICROSTRUCTURE

The integrated areas of these two peaks indicated that this polymer contains approximately 7% of the 3,4-structure, assuming the absence of the 1,2-enchainment. Sivola[15] has reported 10.4-14.7% of the 3,4-microstructure for poly(myrcene) prepared in benzene under various conditions. Other ^{1}H-NMR peaks are observed at $\delta 2.4-1.8$ppm ($-C\underline{H}_2-C=$ and $-C\underline{H}-C=$ protons), at $\delta 1.6$ppm [$=C(C\underline{H}_3)_2$], and a very small, broad absorption in the range of ca. 1.3-0.9 ppm ($C\underline{H} - \overset{|}{C}-$). It was not possible to determine the relative amount of cis- and trans- 1,4-microstructures since there are not satisfactory model compounds available for comparison.

Block Copolymers. A major question regarding the anionic polymerization of myrcene is whether it is a true living polymerization without termination and chain transfer. It was important to establish the living nature of our polymerizations since the myrcene samples used were the materials of commerce (90% purity). One unique characteristic of living polymerization is the ability to prepare block copolymers by sequential monomer addition. In order to determine the living nature of these polymerizations, a low molecular weight sample of poly(myrcenyl)lithium was prepared and then styrene monomer was added to this yellowish solution under high vacuum conditions. GPC chromatograms for the initial poly(myrcene) block ($\underline{A}$, $\bar{M}_n^{stoic}$=9,000), the diblock poly(myrcene-b-styrene) ($\underline{B}$, $\bar{M}_n^{stoic}$ = 110,000, and mixtures of these two polymers ($\underline{C}$) are shown in Figure 1. No peak corresponding to the initial poly(myrcene) block was observed in the chromatogram of the diblock copolymer under conditions where these two polymers are readily separated and observable ($\underline{C}$, Figure 1). This result is consistent with the living nature of the alkyllithium-initiated polymerization of myrcene and also shows that poly(myrenyl)lithium will crossover and polymerize styrene to form block copolymers.

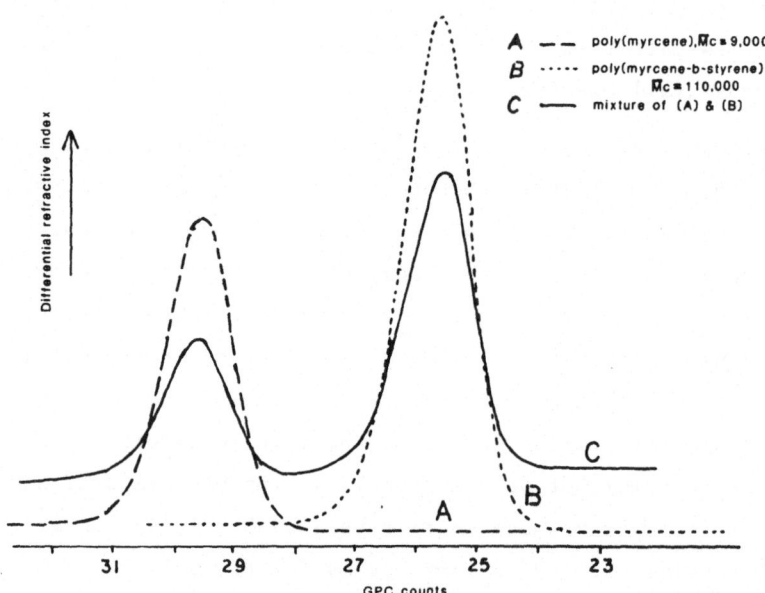

Figure 1.   Gel permeation Chromatograph of homo-poly(myrcene) &
             poly(myrcene-b-styrene)

During the synthesis of SMS triblock copolymers, poly(styrene-b-myrcene-b-styrene), samples of the initial poly(styrene) block and the poly(styrene-b-myrcene) diblock were withdrawn from the reactor and analyzed by GPC. The GPC chromatograms (Figure 2) for a typical triblock copolymer synthesis show distinct, narrow, symmetrical peaks for each initial block, diblock, and final triblock polymer. In some samples, a small, high molecular weight shoulder was observed in the GPC chromatogram of the triblock copolymer; this may have been due to a small amount of oxygen coupling during termination,[24] or to the effect of some of the impurities in myrcene.

Number-average molecular weights ($\overline{M}_n$) and weight average molecular weights ($\overline{M}_n$) of the SMS triblock copolymers were determined by membrane osmometry and GPC methods, respectively. The results are shown in Table I. The weight-average molecular weights of the block copolymers were calculated using the universal GPC calibration method which assumes that the hydrodynamic volume, characterized by $[\eta]M$, determines the retention time.[25] Thus, using monodisperse poly(styrene) samples (Pressure Chemical Co.) as the primary calibration standards, at equal elution volumes the molecular weight can be calculated from eq. 5, where M and $[\eta]$ are the molecular weight and intrinsic

$$M [\eta] \quad = \quad M_s [\eta]_s \qquad (5)$$

viscosity of the block polymer and M_s and $[\eta]_s$ correspond to the same variables for poly(styrene). All dilute-solution viscosity and GPC measurements were performed in THF, a good solvent for the block copolymers and poly(styrene). This universal calibration method has been used previously for the molecular weight characterization of poly(styrene-b-isoprene-b-styrene) (SIS) and poly(styrene-b-butadiene-b-styrene) (SBS) block copolymers,[26,27]

The block copolymers were obtained in relatively low yields (88-91%). This is due at least in part to the presence in myrcene of ca. 10% of impurities, which are not polymerized. Excluding this

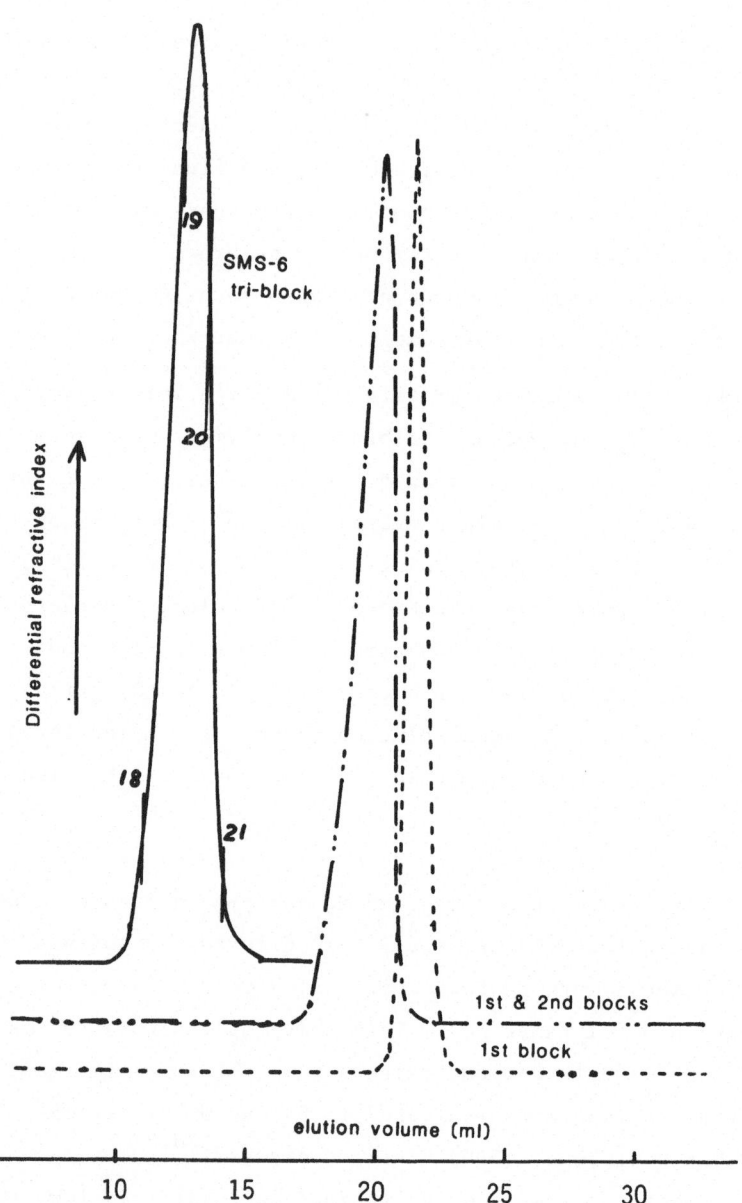

Figure 2. Gel Permeation Chromatograph of SMS-6 tri-Block Copolymer
and Each Segment

Table 1. Characterization of SMS-Block Copolymers

SAMPLE	$\bar{M}_c \times 10^{-3}$ [a] (g/mol)	$\bar{M}_n \times 10^{-3}$ [b] (g/mol)	$\bar{M}_w \times 10^{-3}$ [c] (g/mol)	$[\eta]$ [d] (dl/g)	T_g [e] (°C)	wt. % styrene calc.	wt. % styrene exp. [f]
SMS-6	190 (55-80-55)	187	194	0.815	-63	103	57.9	58.1
SMS-3	150 (30-90-30)	165	157	0.714	-62	95	40.0	41.2
SMS-7	170 (30-110-30)	194	186	0.853	-61	95	35.3	36.0
SMS-9	180 (20-140-20)	190	210	1.065	-60	87	22.2	23.1
SMS-4	110 (17-76-17)	130	125	0.683	-63	85	31.2	32.6
SMS-10	80 (14-52-14)	92	85	0.495	-63	85	35.0	35.5

a   Stoichiometric molecular weight based on the ratio of gm of monomer charged to the moles of initiator after correction for the 10% impurities in myrcene.
b   Determined by membrane osmometry.
c   Determined by GPC using the universal calibration method (ref. 25).
d   Intrinsic viscosity at 25°C in THF.
e   Determined by DSC.
f   Determined by integration of the 1H-NMR spectra (exp.). The calculated value is based on the wt % of the monomers charged into the reactor.

amount of impurity, the apparent polymer yields (92–97%) improve, and the calculated and observed molecular weights are in good agreement (Table 1). The experimental values of the wt% of styrene in the triblock copolymers (see Table 1) were calculated from the integrated areas of the ^{1}H–NMR peaks corresponding to the aromatic protons (δ7.0ppm and δ6.5ppm) and the vinyl protons (δ5.0ppm and δ4.7ppm). These experimental values agree closely with the composition values calculated from the composition of the monomers charged into the polymerization reactor.

By analogy with the theoretical and experimental results for SBS and SIS block copolymers,[9,28-38] it was anticipated that SMS block copolymers would undergo microphase separation because of the thermodynamic incompatability of the poly(myrcene) and poly(styrene) segments.[39] Experimental proof for the heterophase nature of these SMS triblock copolymers was obtained by DSC measurements which showed two discrete glass transition temperatures corresponding to the poly(myrcene) phase [(−60) – (−63)°C] and the poly(styrene) phase (85–103°C) (see Table 1).

The heterophase nature of these SMS triblock copolymers was also confirmed by dynamic mechanical analysis (DMA). Block copolymer samples for DMA study were prepared by solvent casting onto thin brass shim supports.[20] Typical dynamic mechanical spectra for SMS–3 and SMS–9 are shown in Figures 3 and 4, respectively. In general, two loss peaks are observed in these samples, and the maxima provide measures of the glass transition temperatures which are in good agreement with the DSC values (Table 1). The morphology of heterophase polymers can be affected by the casting solvent and the nature of its interaction with the polymer blocks.[9,30,40] The use of selective solvents which are either "good" or "poor" for one block can alter the volume ratios of the phases which determines the morphology of the block copolymers.[30] Selective solvents can determine

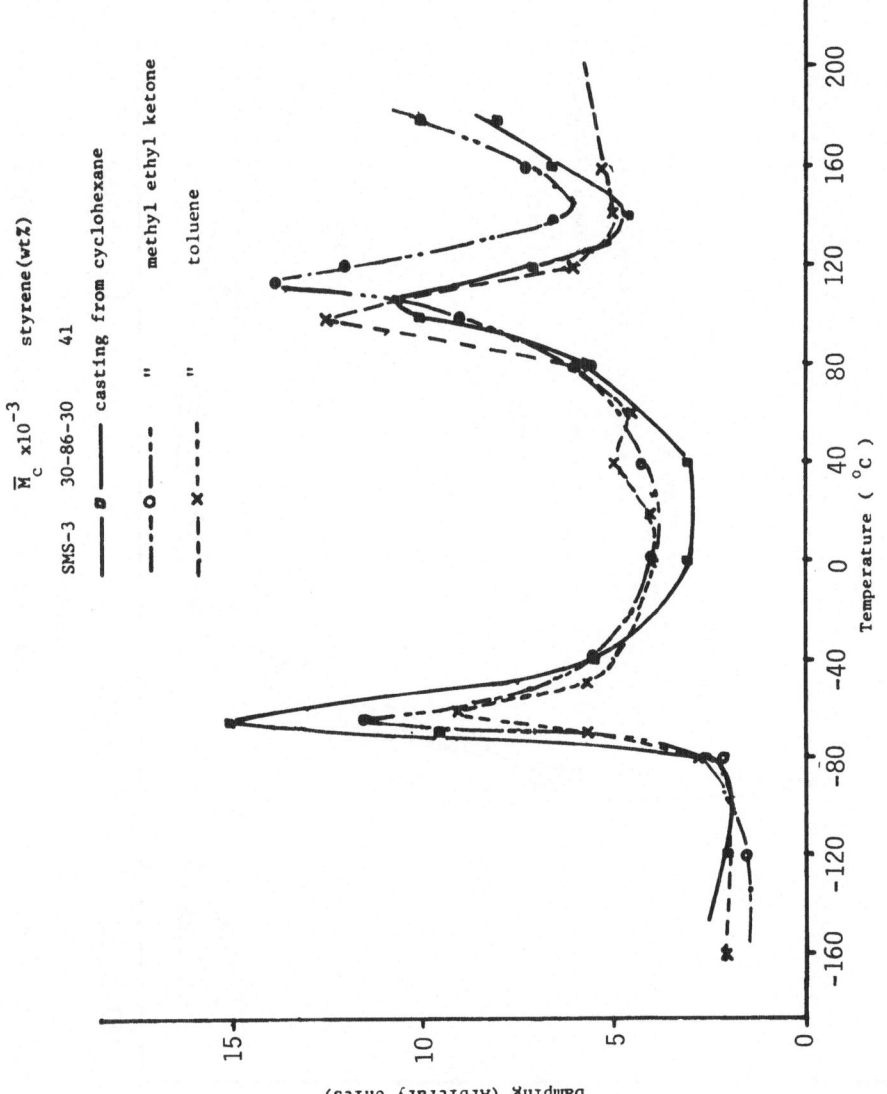

Figure 3. Dynamic Mechanical Analysis of Poly(styrene-myrcene-styrene), SMS-3

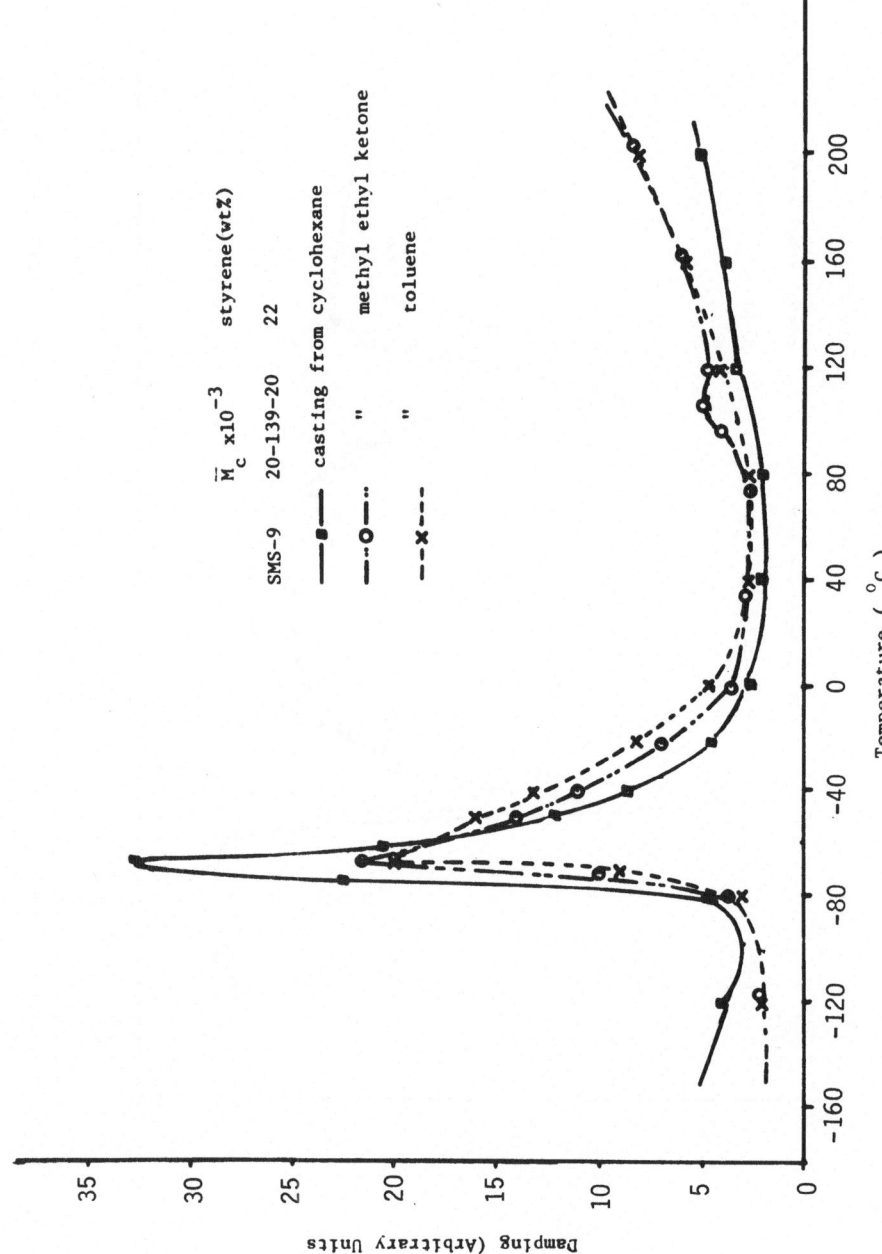

Figure 4. Dynamic Mechanical Analysis of Poly(styrene–myrcene–styrene), SMS–9

which phase will be the first to phase separate during the evaporation process, and the polymer segments which remain swollen with
solvent will then generally form the continuous phase. The solvents
used for DMA film casting were cyclohexane (a good solvent for the
polydiene), methyl ethyl ketone (a good solvent for polystyrene),
and toluene (a good solvent for both block segments).[40,41] In DMA
experiments, the peak areas are proportional to the volume fraction
of the corresponding phase in the block copolymer.[42] From the DMA
energy loss spectra, the peak area corresponding to the poly-
(myrcene) segments is sharper and less diffuse in cyclohexane, while
the styrene peak is most distinct in methyl ethyl ketone (MEK). In
fact, for SMS-9 (22 wt% styrene) no polystyrene DMA peak is observed
except in films cast from MEK. An additional intermediate transition
was observed for SMS-3 (41 wt% styrene) at ca. 40°C in films cast
from toluene, a good solvent for both segments. This intermediate
transition can be assigned tentatively to interphase mixing of the
poly(styrene) and poly(myrcene), as observed for SBS and SIS block
copolymers.[41] These results suggest that SMS films cast from
selective solvents may have different morphologies which should be
observable by electron microscopy and which should affect the mechanical properties of these polymers.

Both theory and experiment show that the morphologies of ABA
triblock copolymers vary with the volume fraction of the block segments.[9,28-38,43] For poly(styrene-b-diene-b-styrene) triblock copolymers, the domain morphologies vary from spherical(polystyrene
spheres in a diene matrix) for polymers containing 0-20% styrene, to
cylinders (polystyrene cyclinders in a diene matrix) for polymers
with ca. 20-30% styrene, to lamellae for polymers with ca. 30-50%
styrene. The same morphologies with inverted phases arise for higher
styrene contents. Preliminary results showing the morphologies of
SMS-3 (Figures 5,6) and SMS-9 (Figures 7,8) block copolymers have
been obtained by electron microscopy.[44] The first conclusion from

Figure 5. Electron Micrograph of SMS-3. Film Cast from Methyl Ethyl Ketone

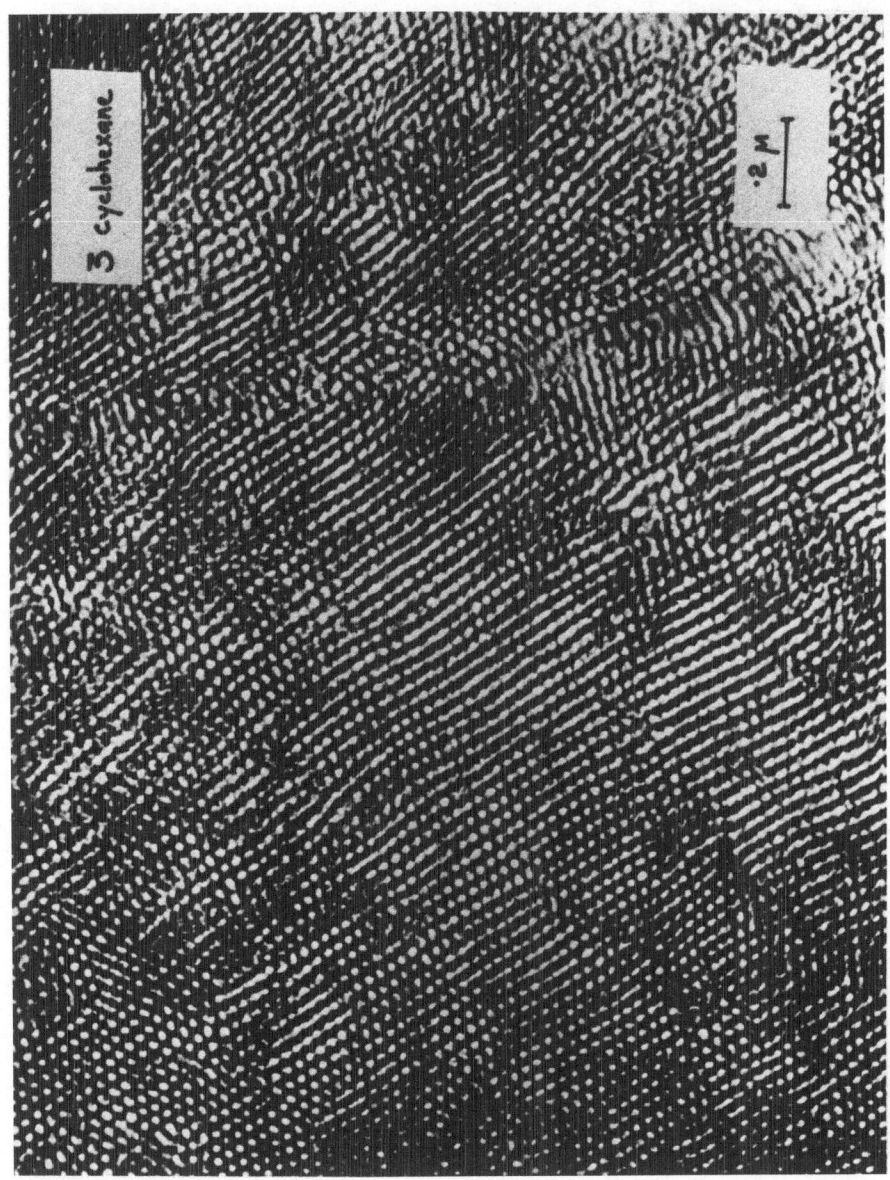

Figure 6. Electron Micrograph of SMS-3. Film Cast from Cyclohexane

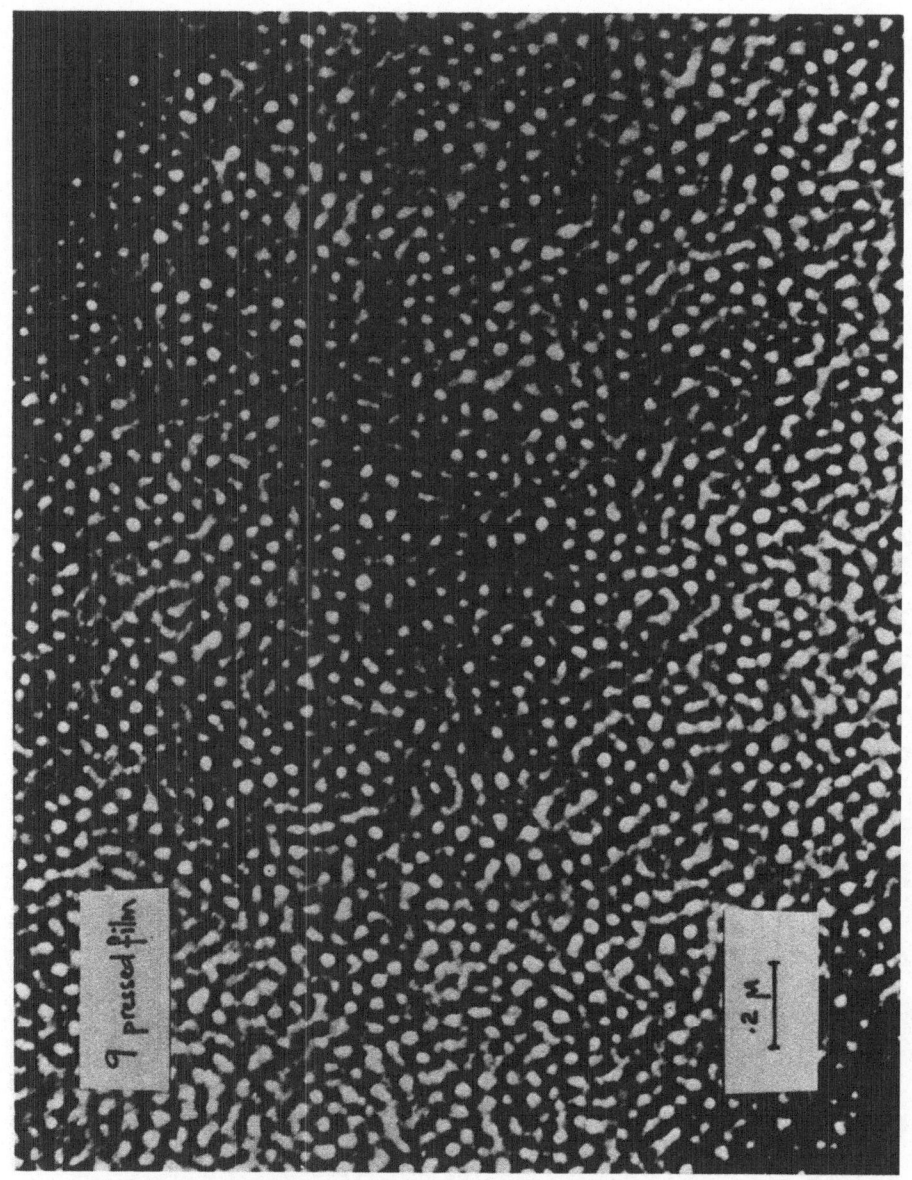

Figure 7. Electron Micrograph of SMS-9. Pressed Film

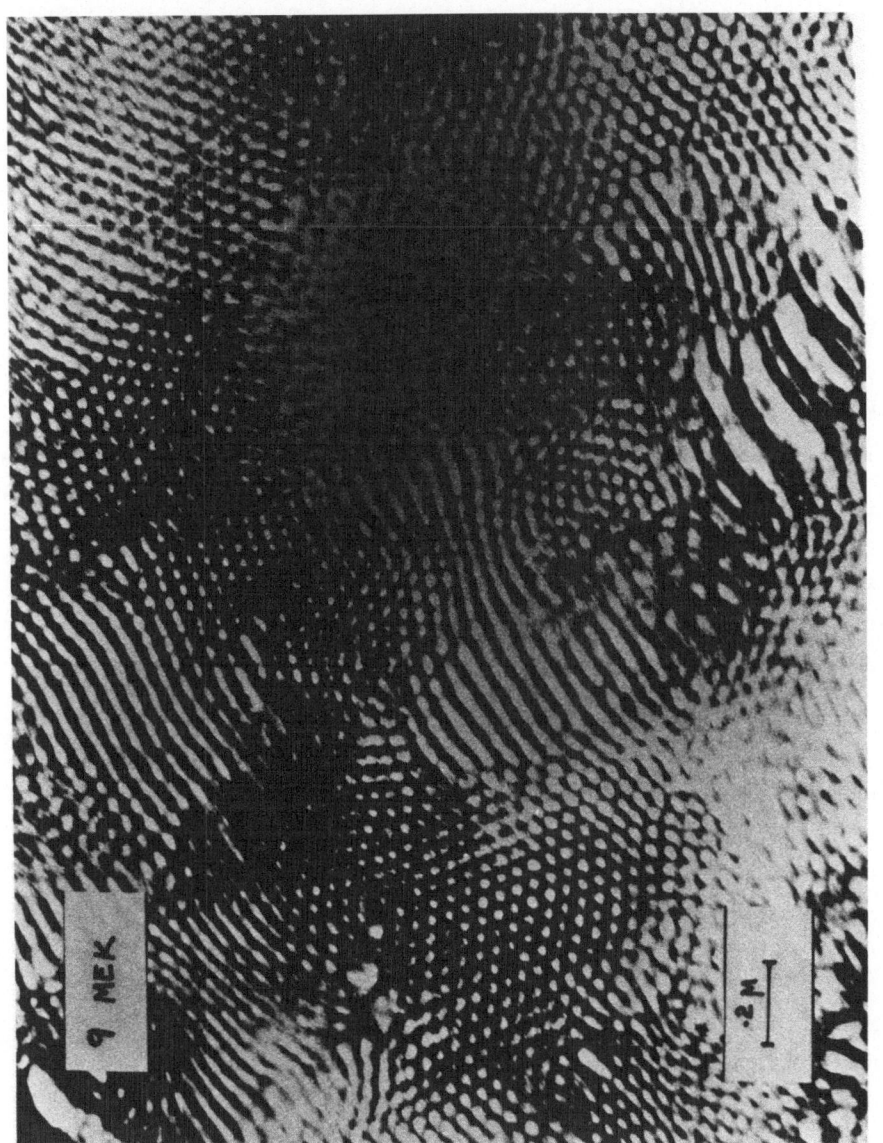

Figure 8. Electron Micrograph of SMS-9. Film Cast from Methyl Ethyl Ketone

these micrographs is that the poly(styrene-b-myrcene-b-styrene) polymers form phase-separated domain structures as expected. For SMS-9
(23 wt% styrene), a spherical domain morphology would be expected
and appears in the electron micrograph (Figure 7); however, since
this pressed film sample was not examined from a variety of angles,
it can only be concluded that the electron micrograph is consistent
with this morphology. MEK is a good solvent for polystyrene and
probably a poor solvent for poly(myrcene); therefore, it would be
expected that solvent casting films from MEK would shift the
morphology of SMS-9 toward cylindrical or lamellar structures.
Figure 8 shows a morpohology consistent with a cylindrical
morphology in that it apparently shows both edge and end-on views.
For SMS-3 (41 wt% styrene) films cast from MEK, one would expect a
lamellar morphology. The electron micrograph of this sample (Figure
5) is consistent with either a lamellar (most probable) or cyclind-
rical (if perfectly oriented axially) morphology. For SMS-3 films
cast from cyclohexane which is a poor solvent for polystyrene and
probably a good solvent for poly(myrcene), a shift of morphology
towards spherical morphology would be expected. The electron micrograph (Figure 6) of this sample is consistent with this prediction.
Thus, the morphological behavior of SMS block copolymers, as deduced
from these preliminary electron microscopic analyses, is similar to
the experimental results for analogous poly(styrene-b-isoprene-
b-styrene) (SIS)and poly(styrene-b-butadiene-b-styrene) (SBS) block
copolymers, and consistent with theoretical predictions.

The tensile stress-strain characterization of the SMS triblock
copolymers is shown in Figure 9. The polymer SMS-6 with a high
styrene content (58 wt%) shows a high tensile modulus at initial
stretching and a high yield point. The morphology of this block
copolymer is probably lamellar. This would explain the plastic-like
behavior, since the hard poly(styrene) phase would be broken on

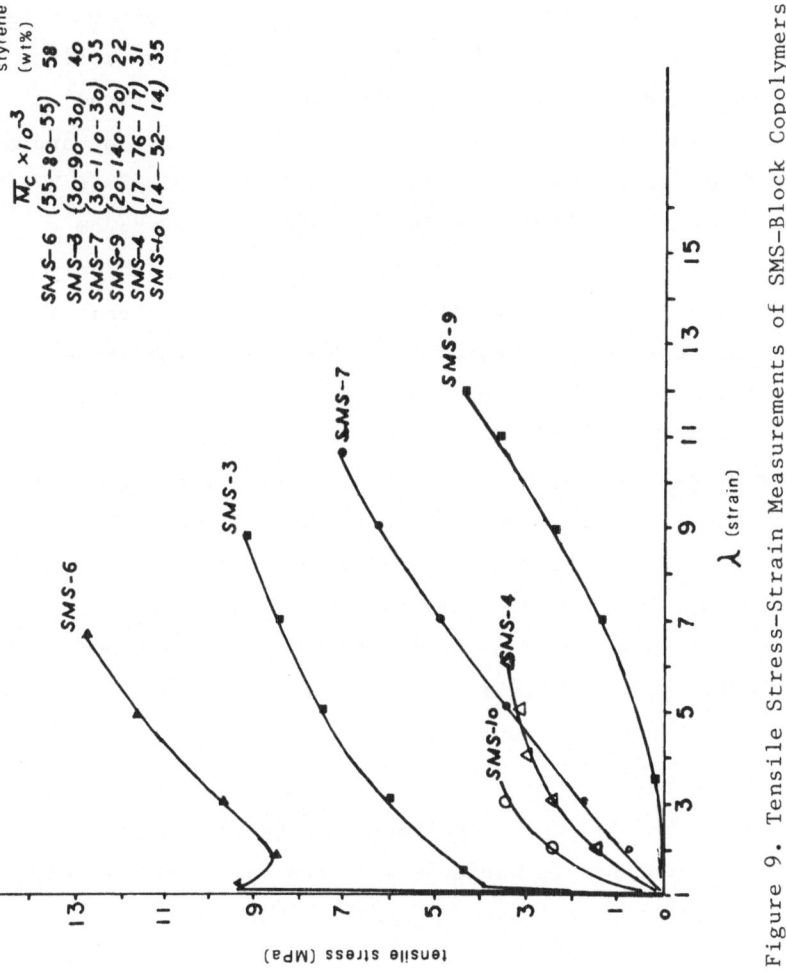

Figure 9. Tensile Stress–Strain Measurements of SMS–Block Copolymers

stretching and then the poly(myrcene) domain could elongate and
provide the relatively high extension behavior observed. Electron
microscopy indicates that SMS-3 has a lamellar morphology (Figure
5). The similarity of the tensile behavior for SMS-3 and SMS-6 is
consistent with a lamellar morphology for both polymers. On the
basis of its composition, polymer SMS-7 would be expected to exhibit
a cylindrical morphology. However, the absence of a distinct yield
point and drawing behavior points to a possible spherical
morphology. The SIS and SBS polymers with 30 wt% styrene and molecular weights of 13,700–100,400–13,700 have extensions at break of
12–13 and tensile strengths in the range of 250–300 kg/cm^2
(25.5–30.6 MPa).[9] The corresponding 22 wt% styrene SMS polymer
(SMS-9) has a high extensibility (ca. 12), but its tensile strength
at break is only about 43 kg/sm^2 (4.4 MPa). Thus the large side
chain attached to the polymer chain backbone in poly(myrcene) lowers
the tensile strength considerably in much the same way as an oil
extender or plasticizer would act in the corresponding SIS or SBS
block polymers. It is noteworthy that polymers SMS-4 (31 wt%
styrene) and SMS-10 (35 wt% styrene) with diene center blocks of
76,000 and 52,000, respectively, have very low extensions at break.
These results indicate that the molecular weight required for long-
range rubber extensibility in poly(myrcene) is much larger than in
the corresponding poly(isoprene) and poly(butadiene) rubbers.

    Preliminary electron microscopic examination indicates that the
morphology of SMS triblock copolymers can be varied by selective
solvent film casting. It would be expected that changes in morphologies would be manifested by changes in the tensile stress-
strain properties also. The effect of selective solvent casting for
SMS-3 is shown in Figure 10. Essentially the same tensile behavior
as for the pressed film is observed for films cast from either the
nonselective solvent toluene or MEK which is good solvent for polystyrene. This is consistent with the predicted lamellar morphology

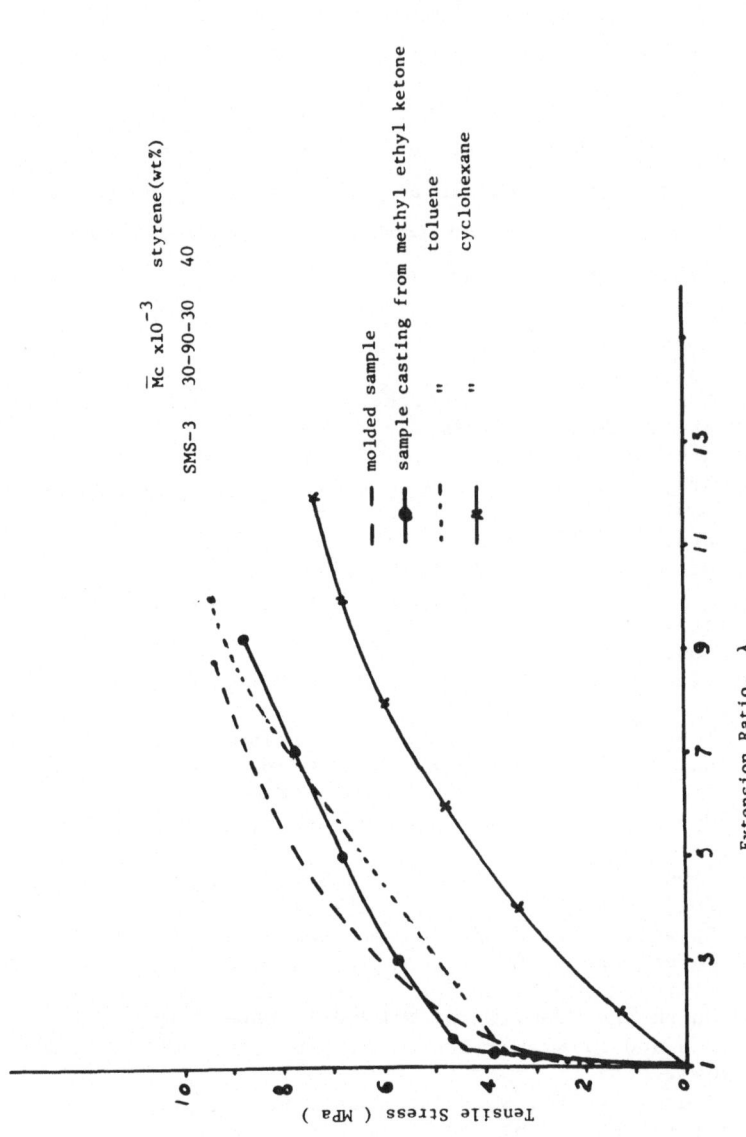

Figure 10. Stress-strain Properties of SMS Block Polymer Films Prepared from Different Solvents

for this polymer (Figure 5). A dramatic change is observed for the
film cast from cyclohexane which is probably a selective solvent for
the poly(myrcene) segments. The stress-strain behavior for this
sample is consistent with a change in morphology toward the
spherical domain structure (Figure 6).

These studies indicate that poly(styrene-b-myrcene-b-styrene)
triblock copolymes exhibit interesting and unique properties
compared with the analogous SIS and SBS triblock copolymers. It
will be interesting to further examine the structure-property relationships for poly(myrcene) and its copolymers.

REFERENCES

1. M. Morton and L. J. Fetters, Rubber Chem. Tech., 48, 359 (1975)
2. S. Bywater, Adv. Polym. Sci., 30, 89 (1979).
3. S. Bywater, Prog. Polym. Sci., 4, 27 (1974).
4. M. Szwarc, "Carbanions, Living Polymers and Electron Transfer
    Processes," Interscience Publishers, New York, 1968.
5. S. Bywater, in "Non-Radical Polymerization," "Comprehensive
    Chemical Kinetics", Vol. 15, Elsevier, Amsterdam, 1976, p. 1.
6. L. J. Fetters and M. Morton, in "Macromolecular Synthesis",
    Collective Vol. 1, J. A. Moore, Ed., Wiley, New York, 1977,
    p. 463.
7. P. Dreyfuss, L. J. Fetters and D. R. Hansen, Rubber Chem.
    Tech., 53, 728 (1980).
8. A. V. Tobolsky and C. E. Rogers, J. Polym. Sci., 40, 73 (1959).
9. A. Noshay and J. E. McGrath, "Block Copolymers. Overview and
    Critical Survey," Academic Press, New York, 1977.
10. G. Holden, E. T. Bishop and N. R. Legge, J. Polym. Sci. Part C,
    26, 37 (1969).
11. D. F. Zinkel, Chem. Technol.,5, 235 (1975).
12. I. S. Goldstein, Science, 189, 847 (1975).
13. R. L. Burwell, Jr., J. Am. Chem. Soc., 73, 4461 (1951).
14. C. S. Marvel and C. C. L. Hwa, J. Polym. Sci., 45, 25 (1960).
15. A. Sivola, Acta Polytechnica Scandinavica, 134, 1 (1977).
16. R. P. Quirk and W.-C. Chen, Makromol. Chem., 183, 2071 (1982).
17. H. Gilman and F. K. Cartledge, J. Organomet. Chem., 2 447
    (1964).
18. C. J. Penther and G. W. Noller, Rev. Sci. Instr., 29, 43 (1958).
19. Z. Grubisic, P. Rempp, and H. Benoit, J. Polym. Sci. B, 5, 753
    (1967).

20. H. W. Starkweather, Jr. and M. R. Giri, J. Appl. Polym. Sci., 27, 1243 (1982).
21. A. D. Corcia, A. Liberti, C. Sambucini, and R. Samperi, J. Chromatography, 152, 63 (1978).
22. J. M. Widmaier and G. C. Meyer, Rubber Chem. Technol. 54, 940 (1981).
23. M. L. Dannis, J. Appl. Polym. Sci., 1, 121 (1959).
24. L. J. Fetters, and E. M. Firer, Polymer, 18, 306 (1977).
25. Z. Grubisic, P. Rempp, and H. Benoit, J. Polym. Sci. B, 5, 753 (1967).
26. G. Kraus, F. E. Naylor, and K. W. Rollmann, J. Polym. Sci. A-2, 9, 1839 (1971).

27. L. H. Tung, J. Appl. Polym. Sci., 24, 953 (1979).
28. S. L. Aggarwal, Polymer, 17, 938 (1976).
29. D. J. Meier, J. Polym. Sci., Part C, 26, 81 (1969).
30. D. J. Meier, in "Block and Graft Copolymers," J. J. Burke and V. Weiss, Eds., Syracuse University Press, Syracuse, New York, 1973, Chapt. 6, p. 105.
31. D. J. Meier, ACS Polym. Preprints, 15(1), 171 (1974).
32. E. Helfand, Macromolecules, 8, 552 (1975).
33. E. Helfand and Z. R. Wasserman, Macromolecules, 9, 879 (1976).
34. E. Helfand, Accounts Chem. Res., 8, 295 (1975).
35. M. Morton in "Encyclopedia of Polymer Science and Technology," John Wiley and Sons, Inc. New York, 1971, vol. 15, p. 508.
36. S. Krause, J. Polym. Sci. A-2, 7, 249 (1969).
37. S. Krause, Macromolecules, 3, 84 (1970).
38. "Block Copolymers: Science and Technology," D. J. Meier, Ed., MMI Symposium Series, Harwood Academic Publishers, 1983.
39. "Polmer Compatibility and Incompatibility, Principles and Practice," K. Solc, Ed, MMI Symposium Series, Harwood Academic Publishers, 1982.
40. R. Seguela and J. Prud'homme, Macromolecules, 11, 1007 (1978).
41. S. L. Aggarwal, R. A. Livigni, L. F. Marker, and T. J. Dudek, in "Block and Graft Copolymers," J. J. Burke and V. Weiss, Eds, Syracuse University Press, Syracuse, New York, 1973, Chapt. 9, p. 157.
42. T. Murayama, "Dynamic Mechanical Analysis of Polymeric Materials," Elsevier, Netherlands, 1978, p. 90.
43. D. J. Meier, ACS Polymer Preprints, 11(2), 400 (1970).
44. Electron micrographic analysis were provided through the generosity of Professor L. J. Fetters, Institute of Polymer Science, University of Akron.

NEW FUNCTIONAL METHACRYLATE

POLYMERS BY ANIONIC POLYMERIZATION

G.D. Andrews and L.R. Melby

E.I. Dupont de Nemours and Company

Central Research and Development Department
Wilmington, Delaware 19898

INTRODUCTION

In the early 1960's, subsequent to the pioneering work of M. Szwarc on the living anionic polymerization of styrene and butadiene, extensive research on the anionic polymerization of methacrylate esters was undertaken by D. L. Glusker, W. E. Goode, R. K. Graham and co-workers of the Rohm and Haas Company laboratories. Their work was augmented by that of D. M. Wiles and S. Bywater of the National Research Council of Canada during the same era, and related studies subsequently issued from many laboratories. (For leading references see Ref. 1.)

Most of the published material has dealt with kinetics, mechanisms and stereochemical aspects of the polymerization. Thus early work by the Rohm and Haas group showed that under proper conditions the anionic polymerization of methacrylates is truly living. That is to say, if monomer, solvent and reagents meet stringent criteria of purity, and the temperature is maintained at -75°C or thereabouts, a carbanion or anion-radical source such as fluorenyllithium (FlLi) or naphthalenesodium will add to methyl methacrylate (MMA) to form an enolate anion($\underline{1}$) which then propagates to form a polymer chain with an anionic terminus ($\underline{2}$).

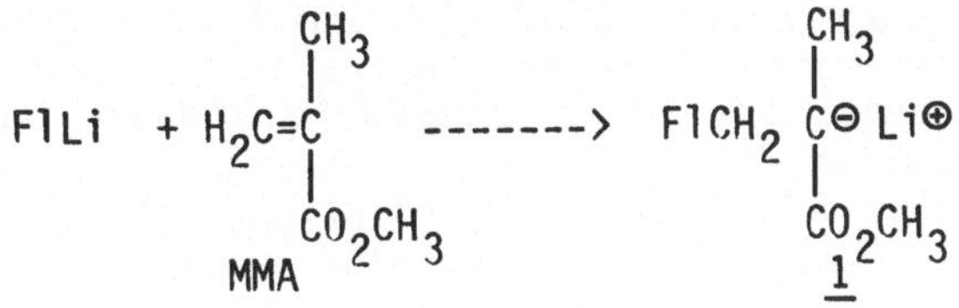

$$\text{FlLi} \quad + \quad \text{H}_2\text{C=C} \overset{\text{CH}_3}{\underset{\text{CO}_2\text{CH}_3}{|}} \quad \text{------->} \quad \text{FlCH}_2 \overset{\text{CH}_3}{\underset{\text{CO}_2\text{CH}_3}{|}} \text{C}^{\ominus} \text{Li}^{\oplus}$$

$$\text{MMA} \qquad\qquad\qquad \underline{1}$$

$$\underline{1} \ + \ MMA \ \ ------> \ \ Fl(MMA)_n^{\ominus} \ Li^{\oplus}$$

The anion $\underline{2}$ has a finite lifetime and propagates further by stepwise addition of more monomer. However, death of the polymer enolate ion can occur by a spontaneous, thermally-induced cyclization to form a terminal, substituted cyclohexanone ring [2] ($\underline{3}$) which results in termination.

PCH$_2$ ... CO$_2$Me ... OMe → PH$_2$C ... CO$_2$Me + MeO$^-$

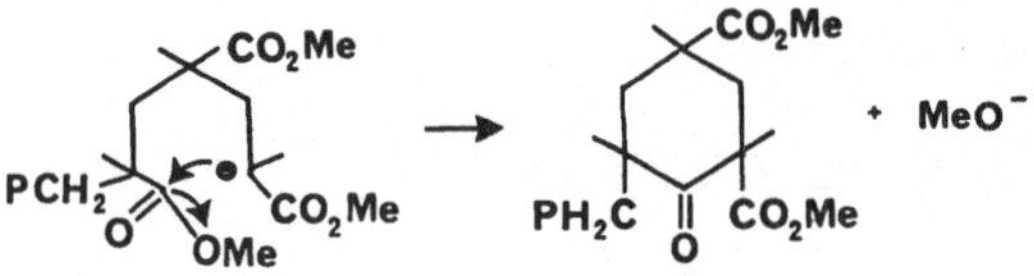

$\underline{3}$

Subsequent work by D. Braun and collaborators dealt with the effect of solvent polarity and countercation on the tacticity of poly(methyl methacrylate) (PMMA) initiated with various organo alkali metal compounds. By using proton NMR analysis they concluded that polar solvents and alkyllithium initiators favored syndiotactic configuration($\underline{4}$) while non-polar solvents favored isotactic placement($\underline{5}$). [3]

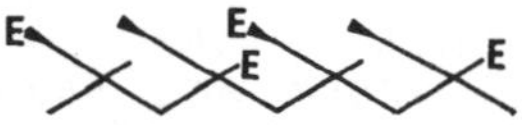

$\underline{4}$ 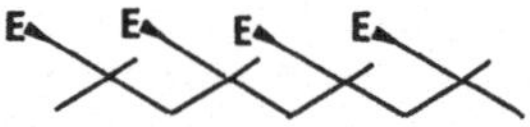 $\underline{5}$

Later, related work by F.A. Bovey and co-workers considered solvation control and stereoselectivity in methacrylate and deuterio-methacrylate polymerizations initiated by fluorenyllithium (FlLi) and 1,1-diphenylhexyllithium (DPHLi). They too used NMR analysis extensively and speculated on the conformation of the propagating enolate ion cavity with its associated counterion and their relationship to the tactic placement of incoming monomer units. [4] Similiar considerations were pursued by G. V. Shulz and collaborators in their kinetic analysis of MMA polymerizations done anionically in tetrahydrofuran (THF) with cumylcesium initiators. Their main concern was to determine the statistics which best described the tactic addition of monomer. By NMR analysis of sequence lengths they concluded that the system obeyed Markovian rather than Bernoullian statistics. [5]

In an ideal living polymerization the degree of polymerization (DP) and thus the molecular weight of the product

are predictable and simply determined by the molar ratio of the
monomer ([M]) to that of the initiator ([In]):

$$DP = \frac{[M]}{[In]}$$

For this condition to prevail the initiator must add
quantitatively to the first monomer unit without destructive side
reactions, and the addition must be faster than the rate of
propagation. We have already remarked on the further necessity
that spontaneous termination must be absent. The molecular weight
of the product will of course be augmented by the attached
initiator moiety. At very high DP's the chemical and physical
characteristics of that moiety will generally not materially
affect the gross properties of the product polymer, but at low
DP's it may.

Another consequence of such ideality is that the molecular
weight distribution (Mw/Mn) will be very narrow and at high DP[6]
will approach unity according to the relation:

$$\frac{Mw}{Mn} = 1 + \frac{1}{DP}$$

That is to say, the polymer will be essentially monodisperse.
The ability to produce polymer with predictable and narrow
molecular weight distribution is highly desirable from the
standpoint of characterization, of correlating physical
properties with chemical constitution, and of preparing
well-defined compositions or formulations for practical uses.
Absolute ideality cannot of course be achieved experimentally
but, as we shall see, appropriate manipulation of conditions can
for practical purposes approach the ideal.

Our main interest in anionic polymerization of methacrylates
has been to develop the procedure into a simple, reliable
synthetic method for polymethacrylates with controlled,
predictable molecular weights and with functional groups
introduced either by manipulating the initiator, the polymer
enolate ion, or by introducing specific reactive comonomers. Thus
we have not been greatly concerned with pursuing physico-chemical
aspects which in any case are under continued capable
investigation in other laboratories.

What follows is derived mainly from our own research since
it has been extensive enough to fulfill the charter of this
chapter. Some of the work we describe here has been the subject
of a brief written communication,[7] and some has been discussed at
length in various forums,[8,9] but much remains to be published in
detail.

PREPARATIVE PROCEDURES

General Considerations

Anionic polymerizations are notoriously susceptible to disruption by protic contaminants or oxygen. Such contaminants must therefore be rigorously excluded from the polymerization atmosphere, media and monomer to achieve finest control. Although much of the previously published work on methacrylates involved the use of vacuum trains, break-seal ampules and the like, such precautions are unsuitable for preparative purposes and indeed are not necessary. Ordinary organic synthesis glassware may be used if it is appropriately dried. Purifications of monomers and solvents are straightforward (distillation followed by filtration through alumina), and the polymerizations can be carried out in atmospheres of standard commercially-available, high-purity nitrogen or argon.

Our preferred initiators were generally prepared from 1,1-diphenylethylene by addition of n-butyllithium or specially synthesized substituted alkyllithium reagents (see below). The ready availability of diphenylethylene and commercial standard n-butyllithium solutions precludes the need for elaborate apparatus for initiator generation such as described by Galluccio and Glusker in Macromolecular Synthesis.[10]

Solvents and Temperature

In their work on solvent effects on PMMA configuration, Braun et al used among others, mixtures of toluene with pyridine and initiated with n-butyllithium at -30 to -70°C. Our reexamination of this system showed it to be truly living and controllable as specified above.[11] Toluene alone is not a suitable solvent; the pyridine is required to meet living polymerization criteria. However, such solvent mixtures have both practical and chemical shortcomings. Polymer isolation is inconvenient, solvent disposal becomes a problem, and the chemical reactivity of the excess pyridine severely limits the possibilities for post-reaction of the living polymer. Other solvents or mixtures including aliphatic hydrocarbons, diethyl ether, esters, etc., are unsuitable for a variety of reasons which need not be specified here.

Summarily, the solvent of choice is tetrahydrofuran (THF). From the standpoint of availability, cost, solvent power, solubility in polymer precipitants as diverse as hexane or water, relative inertness, easy purification and so on, it cannot be matched for these polymerizations. Critically important is its donor character which influences the polymer enolate ion pair in such a way as to ensure its orderly, stepwise propagation and maintain its living character. In most respects dimethoxyethane functions as well but its higher cost and easier peroxidizability make it less desirable than THF. We distill THF from benzophenone/sodium in a Soxhlet extactor modified to allow THF removal with a syringe.

In toluene-pyridine mixtures the polymerization maintains living character at temperatures as high as -10°C, but THF does not allow this latitude. In THF termination becomes significant at -30°C and so the temperature must be kept as low as possible short of freezing the medium. In practice, temperatures between -60 and -78°C are adequate and these can be conveniently attained with solid carbon dioxide-acetone cooling baths.

Maintaining the low temperature is of course countered by the exothermicity of the polymerization reaction (12.9 kcal/mole for MMA[12]). Thus when a large scale monomer charge is added rapidly in bulk to an appropriately cooled solution of anionic initiator, polymerization proceeds to completion in a matter of a few seconds and heat evolution is so rapid that low temperature can not be maintained, therefore termination becomes significant. This difficulty is circumvented by adding monomer dropwise at a rate consistent with keeping the temperature within the desired limits. Addition of the initiating anion to monomer is so rapid that the 1:1 adduct is completely formed before propagation proceeds. As monomer addition continues it is necessary only that the polymerizing mixture be adequately mixed. Mechanical stirring at a reasonable rate is sufficient. High speed or other sophisticated mixing devices give only a marginal narrowing of the molecular weight distribution compared with standard stirring and need only be used in special circumstances such as in large-scale runs (see below).

Termination and Isolation

Completed polymerizations can be terminated at -78°C by introducing a protic material such as methanol or acetic acid in which case the product will have isobutyrate ester termini. Alternatively the mixture can be warmed to room temperature whereupon spontaneous termination occurs to form cyclic ketone ends (3). With THF solvent, isolation of polymer is accomplished by precipitation in water, hexane or methanol depending upon the sorts of contaminants or by-products one wishes to remove; or, the polymer solution can simply be stripped free of solvent.

Polymer Analysis and Characterization

Although the polymers from these anionic systems have been as carefully constructed as is practically possible, their analysis and characterization has been a challenging task.

Despite the control possible in these syntheses, the products are nevertheless still statistical mixtures of oligomers complicated by positional and stereo isomerism. We have of course employed the classical techniques such as NMR and GPC analysis which measure properties averaged over the total compositional range of the sample. We have also made good use of the newer analytical technique of High Performance Liquid Chromatography (HPLC). HPLC is an ideal method for characterizing the functionalized polymers that we have made, especially for separation of species which differ in the number of hydroxyl groups per molecule, and for separation of PMMA oligomers and stereoisomers.[13] Details of the method are discussed in the pertinent sections below.

INITIATORS

Nonfunctional Initiators

Numerous anionic initiators for methacrylates have been described in the literature including naphthalenesodium,[2] fluorenyllithium,[2] biphenylsodium,[14] and 1,1-diphenyhexyl-lithium (DPHLi,6).[4,15] Of these we find the latter to be most convenient and dependable. It is rapidly and quantitatively generated by adding n-butyllithium to 1,1-diphenylethylene in THF at room temperature.

$$n-C_4H_9Li + H_2C=CPh_2 \quad ------\!> \quad C_5H_{11}\overset{\displaystyle Ph}{\underset{\displaystyle Ph}{C^{\ominus}}}Li^{\oplus}$$

<u>6</u>

At room temperature this reaction is almost instantaneous and forms the blood-red adduct. The solution is then cooled to -78°C at which temperature the initiator remains soluble even at concentrations high enough to give oligomers with DP as low as two or three. It is important the the DPHLi be formed at room temperature because at -78°C its formation is so slow that one cannot be certain of just when it is complete. When monomer (e.g. MMA) is added to the initiator solution at -78°C the red color fades and is discharged to light yellow at an equivalence point corresponding to the addition of one mole of monomer per mole of DPHLi and to the formation of the enolate ion <u>7</u>. Further addition of monomer results in chain propagation.

$$C_5H_{11}\overset{\displaystyle Ph}{\underset{\displaystyle Ph}{C}}-CH_2\overset{\displaystyle CH_3}{\underset{\displaystyle CO_2Me}{C^{\ominus}}}Li^{\oplus}$$

<u>7</u>

Besides being easy to prepare and efficient in function, DPHLi affords a convenience in polymer characterization. The aromatic protons at the initiatory end of the polymer are well resolved in NMR spectra and their intensity can be related to the intensity of other resolved bands to give compositional information. For example, the aromatic resonance is necessarily assigned a value of ten protons and integration of its intensity vs that of the ester methyl group resonance in an MMA homopolymer gives a direct measure of degree of polymerization. At molecular weights of about 5000 or less this method is more reliable than gel permeation chromatography (GPC) for determining M_n, but the NMR method tells nothing about molecular weight distribution.

Functional and Masked Functional Initiators

Anionic polymerization has been used to its fullest advantage for preparation of polymers with functional groups at the chain ends. All previous work has been on butadiene and styrene polymers because the living ends are sturdier and there are fewer side reactions than with the methacrylates. Our early work on prolonging the life of the living ends and elimination of side reactions allowed us to bring the methacrylates under control. This synthetic control, coupled with HPLC analytical techniques, has enabled us to prepare and characterize a new series of functionalized methacrylate polymers.

The initiator preparation chemistry described above was readily modified to provide functionalized initiators. Ethyl 3-lithiopropyl acetaldehyde ethyl acetal (8) reacts with 1,1-diphenylethylene to provide the blocked hydroxyl-containing initiator 9. Lithium reagent 8 is easy to prepare in large scale. It was originally developed by Eaton for hydroxypropylation [16] and has been used by Schultz to prepare hydroxy-functional polybutadiene. [17]

$$
\begin{array}{cc}
\overset{\displaystyle H}{\underset{\displaystyle CH_3}{C_2H_5OCO-(CH_2)_3Li}} + H_2C=CPh_2 \xrightarrow{\hspace{1cm}} & \overset{\displaystyle H \qquad Ph}{\underset{\displaystyle CH_3 \quad Ph}{C_2H_5OCO(CH_2)_4C^{\ominus}Li^{\oplus}}}
\end{array}
$$

$$\underline{8} \qquad\qquad\qquad\qquad \underline{9}$$

Initiator 9 reacts with methyl methacrylate at -78°C in THF to produce a quantitative yield of acetal-ended polymer 10 with n from 1 to 500 and $M_w/M_n < 1.2$. DP's determined by NMR agree well with GPC molecular weights and hydroxyl equivalents. The presence of the blocked hydroxyl group in the polymer was verified by the presence of the phenyl protons in the correct proportion, and by the weak signals from the methine proton of the acetal group. This methine signal is too weak, however, to accurately estimate its amount.

$$9 \ + \ MMA \ -----> \ C_2H_5OCO(CH_2)_4\overset{\overset{\displaystyle H}{|}}{\underset{\underset{\displaystyle CH_3}{|}}{C}}\overset{\overset{\displaystyle Ph}{|}}{\underset{\underset{\displaystyle Ph}{|}}{C}}(MMA)_nH$$

10

That the acetal group is present in oligomers such as 10 and can be removed to produce the hydroxy derivative 11 was demonstrated in the simple case where n=1. The initiator 9 was titrated with one equivalent of methyl methacrylate and the protecting group removed with dilute aqueous HCl in THF. Preparative HPLC separated the product into pure samples of alcohol 11 with n=1 (71%) and n=2 (25%). This experiment demonstrates that the basic chemistry works for low molecular weight materials.

$$10 \ ----------> \ HO(CH_2)_4C(Ph)_2(MMA)_nH$$

11

Proof that the chemistry also works for higher molecular weight materials rests on HPLC evidence. When the acetal-ended PMMA 10 (n=10) is eluted through a silica gel column with ethyl acetate, a single peak is produced at the solvent front (Figure 1). Ethyl acetate is more polar than the polymer molecules and they are not retained. The polar hydroxyl group of 11, (n=10) however, interacts strongly with the silica gel and the polymer is retained (Figure 1). The peak is much broader than a single component would be, but is as narrow as it is because the sample is nearly monodisperse. Broader distribution samples give broader peaks. The separation is not limited to the low molecular weight material shown. If one uses a column with pores large enough to eliminate size-exclusion effects (10nm), hydroxy PMMA with M_n to 10,000 (n=100) can be separated from its acetal precursor with almost baseline resolution.

Hydroxy-terminal PMMA 11 can be prepared in a different way. Allyllithium adds to 1,1-diphenylethylene to give initiator 12. Again, 12 readily polymerizes MMA to give 13. 9-Borabicyclo-[3,3,1]nonane (9-BBN) reacts selectively with the carbon-carbon double bond of 13 to produce alcohol 11 after oxidation with basic hydrogen peroxide. HPLC analysis as above shows that hydroxy PMMA 11 is produced cleanly. This method is the easiest way to prepare hydroxy PMMA 11 because allyllithium is prepared from commercially available allyl phenyl ether.[38]

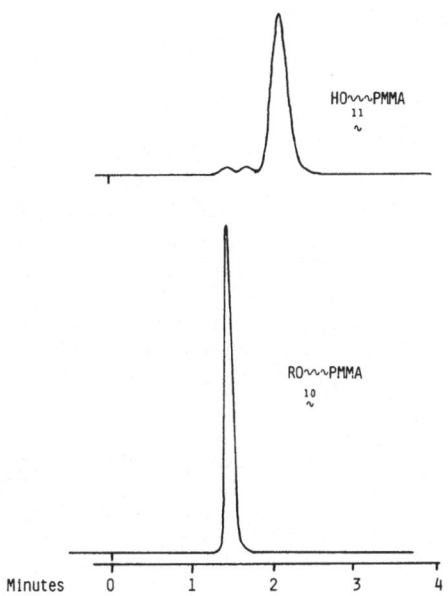

Figure 1: HPLC of functional PMMA 10 and 11 on a silica gel column with ethyl acetate elvent (Ref. 13).

$$H_2C=CHCH_2CH_2C(Ph)_2Li \quad --------> \quad H_2C=CHCH_2CH_2C(Ph)_2(MMA)_nH$$

$$\underline{12} \hspace{7cm} \underline{13}$$

This simple and rapid HPLC analysis provides direct proof of the structure of the polymers. The classical approach which compares measured M_n to measured hydroxyl equivalent shows only that on average there is one hydroxyl group per molecule. Chromatography has been used to do similar analyses of hydrocarbon polymers such as polystyrene and polybutadiene.[18] Our work shows that this technique is general and can be applied to more polar polymers.

REACTIONS OF THE LIVING END

Capping Reactions

Although the methacrylate polymer enolate ion is centered on a hindered tertiary carbon, it is reactive enough to participate in nucleophilic substitution reactions with suitably reactive electrophiles. Such reactions must necessarily be carried out at or near -78°C to ensure that the anion is not inactivated by the thermal cyclization reaction. Typical of such reactions is that with vinylbenzyl[19] iodide which gives the styryl-capped oligomer 14.

$$DPH(MMA)_n^\ominus + ICH_2C_6H_4CH=CH_2 \quad -----> \quad DPH(MMA)_nCH_2C_6H_4CH=CH_2 + I^\ominus$$

$$\underline{14}$$

Ultraviolet spectral and NMR analysis showed that the reaction proceeds to about 50% conversion in ten or fifteen minutes and reaches a maximum conversion of 92% in several hours using a two fold excess of the halide. The bromo analog is also effective, but the chloride fails to react. Similarly capped products are obtained from allyl bromide ($\underline{15}$), -bromo esters($\underline{16}$) and bromomethyl ketones ($\underline{17}$), conversions again being in excess

$$DPH(MMA)_nCH_2CH=CH_2 \quad DPH(MMA)_nCH_2CO_2tBu \quad DPH(MMA)_nCH_2-COAr$$

$$\underline{15} \hspace{4cm} \underline{16} \hspace{4cm} \underline{17}$$

of 90%. Typically these reactions were carried out on MMA oligomers with DPs of 10 to 30 to facilitate characterization but similar reactions proceed with higher molecular weight materials.

Coupling Reactions

In a manner analogous to the capping reaction, polyfunctional electrophiles react with the living PMMA anion to multiply the polymer molecular weight. Among the simplest and most effective of such reagents are the poly(-bromomethyl)benzenes.

In our early work on the coupling reaction we used equimolar amounts of pyridine and sec-butyllithium as the initiator[11] in THF solvent at -78°C. When polymerization was complete an aliquot was removed to isolate a sample of uncoupled control polymer, and to the remainder was added a stoichiometric amount of 1,4-bis-(α-bromomethyl)benzene to effect the transformation.

$$2In(MMA)_n^{\ominus} + p\text{-}BrCH_2C_6H_4CH_2Br \quad \text{--------}>$$

$$In(MMA)_nCH_2C_6H_4CH_2(MMA)_nIn \quad + 2\ Br^{\ominus}$$

In = initiator fragment

The molecular weight distribution of the control and the product were almost identical (Mw/Mn ~ 1 by GPC), but the Mn of the product was almost twice that of the control (11,000 vs 5,800). The coupling efficiency was thus 95% as calculated from the relationship:

% Coupling=200(1-Mc/Mp)

where Mc=control Mn and Mp=product Mn

There was some concern that the coupling might have occured by bis-quaternization of the putative dihydropyridine initiatory end groups. This was ruled out by quenching the living anion prior to adding the coupling agent. Control and product were identical. Similar results were obtained with DPHLi-initiated polymer and in this case it was shown that no coupling occurs if the polymer anion is allowed to die by warming to 0°C before adding the coupler.

Coupling efficiency with the meta,bis(bromomethyl) benzene was essentially the same as with the para, but considerably less with the ortho presumably because of steric hindrance. In an isolated experiment with 1,4-diiodobutane coupling efficiency was only about 67%.

Limitations of the GPC analytical method necessarily limit the accuracy with which coupling efficiency and reaction rate can be determined. This limitation is largely overcome by using HPLC on polar functional polymers.

The functional polymers discussed above can also be coupled

to produce linear and star polymers with hydroxyl groups at the chain ends. The HPLC technique described above allows structure proof of these polymers. Figure 2 shows the chromatographs of products from coupling with 1,4-bis(bromomethyl)benzene, 1,3,5-tri(bromomethyl)benzene, and 1,2,4,5-tetrakis(bromomethyl)-benzene. As the number of hydroxyl groups per molecule increases, the retention times increase so that polymers with 1,2,3, and 4 hydroxyl groups can be separated regardless of molecular weight. Proof of this was found by GPC analysis of peak samples isolated by preparative HPLC. The number-average molecular weights of the peaks are shown above them in Figure 2.

The HPLC analysis can also be used to monitor the coupling reactions to indicate the best reaction conditions. Figure 3 shows the coupling yield as a function of reaction time for coupling with 1,4-bis(bromomethyl)benzene. This shows that for this reaction, a second order reaction with equimolar reactants, complete conversion requires a very long reaction time. During this time the living ends are dying so that the maximum coupling yield attainable is about 90-95%.

More information can be obtained by further examination of the monofunctional material produced. It is a mixture of material capped with a 4-(bromomethyl)benzyl group and of dead polymer. The bromomethyl group can be hydrolyzed to the alcohol with aqueous silver nitrate in THF to convert it into dihydroxy material. HPLC analysis shows that the monofunctional material produced in the coupling reaction is about 1/3 capped with (bromomethyl)benzyl groups, the remainder being uncapped dead material.

ALKYL METHACRYLATE POLYMERS AND COPOLYMERS

Block Copolymers

Early work on alkyl methacrylate anionic copolymerizations by Graham and co-workers showed that the polystyrene living anion could initiate MMA monomer to incorporate it and form polystyrene-b-PMMA block copolymer, but that styrene is inert to the PMMA anion. Later they demonstrated the incorporation of various other alkyl methacrylates into block copolymers with PMMA.[20,21]

Subsequent studies by Ailhaud, Gallot and Skoulios dealt with block copolymers of MMA with hexyl methacrylate (HMA), lauryl methacrylate (LMA) and octadecyl methacrylate initiated with diphenylmethylsodium at -70°C in THF. By solvent-nonsolvent fractionation of polymer pairs such as PMMA-b-PHMA and PHMA-b-PMMA they were able to distinguish differences in composition depending on which block was formed first. Thus when

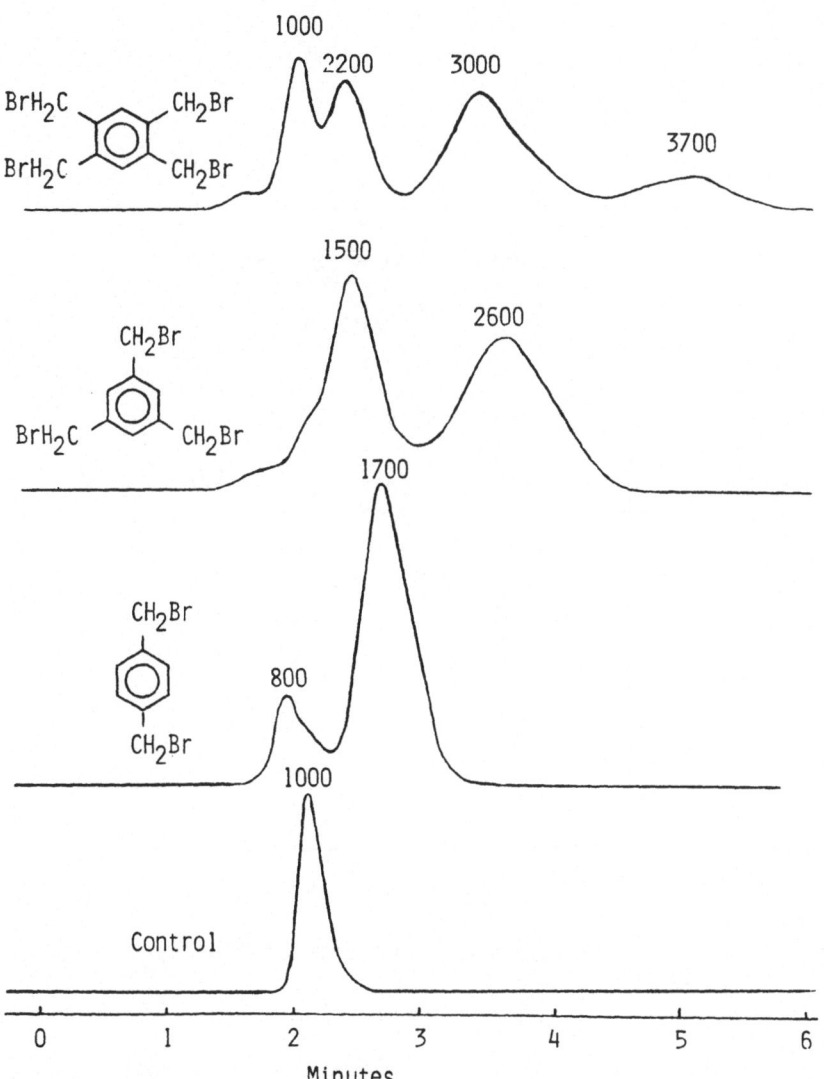

Figure 2: HPLC separations of hydrolized products from coupling of living, functional PMMA with the indicated bromomethyl benzenes. The numbers above the peaks are the GPC - determined number average molecular weights of preparatively isolated peaks.

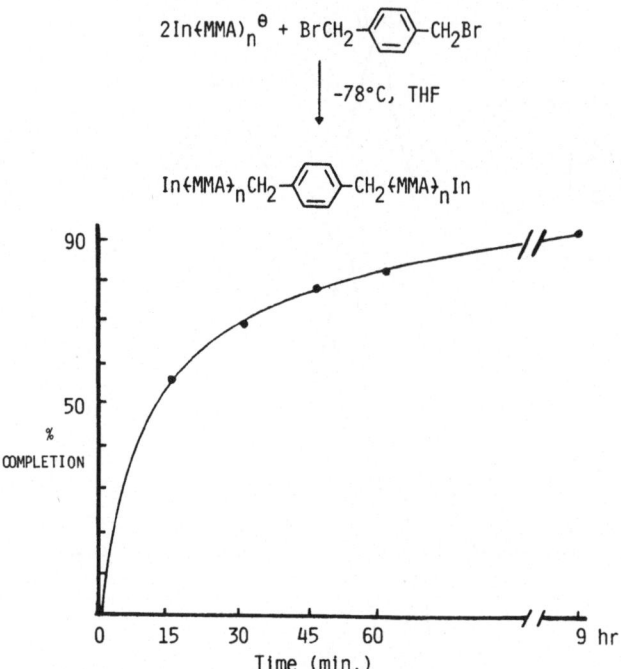

Figure 3: Coupling yield as a function of time as determined
           by HPLC Analysis.

the PHMA block was formed first and MMA was added to it, all fractions of the resultant PHMA-b-PMMA were virtually indistinguishable. However in the obverse case of PMMA-b-PHMA the molecular weight distribution was significantly broader and the more soluble fractions contained systematically decreasing amounts of PHMA.[22] Similar fractionation studies were done by Seow and co-workers on block terpolymers of ethyl MA with HMA and LMA.[23]

Later work on copolymers of methyl α-phenylacrylate, and benzyl and trityl-type methacrylates by Hatada and collaborators was primarily concerned with NMR and optical activity studies of the products.[24]

Our efforts on non-functional alkyl copolymers have been directed toward optimizing our system for preparative scale syntheses, and to prepare copolymers of MMA with Tg's adjusted by variation of the kind and amount of alkyl MA comonomers.

For example, for certain applications we prepared random copolymers of MMA with n-butyl MA (typically MMA/nBMA, 75/25 mole %) with molecular weights in the 30,000 range and with molecular weight distributions Mw/Mn=1.2-1.4. However to achieve such narrow distribution on a 300 gram polymer scale it was necessary to resort to a cylindrical reactor with a high surface/volume ratio and very efficient mixing such as provided by a high-frequency vibrating device;[25] these were needed to achieve adequate mixing and to maintain the required low temperature.

In another investigation relating copolymer composition to Tg and other lower order thermal transitions we prepared a series of MMA-laurylMA (LMA) block copolymers of well-defined sequence.[26] A portion isolated from a living initial block showed it to have a nominal composition corresponding to DPH(MMA)$_{200}$ with Mw/Mn=1.32 and a Tg of 109°C. LMA monomer was then added to the living initial block to form an intermediate diblock copolymer DPH(MMA)$_{200}$-b-(LMA)$_{67}$ having Mw/Mn=1.31 with Tg=105°C and a presumed β-transition at -5°C. Subsequent block building gave a further intermediate diblock DPH(MMA)$_{200}$-b-(LMA)$_{134}$, Mw/Mn=1.72, Tg 104°C, β-transition 0°C and a third, γ-transition at -44°C. The final composition corresponding to the triblock polymer DPH(MMA)$_{200}$-b-(LMA)$_{134}$-b-(MMA)$_{200}$ had a greatly broadened molecular weight distribution (Mw/Mn=3.56) and showed the same three thermal transitions as the previous diblock. This extensive distribution broadening is noteworthy and relates, among other things, to the low solubility of LMA in THF at -78°C., to the slow death of living ends during the protracted reaction period and probably to a kinetic factor involving poor accessibility of living ends. In any event the general controllability of polymerizations of this sort provide a superior way of tailoring polymer compositions for correlation of physical properties with composition or structure.

Poly tert-Butyl Methacrylate

The high reactivity of acrylates in general prevents them from polymerizing by the anionic mechanism. However, Kitano and co-workers have reported the anionic polymerization of t-butyl acrylate at -78°C, and this success is ascribed to the bulky t-butyl groups which inhibit destructive side reactions but allow the Michael addition by which initiation and propagation proceed.[27] We thus reasoned that anionic polymerization of t-butyl methacrylate might proceed at room temperature since the t-butyl groups could inhibit the cyclization termination reaction. The polymerization did indeed occur in THF with DPHLi initiator. The initial temperature was 26°C which rose to 37°C during monomer addition and remained there until addition was complete.[28] The Pt-BMA was obtained in quantitative yield by precipitation in water[29] and it had a molecular weight distribution considerably broader than that for PMMA from an analogous run at -78°C. (Mw/Mn 2.3 for Pt-BMA vs. 1.2 for PMMA). In several runs designed to give a product with Mn 65,000, the GPC values found were about 20,000 whereas that for a comparable PMMA sample were only 15% lower than theory. The large discrepancy for Pt-BMA probably relates to the difference in its effective hydrodynamic volume compared to that of the materials used in GPC column standardization. We had expected the Tg of Pt-BMA to be rather low because of the hindrance to chain ordering by the bulky t-butyl groups, but, surprisingly, its Tg was 102°C, while that for the analogous PMMA was 118°C.

FUNCTIONAL POLYMERS

Enyl and Dienyl Methacrylates

Anionic polymerization can be used to prepare methacrylate polymers from monomers other than alkyl methacrylates. Some of these monomers are also copolymerizable by free radical techniques, however, anionic polymerization provides the advantage of controlled molecular weight and distribution in these polymers. Monomers which cannot be readily polymerized by free radical initiators are of special interest. Methacrylates which contain monoolefins and dienes are readily polymerized by anionic polymerization. The resulting copolymers have useful reactive groups attached for crosslinking, grafting, etc., in addition to controlled molecular weights and distributions.

Our interest in allyl methacrylate copolymers stemmed from a need for low molecular weight copolymers with pendant air-drying

groups.   Two literature reports documented the air-induced
crosslinking of poly (allyl methacrylate).[30]   We prepared some
allyl   methacrylate   and   related   mono-olefinic   methacrylate
copolymers as shown in Table I.   The standard methods discussed
earlier provided the polymers without difficulty and NMR analysis
confirmed the indicated compositions.

TABLE I

MONO-OLEFINIC METHACRYLATE COPOLYMERS
(10% olefinic methacrylate, 45% n-butyl methacrylate, 45% methyl
methacrylate prepared as in Ref. 31 in THF at -78°C with DPHLi
initiator at ca. 20 wt % solids.)

Methacrylate	Mn (calc)	Mn (obs)	Mw/Mn	Tg (°C)
Allyl	10,000	7,300	1.25	55
Methallyl	10,100	8,900	1.11	55
Crotyl	10,100	11,000	2.01	50
3-methyl-2-butenyl	10,200	8,800	1.20	51

These mono-olefinic methacrylate copolymers do not cure fast
enough to make good air-drying finishes.   The allyl methacrylate
copolymers can be heated for several days at 80°C with cobalt
naphthenate   accelerator   without   change   in   solubility,   NMR
spectrum, or molecular weight.   An earlier report showed that
peroxide initiators might solve this problem,[30b] but we chose to
proceed by modifying the olefin to improve the crosslinking rate.
It is well known from drying oil chemistry that fatty esters with
high proportions of mono-unsaturated acids do not dry and that
fatty esters with 1,3- and 1,4-dienes do.   This suggested that we
prepare and polymerize dienyl methacrylates.

Methacrylate esters with 1,3- and 1,4-diene groups were not
known compounds when we began this work.   Special techniques must
be used in their synthesis because both the starting alcohols and
product esters are sensitive to acid and base-catalyzed degradation. Titanate esters, such as tetraispropyl titanate, are useful
catalysts in such reactions and provide dienyl methacrylates in
good yield.   Table II shows the variety of dienyl methacrylates
that we have made in this way.[31]

The dienyl methacrylates in Table II copolymerize readily by
our standard anionic techniques.   The polymers listed in Table
III were prepared with diphenylhexylithium initiator in THF at
-78°C.   NMR and GPC analyses provided the characterization
included in Table III.   We have not seen any evidence that the
diene groups either copolymerize with the methacrylates or
interfere with the polymerizations by providing acidic hydrogens.
These observations are to be expected since the acidity
difference between the ester enolate living end and any potential
side products based on hydrocarbon anions is very large.

Consistent with our prediction, the dienyl methacrylate copolymers crosslinked rapidly on exposure to air[31] to form coatings with excellent appearance and durability. The 2,5-cyclohexadienylmethyl methacrylate copolymer with butyl and methyl methacrylate is an interesting exception. The only reaction of the cyclohexadiene groups which we could see was their oxidative conversion to benzene rings.

Table II

DIENYL METHACRYLATES Prepared By
Transesterification of the Corresponding Alcohol

METHACRYLATE	YIELD
2,4-Hexadienyl	71
2,4-Octadienyl	78
2,4-Decadienyl	65
4,7-Octadienyl	87
1,3-Cyclohexadienylmethyl	72
2,5-Cyclohexadienylmethyl	87
9,12-Octadecadienyl	100

TABLE III

DIENYL METHACRYLATE POLYMERS AND
COPOLYMERS PREPARED AS DESCRIBED IN TEXT

METHACRYLATES	MONOMER MOL % FEED	POLYMER	DP CALC	OBS	Mn 1000	Mw/Mn	Tg(°C)
2,4-HEXADIENYL	100	100	10	12	1.5	1.8	--
2,4-HEXADIENYL	15	15					
BUTYL	57	53	100	94	13	1.1	23
METHYL	28	32					
2,4-OCTADIENYL	15	15					
BUTYL	53	54	110	136	16	1.6	39
METHYL	32	30					
2,4-DECADIENYL	15	16					
BUTYL	34	38	110	109	15	1.4	42
METHYL	51	47					
4,7-OCTADIENYL	15	17					
BUTYL	53	54	110	106	13	1.4	31
METHYL	32	29					
1,3-CYCLOHEXA-DIENYLMETHYL	15	16					
BUTYL	53	55	110	121	13	1.5	47
METHYL	32	29					
2,5-CYCLOHEXA-DIENYLMETHYL	15	13					
BUTYL	53	55	110	149	19	1.7	54
METHYL	32	31					

This chemistry provides an attractive route to a family of highly reactive methacrylate polymers with many possible commercial applications.

Glycidyl Methacrylate (GMA) Polymers

Discussion

The chemical reactivity of the epoxide function in glycidyl methacrylate (GMA) (18) commends it as a comonomer for the introduction of functionality into acrylic copolymers.

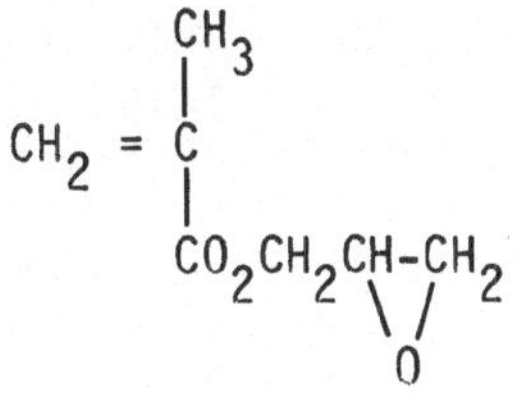

$$CH_2 = \underset{\underset{CO_2CH_2CH-CH_2}{|}}{\overset{\overset{CH_3}{|}}{C}}$$

$$\underset{O}{\diagdown\diagup}$$

18

We have encountered only two reports of the anionic polymerization of glycidyl methacrylate, both with n-butyllithium as initiator. The first, by Arbusova and co-workers, gives only a brief statement that the yield of the polymer was only 11% with 80% of the epoxide groups intact, and conclude that the mechanism was by vinyl polymerization rather than by epoxide ring opening.[33] No experimental details or characterization data were given. The second report was by the Tokyo group of Iwakura et al who specified n-butyllithium initiator in THF solvent at -78°C., and presented tacticity data,[34] but in our hands this system failed to produce polymer. It is pertinent that a subsequent more detailed paper by the Tokyo group gives extensive comparative data on PGMA's obtained with various combinations of organometallic initiators and solvents, but specifically exclude mention of the n-BuLi/THF product in their comparisons.[35]

In our work on PGMA we again expoited DPHLi as the initiator also in THF at -78°C. In an early experiment we first added GMA monomer to the initiator followed by a mixture of GMA and MMA, anticipating the formation first of a short PGMA block to which a random P(MMA/GMA) segment would be attached. Propagation proceeded to give a quantitative yield of polymer with all epoxide groups intact as inferred from the NMR spectrum. However, the Mn was almost ten-fold larger than anticipated from the monomer/initiator ratio. Calculation showed that about 85% of the original initiator had been destroyed presumably by reaction with the epoxy groups. Moreover, the molecular weight distribution was much broader than expected for an orderly living polymerization (Mn/Mw 1.9 vs 1.1).

However, it is clear that despite the presence of the reactive epoxide group, GMA is not inherently insusceptible to vinyl anionic propagation. And since propagation proceeds through the methacrylate enolate ion we reasoned that the substituted isobutyrate ion generated by addition of DPHLi to MMA (7) which in effect is an initiator itself, might be too weak a nucleophile to react with the epoxy group and thus avoid destruction. This turned out to be the case.

To generate the initiator, three MMA units were added to DPHLi to ensure that statistically all fragments carried at least one MMA unit, therefore all were of the enolate ion type. Thus the "initiator" concentration is predetermined by the original concentration of DPHLi. Alternatively lithium diisopropylamide (generated from n-BuLi and diisopropylamine) can be used instead of DPHLi to form the "initiator" ion (19).

$$(\underline{i}\text{-Pr})_2\text{N-CH}_2\text{-}\overset{\overset{\displaystyle\text{CH}_3}{|}}{\underset{\underset{\displaystyle\text{CO}_2\text{Me}}{|}}{\text{C}}}{}^{\ominus}\text{Li}^{\oplus}$$

$$\underline{19}$$

Using such systems we obtained random and block copolymers of GMA with narrow molecular weight distributions (Mw/Mn 1.1-1.3) and Mn's closely predicted by the M/I ratio. Conversions to polymer were virtually quantitative.

One should note that to form polymers of low DP, where large amounts of initiator are needed and to ensure narrow molecular weight distribution, fluorenyllithium is not satisfactory because of its low solubility in THF at -78°C.

"Homo" PGMA

Representative of what is tantamount to homoPGMA is the product obtained by forming initiator from three equivalents of MMA per DPHLi followed by fifteen equivalents of GMA. The predicted composition corresponds to formula 20 for which the theoretical Mn is 2627 and calculated epoxide oxygen is 9.1%.

$$\text{DPH(MMA)}_3\text{-b-(GMA)}_{15}$$

$$\underline{20}$$

The Mn as determined by NMR integration was 2500, and epoxide oxygen was 9.1% by potentiometric titration. The molecular weight distribution Mw/Mn was 1.14 as shown by GPC.[36,37]

Vinyl Methacrylate (VMA) Copolymers with GMA

Introduction of other functionality is possible by copolymerization with, for example, vinyl methacrylate. Addition of a mixture of nBMA, GMA, MMA, and VMA monomers to the initiator fragment was calculated to give a random copolymer of composition 21 with Mn 5711.

random

$$DPH(MMA)_3-b-(BMA_{12}/GMA_{15}/MMA_{22}/VMA_5)$$

21

The experimentally determined Mn was 5500 (GPC) but the distribution was greatly broadened (Mw/Mn 2.21) probably because of a significant difference in the kinetics of polymerization of VMA vs the other comonomers. The temperature characteristics of this polymerization were unusual; when addition of the comonomer mixture was complete the temperature remained at -70°C for several minutes then abruptly rose to -62°C before subsiding to the usual completion temperature of -74°C. Other simpler systems simply subsided rather quickly from -70°C to -74°C on completion. One may speculate that these polymers have two kinds of living ends, "normal" ends associated with simple methacrylates and those with vinyl methacrylate enolate ion ends which coordinate the lithium counterion differently and thus propagate differently.

However if the VMA was added last as a separate block to the previously constituted random block, the isomeric product 22 now had the customary narrow molecular weight distribution, (Mw/Mn=1.19).

random

$$DPH(MMA)_3b-(BMA_{12}/GMA_5/MMA\ _{22})-b-(VMA)_5$$

22

CONCLUSION

The anionic polymerization of methacrylate esters is a versatile method for preparing polymers with well-defined compositions and a large variety of chemically tractable functional groups. The method provides the ability to prepare oligomers and polymers over a wide range of molecular weights. Thus, it provides materials for study of functional group reactivity as related to polymer structure and size, and allows optimization of molecular weight and functionality for whatever may be the practical applications of choice.

The opportunities for further work are evident from the sampling of materials which we have described here.

ACKNOWLEDGMENTS

We wish to acknowledge contributions by our colleagues B. C. Anderson, P. Arthur, Jr., H. W. Jacobson, C. H. Park, A. J. Playtis and W. H. Sharkey, and the co-workers of our physical characterization and analytical divisions, especially A. Vatvars and M. C. Han. Finally, we thank Mr. J. R. Butera for help in preparing the manuscript.

REFERENCES

1. D. M. Wiles in "Structure And Mechanism In Vinyl Polymerizations", Eds., T. Tsuruta and K. F. O'Driscoll, Marcel Dekker, Inc., New York, 1960, p. 223.

2. W. E. Goode, F. H. Owens, and W. L. Myers, J. Polym. Sci., 47, 75 (1960).

3. D. Braun, M. Herner, U. Johnson, and W. Kern, Makromol. Chemie, 51, 15 (1962).

4. W. Fowells, C. Schuerch, F. A. Bovey, and F. P. Hood, J. Amer. Chem. Soc., 89, 1396 (1967).

5. A. H. E. Müller, H. Höcker, and G. V. Schulz, Macromolecules, 10, 1086 (1977).

6. "Principles of Polymer Chemistry", P. J. Flory, Cornell University Press, 1953, p. 337.

7. (a) B. C. Anderson, G. D. Andrews, P. Arthur,Jr., H. W. Jacobson, L. R. Melby, A. J. Playtis, and W. H. Sharkey, Macromolecules, 14, 1599 (1981). (b) G. D. Andrews, U. S. Patent 4,351,924, Sept. 28, 1982.

8. G. D. Andrews, IUPAC Meeting, Amherst, 1982.

9. B. C. Anderson, Polymer Synthesis - The 1980's, Polytechnic Institute of New York symposium, March 11, 1983.

10. R. A. Galluccio and D. L. Glusker, Macromolecular Syntheses, 7, (1979).

11. We further showed that in toluene/pyridine with n-BuLi, the initiator is in fact not the alkyllithium, but a pyridine alkyl- lithium adduct; L. R. Melby, P. Arthur, Jr. , and W. H. Sharkey, work to be published.

12. C. E. Schildknecht, "Vinyl and Related Polymers", John Wiley and Sons, New York, 1952, p. 245.

13. G. D. Andrews and A. Vatvars, Macromolecules, 14, 1603 (1981).

14. A. Roig, J. E. Figueruelo, and E. Llana, J. Polm. Sci., B3, 171 (1965).

15. (a) D. M. Wiles and S. Bywater J. Polym. Sci., B2, 1175 (1964). (b) idem, Trans. Faraday Soc., 61, 150 (1965).

16. P. E. Eaton, G. F. Cooper, R. C. Johnson, and R. H. Mueller, J. Org. Chem., 37, 1947 (1972).

17. D. N. Schulz, A. F. Halasa, and A. E. Oberster, J. Pol. Sci., Pol. Chem. Ed., 12, 153 (1974).

18. (a) H. Inagaki, Adv. Polym. Sci., 24, 207 (1977). (b) T. Min and H. Inagaki, Polymer, 21, 309 (1980). (c) P. Mansson, J. Pol., Sci., Polym. Chem. Ed., 18, 1945 (1980).

19. L. R. Melby and H. W. Jacobson, work to be published.

20. R. K. Graham, D. L. Dunkelberger, and E. S. Cohn, J. Polym. Sci., 42, 501 (1960).

21. R. K. Graham, J. R. Panchak, and M. J. Kampf, ibid., 44, 411 (1960).

22. H. Ailhaud, Y. Gallot, and A. Skoulios, Makromol. Chemie, 140 179 (1970).

23. P. K. Seow, J.-P. Lingelser, and Y. Gallot, ibid., 178, 107 (1977).

24. H. Yuki, K. Ohta, K. Hatada, and H. Ishikawa, Polymer Journal, 11, 323 (1979).

25. An elongated, creased resin kettle and Chemapec Vibro Mixer were used.

26. The subscripts which follow refer to the approximate DP corresponding to the amount of monomer actually added. GPC peak molecular weight measurements were in good agreement.

27. T. Kitano, T. Fujimoto, and M. Nagasawa, Polymer Journal, 9, No. 2, 153 (1977).

28. The addition rate was the same as for an analogous PMMA run in which the temperature was maintained at -70°C.

29. The polymer is soluble in hexane.

30. (a) D. M. Wiles and S. Brownstein, Polymer Letters, 3, 951 (9165). (b) G. F. D'Alelio and T. R. Hoffend, J. Pol, Sci., pt. A1, 5, 323 (1967).

31. G. D. Andrews, U. S. Patent 4,293,674, Oct. 6, 1981.

32. "Block Copolymers", A. Noshay and J. E. McGrath, Academic press, New York, 1977, p. 31.

33. I. A. Arbusova, V. N. Yefremova, A. G. Eliseyeva, and M. F. Zinder, Vyskomel. soyed., 5, No. 12, 1819 (1963).

34. Y Iwakura, F. Toda, T. Ito, and K. Aoshima, J. Polym. Sci., B5, 29, (1967).

35. T. Ito, K. Aoshima, F. Toda, K. Uno, and Y. Iwakura, Polymer Journal, 1, No. 3, 278 (1970).

36. In this example the GPC value for Mn was 1600. At these low DPs GPC values are consistently lower than those from NMR analysis. The discrepency relates to vagaries in GPC calibration and standards for low molecular weight materials.

37. The NMR spectrum shows a characteristic epoxide proton multiplet at 2.5-3.3 ppm., and glycidyl ester proton resonances at 3.55 ppm, 4.25 ppm, and 4.48 ppm.

38. J. J. Eisch and A. M. Jacobs, J. Org. Chem., 28, 2145 (1963).

CONDUCTIVE POLYMERS: AN OPPORTUNITY FOR NEW MONOMERS AND POLYMERS

Harry W. Gibson

Webster Research Center
Xerox Corporation
800 Phillips Road
Webster, N.Y. 14580

INTRODUCTION

Electrically conductive polymers have been the subject of sporadic interest to chemists and physicists over the past twenty-five years.[1,2] This interest arises from the intriguing prospect of being able to combine in a single material the electrical properties of a metal (high conductivity, greater than 1 s/cm) or semiconductor (moderate conductivity, 10^{-7} to 1 s/cm) with those of a polymer (mechanical strength, flexibility, lighter weight, low preparation and fabrication costs, etc.). Part of the reason for the sporadic nature of the interest and progress in this field until recently was the fact that the interdisciplinary component necessary in such endeavours was not in place. A second cause was the physical nature of the polymers.

Thus, even though polyacetylene had been known since 1958[3] and had been oxidized with Lewis acids (doped) to relatively high conductivity,[4] since there was little interdisciplinary communication apparently and since polyacetylene was an intractable powder, it had aroused little interest. The field was relatively dormant until the 1970's when two significant events occurred. First Ito, Shirakawa and Ikeda reported the synthesis of continuous free-standing polyacetylene films in situ using a very concentrated catalyst.[5] Second this discovery came to the attention of an interdisciplinary team of chemists and physicists led by A.G. MacDiarmid and A.J. Heeger at the University of Pennsylvania. The combination of a useful film form and interdisciplinary attack led quickly to demonstration of metallic conductivity in cases of oxidation or reduction (p- or n-doping) by suitable

Lewis acids or bases.[6] These results in turn piqued the interest of laboratories around the world.

DISCUSSION

Our Work – Cyclopolymerization

As with any prospective new application we reasoned that optimization of physical and chemical properties would be required in order to generate practically useful electrically conductive polymers. We were concerned about mechanical properties, flexibility, conductivity levels, solubility, processability, oxidative stability, etc. Based upon the perceived requirement of a conjugated polyene structure, substituted polyacetylenes were the obvious way to introduce substituents for the purpose of tailoring these characteristics. Unfortunately the literature provided ample evidence of the sluggish nature of substituted polyacetylenes toward polymerization.[7]

Poly (1,6-heptadiyne)

However, we were we were aware of the utility of cyclopolymerization in the area of polyolefins[8] and also were aware of the work of Stille and Frey on the cyclopolymerization of 1,6-heptadiyne to a soluble polymer of proposed structure 1 using a heterogeneous Ziegler-Natta catalyst.[9] With the initial objective of determining if 1) substituted polyacetylenes could be polymerized to free-standing films and 2) then converted to a highly conducting state, we began our investigation of this cyclopolymerization in early 1979.

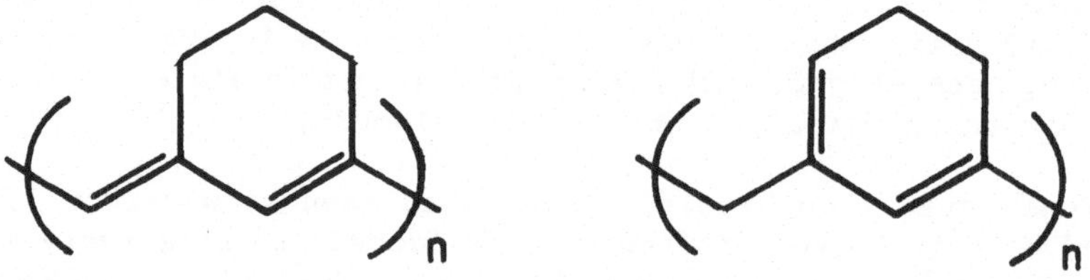

$\underline{1}$ $\underline{2}$

Free-standing films of poly(1,6-heptadiyne) were indeed prepared using a number of homogeneous Ziegler-Natta catalysts. Inasmuch as the details are reported elsewhere,[10] only a synopsis of the results will be presented here. Spectroscopic studies (infrared, nuclear magnetic resonance) using model monomers and the polymer film prepared from 1,6- heptadiyne-1,7-d_2 supported structure 1 for the polymer, which is believed to be a mixture of the syn or helical form and the anti or linear, planar form; an exotherm at 107 °C in differential scanning calorimetry is believed to result from conversion of the former to the latter. All experiments were done on as-prepared films stored at 25 °C under argon. As reported by Stille and Frey for the soluble polymer, the insoluble "poly-1,6" also underwent a thermal rearrangement (exothermic peak at 275 °C in differential scanning calorimetry at 20 °C per minute) which converted it from a lustrous golden-green opaque film to a transparent red-orange film upon heating at 203 °C for 18 hours. By spectroscopic studies using model compounds, unconjugated structure 2 was indicated for the rearranged polymer.

Poly-1,6 is distinct from polyacetylene in its morphology. The latter is very porous, being comprised of fibrils of 100-800Å diameter,[11] and has an apparent density[6] of about 0.4 g/cm^3. Poly-1,6 has very little void space and displays a number of morphologies depending upon polymerization conditions. While polyacetylene prepared by the Shirakawa catalyst is 80-90% crystalline,[12] poly-1,6 displays no crystallinity to x-ray diffraction, although differential scanning calorimetry of some samples shows a sharp transition (endotherm) in the range 164-184 °C, perhaps indicative of crystalline domains.

With these contrasts in molecular structure, morphology and crystallinity between poly-1,6 and polyacetylene, we undertook a study of the oxidation (doping) of poly-1,6 to a conductive state. Indeed exposure to iodine vapor in vacuo at room temperature successfully raised the conductivity from an initial value of 10^{-12} s/cm to 10^{-2} to 10^{-1} s/cm over the period of one hour, a rate similar to that of polyacetylene. However, the conductivity-time curve proceeded through this maximum value and then decreased by about a factor of 10 before plateauing (Figure 1). No dependence of conductivity upon film thickness was observed. If the iodine were removed at the maximum, the conductivity fell by only a factor of about four. This was hypothesized to be due to iodine catalyzed rearrangement of 1 to 2, resulting in loss of conjugation. To test this hypothesis, the "doping" was carried out at -78 °C in toluene saturated with iodine; a value of 0.4 s/cm was reached after ten days with no sign of loss of conductivity, until the temperature was raised (Figure 2). This corresponds to a room temperature value of 5 to 10 s/cm based upon the activation energy of 3 kcal/mole (0.1 eV); this value may be compared to the 10-20 s/cm we observed for iodine treatment of polyacetylene. AsF$_5$ treatment of poly-1,6 led to a conductivity of 10^{-2} s/cm at 25 °C, in comparison to a reported value of 10^3 s/cm for polyacetylene.[6]

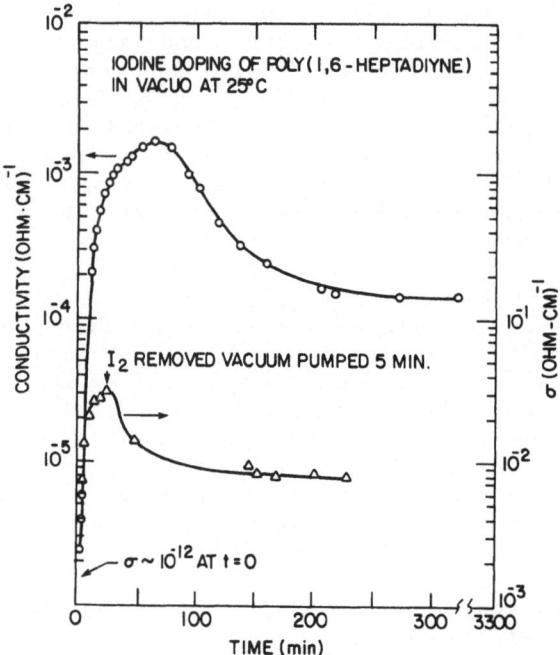

Fig. 1   Time dependence of conductivity of poly(1,6-heptadiyne) upon
         exposure to I_2. Note factor of ten drop after maximum in top
         curve, while only factor of four drop when I_2 removed (bottom
         curve).

         Iodine content on an I atom per double bond basis is the same
(0.50) for poly-1,6 and polyacetylene prior to vacuum treatment.
Vacuum treatment (26 hrs., 25°C, 10^{-6} Torr) leaves only 0.11 I
atom per double bond, compared to 0.30 for polyacetylene. Similarly poly-1,6 after vacuum treatment retains 0.11 mole AsF_5 per
double bond as compared to 0.28 for polyacetylene.

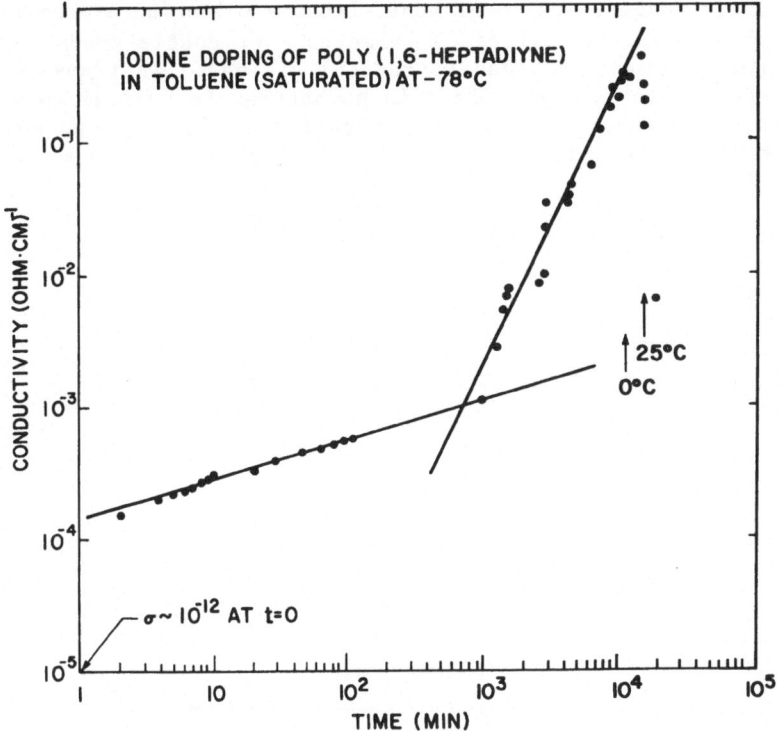

Fig. 2   Time dependence of conductivity of poly(1,6-heptadiyne)
exposed to saturated toluene solution of iodine at $-78°C$.

The oxidation (doping) process in the case of iodine has been
monitored by electron spin resonance (esr) and optical spectroscopies. Esr reveals a narrowing of the initially broad signal at
g ~ 2.0 (Δ H $_{pp}$ $\sim 20G$ vs.   9G for cis- and 0.8G for trans-
polyacetylene) to   10G upon iodine vapor treatment. The number
of spins increases with first order kinetics. In contrast, with
polyacetylene there is no change in linewidth and the radical
population decreases with first order kinetics.[13] Optically,
oxidation with iodine causes the peak at 2eV (cf. 1.9eV for
polyacetylene)[6] to completely disappear and a new peak to appear
at 0.9eV (Figure 3) (cf. 0.8eV for polyacetylene).[6]

These results are consistent with the following explanation. In both systems removal of π -electrons of double bonds by the acceptor (iodine) can lead to radical cations. In trans-polyacetylene due to symmetry there is no barrier to radical coupling to reform a bond and leave two cations, whereas due to the assymmetry of poly-1,6 there is a finite barrier to such a process

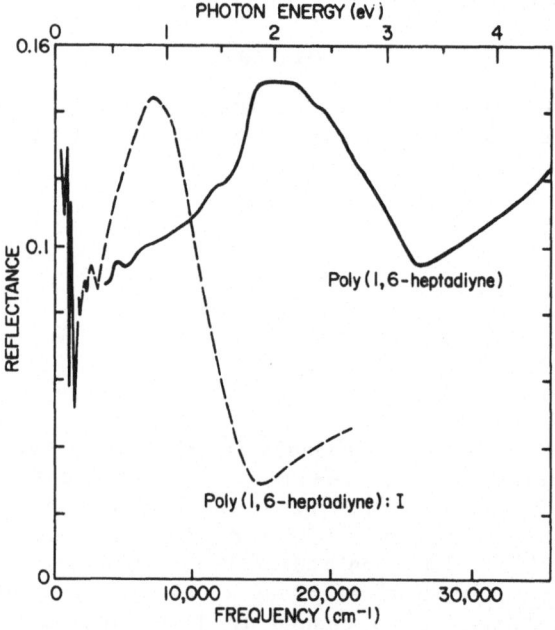

Fig. 3   Infrared to visible reflectance spectra of "undoped" and iodine
         treated (conductivity 10^{-3} s/cm) poly(1,6-heptadiyne).

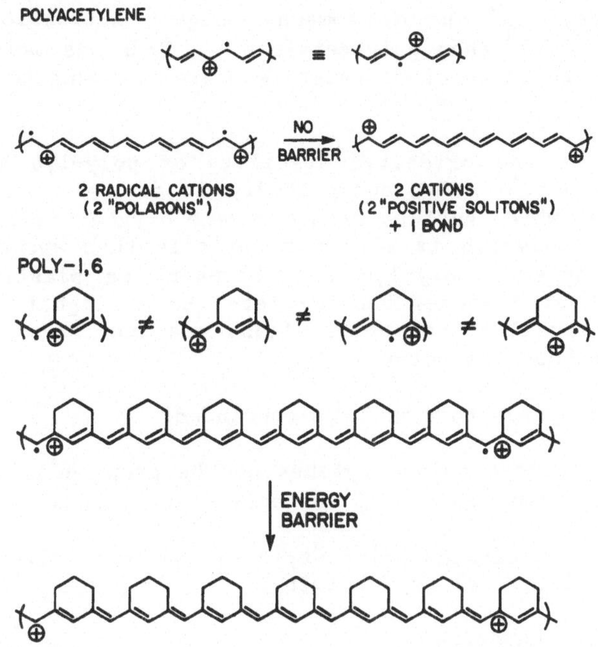

Fig. 4 Difference in symmetry and radical caption coupling in polyacetylene vs. poly(1,6-heptadiyne).

(Figure 4). This results in a buildup of spins in the esr of poly-1,6 similar in mechanism and rate to the loss of spins in polyacetylene. The optical transition in poly-1,6 is most likely then due to a transition between the radical state and the empty π -orbital of the radical cation, while that in polyacetylene is a transition from the valence band to the empty free-radical orbital midway between valence and conduction bands.[6]

Thus a number of contrasts exist between poly-1,6 and polyacetylene: morphology, crystallinity, undoubtedly molecular weight, the dopant content and the final state after doping. Yet in the absence of the rearrangement poly-1,6 is capable of similar conductivities, albeit at somewhat higher activation energy (0.1eV vs. 0.04eV)[6] than polyacetylene. Thus one may question whether these features of polyacetylene are relevant to its high conductivity.

Studies of the oxidative stability of poly-1,6 have also been carried out.[14] Not unexpectedly it is less stable than polyacetylene, which suffers from a severe oxygen sensitivity.[15] Mechanically, poly-1,6 is much more brittle than either cis-or trans-polyacetylene; poly-1,6 can, however, be plasticized by solvents such as toluene, benzene, etc. to a slightly flexible state. These qualitative observations are consistent with its apparent amorphous character.

In summary from this work we concluded:

1.   substituted polyacetylenes can be prepared effectively by cyclopolymerization of suitable diacetylenes
and
2.   such substituted polyacetylenes can be oxidized (doped) to high electrical conductivity.

Poly(propiolic anhydride)

In an attempt to alleviate the oxidative problem encountered with polyacetylene and poly-1,6, we have examined the cyclopoly-merization of propiolic anhydride (3) to polymer 4. This process was reported to yield polymer of molecular weights up to 25000 (200 double bonds) by anionic cyclopolymerization using mild nucleophiles such as I^-, CNS^-, Cl^- in dimethylformamide(DMF)[16] and by coordination polymerization using $PdCl_2$ in DMF.[17]

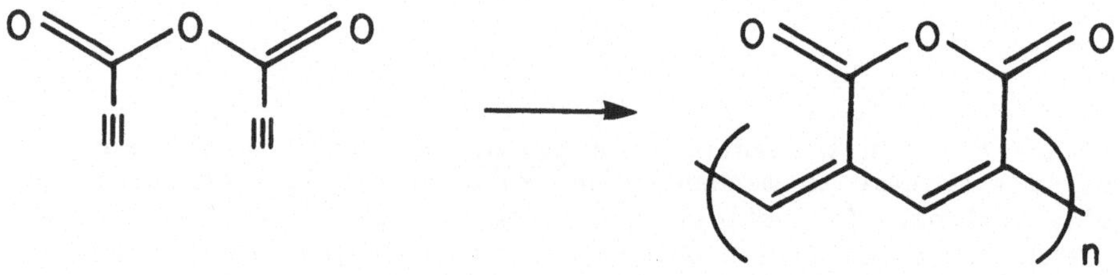

3   4

Our preliminary investigation shows that it is possible to prepare insoluble, crosslinked polymer or soluble uncrosslinked cyclopolymer by control of the initiator: monomer ratio. In DMF at 78 ° C using KI as the initiator the molecular weight by end group analysis is 4600-6000 (37-50 double bonds), regardless of initiator/monomer ratio.

The value of cyclopolymerization is demonstrated by the fact that neither ethyl propiolate (5) nor propiolic acid (6) yield any polymer under these conditions.

The polymer prepared by this process contains DMF as solvent and probably also in the form of structure 8 formed from 4 via decarbonylation of the formyl anhydride intermediate 7. This side reaction precludes meaningful doping studies of this system in view of the potential for reaction of the carboxyl and amido functionalities. Nonetheless, treatment with iodine or AsF_5 at 25 °C raised the conductivity of pressed pellets from 2×10^{-14} to 5×10^{-4} and 2×10^{-6} s/cm, respectively. Moreover, as anticipated the undoped polymer was stable in dry air.

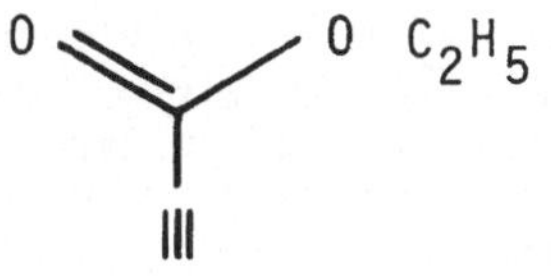

$O = \cdot O \ C_2H_5$

|||

5

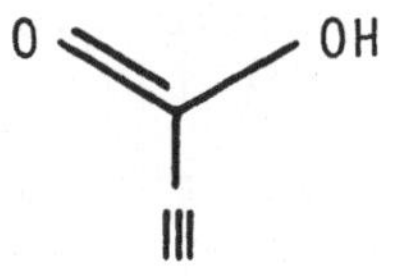

$O = \cdot OH$

|||

6

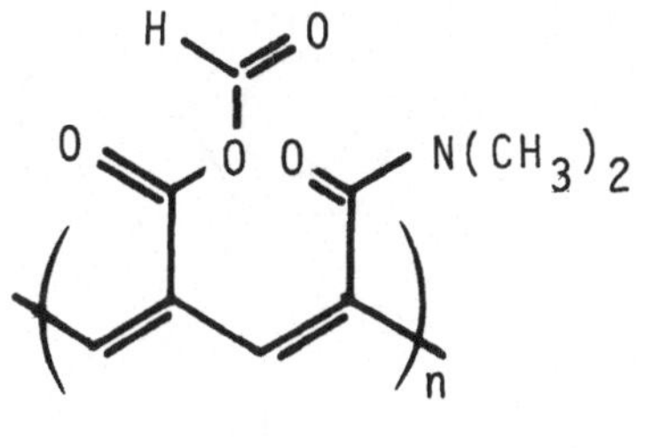

$H \cdot O$ / $O = \cdot O \ O = \cdot N(CH_3)_2$

$\left(\right)_n$

7

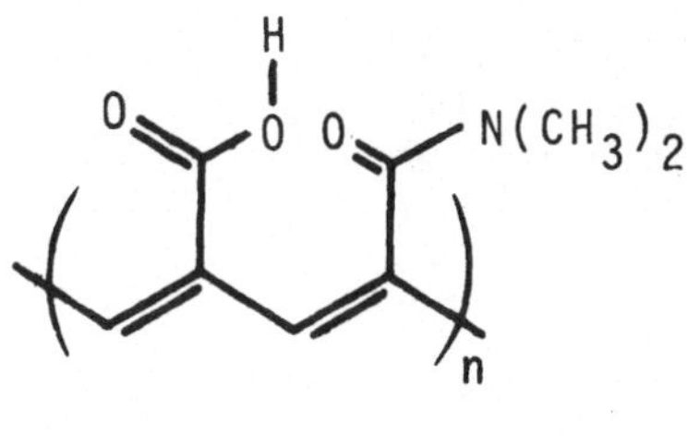

H / $O = \cdot O \ O = \cdot N(CH_3)_2$

$\left(\right)_n$

8

$COOH$

$\left[\left(\right)_{3.3} \left(\right)_1 \right]_n$

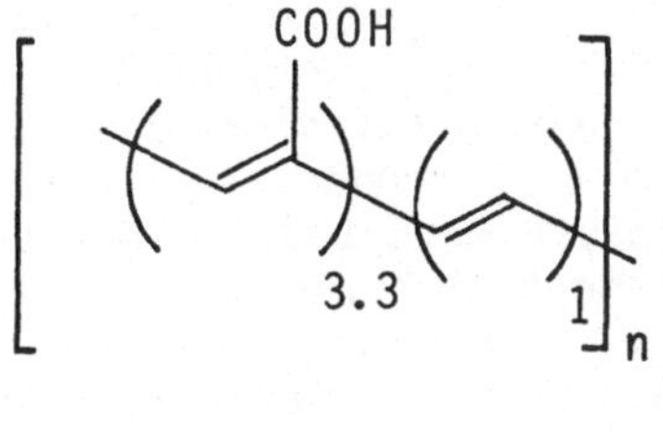

9

Hydrolysis of the polymer with aqueous NaOH at 25 °C was accompanied by decarboxylation to yield a copolymer (9) of acetylene and propiolic acid.   This polymer after purification by reprecipitation (fractionation) had MW ⫫ 5400 (90 double bonds). Upon doping it gave conductivity values comparable to polymer 4.

These preliminary results and the possibility for facile modification of starting diacetylenes and the resultant cyclopolymers make this system attractive for further investigation.

Other Recent Work

Oxidation of Polyaromatic Polymers to High Conductivity

A number of well-known polymers have been converted to a highly conductive state by AsF_5 treatment.[1,2]   These include poly(p-phenylene)(10) [5×10^2 s/cm] , poly(m-phenylene)(11) [10^{-3} s/cm] , poly(p-phenylene sulfide)(12) [1 s/cm] , poly(m-phenylene sulfide)(13) [10^{-2} s/cm] and poly(p-phenylene oxide) (14) [10^{-3} s/cm][19]. In all of these cases studied to date there is evidence for formation of condensed ring products by Friedel-Crafts types of reactions and probably chain extension also occurs.[19]

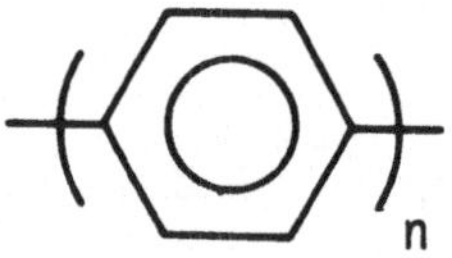

10 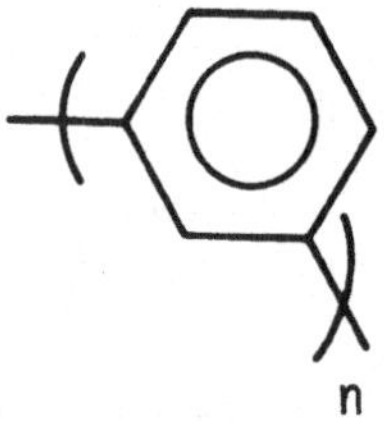   11 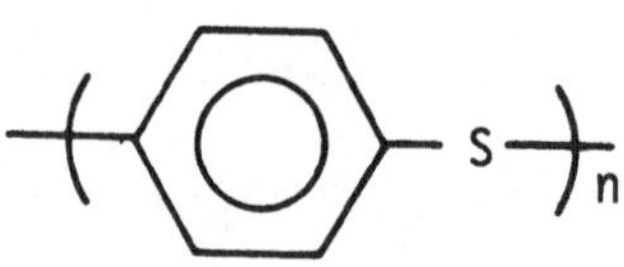   12

A second group of polyaromatic polymers have also been studied.   These include poly(phenylene vinylene) (15), which may be considered an acetylene copolymer, and its derivatives; these polymers are dopable to conductivities up to 3 s/cm.[20]   Poly(2,5-thienylene) (16) yields a conductivity of 10^{-1} s/cm with iodine.[21]   Recently poly(N-methyl-3,3'-carbazolyl (17) was synthesized and upon treatment with iodine yielded a conductivity of 1 s/cm.[22]   Polymers (18) containing phenylene and sulfur nitride linkages are also able to yield relatively high conductivities upon oxidation.[23]

Pyrrole, its derivatives, aromatic amines, thiophene and its derivatives can be electrochemically polymerized to an oxidized state that is conductive, up to 40 s/cm for polypyrrole(19), and less for the others.[24]   Similar polymers are preparable with conductivities in the range 10^{-10} to 10^{-1} s/cm by treatment of thiophene, pyrrole, furan, stilbene, etc. with AsF_5.[25]

Other Substituted Polyacetylenes

Polymerization of methylacetylene[26] with the Shirakawa catalyst led to a soluble orange polymer.   Iodine doping led to a conductivity of only 10^{-3} s/cm and the iodine was readily removed in vacuo. Moreover, AsF_5 had no effect on the conductivity. The low conductivity and the large band gap (orange instead of the black) of polyene 20 were attributed to short conjugation lengths due to steric effects of the methyl substituent. Thus another apparent advantage of cyclopolymerization as in poly(1,6-heptadiyne) is that the resultant ring allows for coplanarity of the carbons of the double bonds, thereby affording extended conjugated sequences.

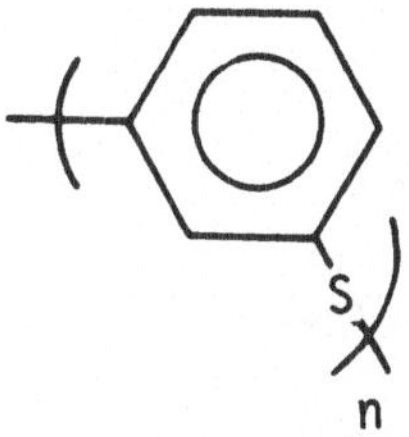

13

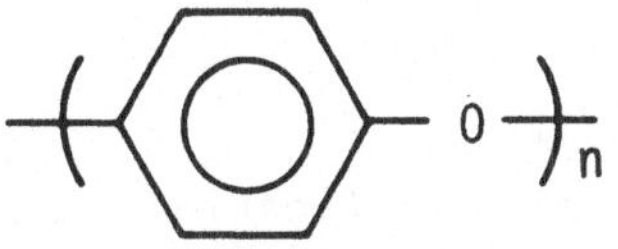

14

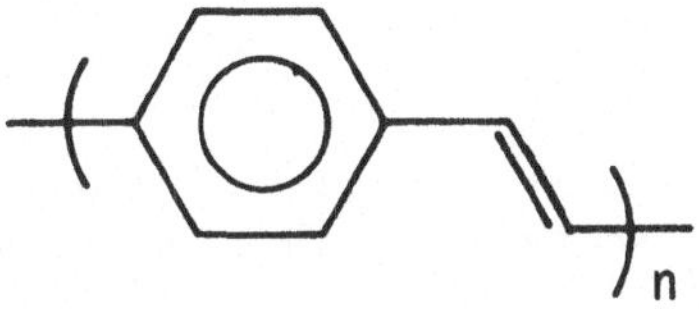

15

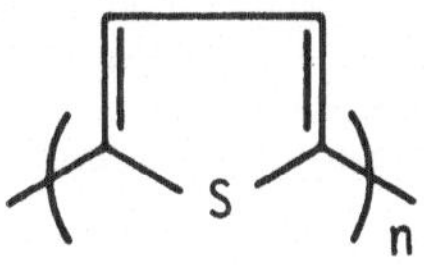

16

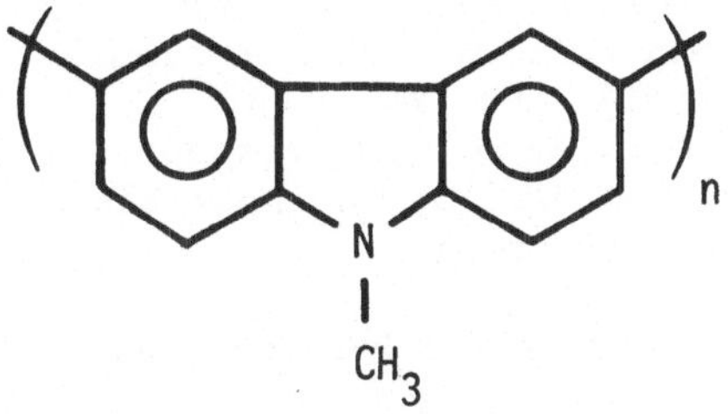

17

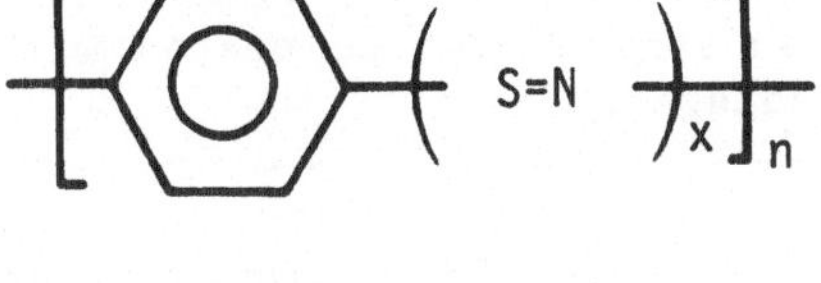

18   x=1 to 4

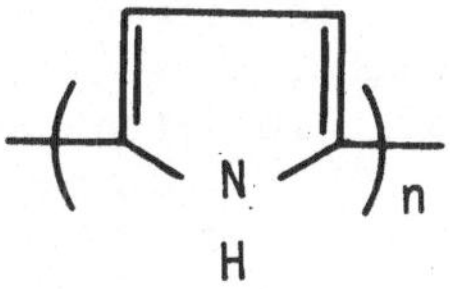

19

Other substituted polyacetylenes (21) have also been synthesized.[27] In all of these cases iodine treatment led to a
conductivity increase: 21a to 10^{-7} s/cm, 21b to 10^{-3} s/cm and 21c
to 10^{-4} s/cm, but the iodine was readily removed in vacuo. The
authors suggested that the conductivity was due to the iodine
itself and not to oxidation of the polyene.

Through the use of transition metal (W and Mo) catalysts a
large number of mono- and di- substituted acetylenes have been
polymerized to molecular weights of 10^5 to 10^6.[28] These polymers
were nearly colorless, indicating lack of conjugation, presumably
because of steric effects. Apparently no attempt was made to
oxidize (dope) them.

d. X=CN

b. X=CH₂Cl

c. X=CF₃

20   21   22

A rather unique approach to the preparation of substituted
polyacetylenes is via chemical modification of polyacetylene
itself. The preparation of copolymers (22) of acetylene and
bromoacetylene by this means has been reported.[29] Thermolysis of
the bromine charge transfer complex of cis-polyacetylene at
150 °C results in bromine addition to the double bonds, followed
by elimination of hydrogen bromide. Bromoacetylene contents as
high as 23% were achieved. No loss of conjugation length was
detected by Raman spectroscopy. Iodine treatment leads to conductivities similar to those observed with polyacetylene itself
upon such treatment.

Other Approaches to Polyacetylene

One of the problems with polyacetylene as prepared via the
Shirakawa catalyst is that the process is not easily amenable to
commercial scale operation. Moreover, if it is desired to coat a
substrate with polyacetylene, the substrate must be coated with
the catalyst. This too has disadvantages.

Recently a very clever solution to these problems was reported.[30] Precursor polymer 24 is prepared by a ring opening
metathesis reaction of a monomer(23) preparable by a Diels-Alder
reaction from commercially available starting materials. 24 is
soluble and allows substrates to be coated. Thermolysis of 24

results in formation of polyacetylene via elimination of o-
bis(trifluoromethyl)benzene(25). The resultant polyacetylene, in
spite of being of shorter conjugation length and amorphous, is
dopable to the same conductivity levels as Shirakawa
material.[31,32] Control of the pyrolysis temperature by structural
variations would seem to offer a broad range of application of
this process.

23   24   25

Rationales for Future Work

Up to this point in time the recent work has dealt solely
with conjugated polyenes. Note that in some cases the conjugation
is via heteroatoms such as S or N. This is in accord with the
generally held view that long conjugated chains are required for
high conductivity. This view emphasizes the importance of delocalization of the charge carriers (positive or negative
charges, i.e., cations or anions) along the chain. From a
theoretical point of view this one-dimensional approach is very
attractive in terms of analysis.[33]

Another view has recently been proposed by Wegner.[34] Naph-
thalene and other simple aromarics can be oxidized electrochemically to form monomeric radical cation salts ($Ar_2 \cdot X^-$) which have
conductivities of 10^2 to 10^3 s/cm.[35] The crystal structures of
these reveal that the aromatic moieties form stacks, along which
the charges and the electrons are presumably delocalized. The
structure is formally analogous to that deduced for oxidized
(doped) polyacetylene in which the polyene chains are arranged in
stacks. This leads to the idea that intermolecular delocalization is the important feature which leads to high conductivity.
Other data are consistent with this rationale. Biphenyl and
terphenyl radical cation salts have crystal structures very similar to that of oxidized (doped) poly(p-phenylene)(10). In the
older literature oligoanilines (26) are reported[36] upon iodine
treatment to yield conductivities up to 1 s/cm; the aniline
moieties are stacked in these materials as well. Poly(N-vinyl-
carbazole) (27) forms radical cation structures by oxidation with

$SbCl_5$ with conductivities as high as 10^{-5} s/cm.[37] Furthermore there are a large number of TCNQ salts of non-conjugated polymers which exhibit relatively high conductivities.[1,2] Raman spectroscopic evidence for <u>trans</u>-polyacetylene has been interpreted to argue for the presence of a large number of short conjugation lengths,[38] which would disfavor the importance of intrachain charge transport.

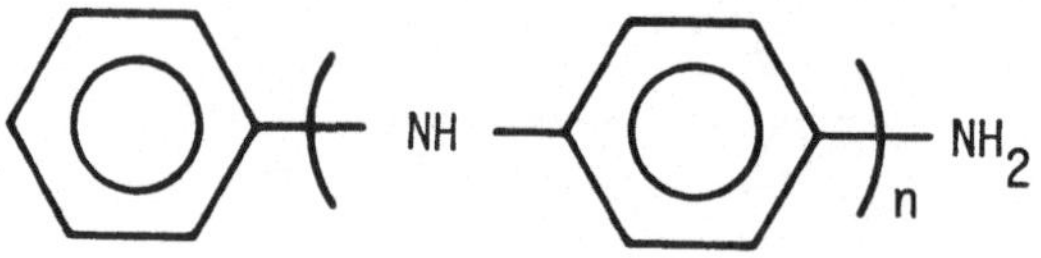

$0 < n < 5$

<u>26</u>

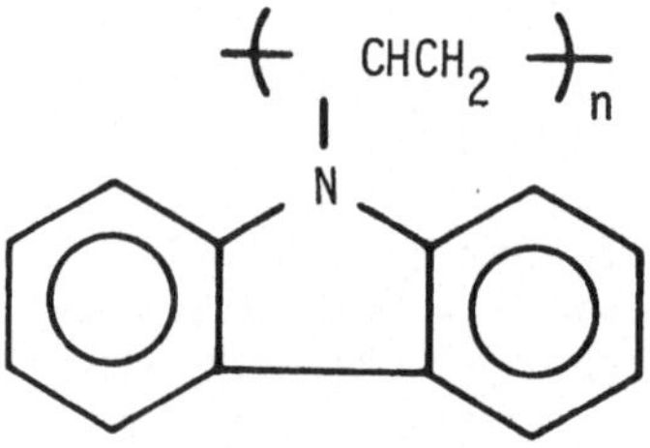

<u>27</u>

While Wegner's hypothesis is currently controversial, it is one that deserves the synthetic chemists' attention, because it opens a second potential avenue of approach to conducting polymers, namely the use of non-conjugated polymers containing aromatic moieties. Such polymers should have improved processability, mechanical properties and oxidative stability with respect to conjugated polymers.

CONCLUSIONS

Conductive polymers offer the prospect of a significant number of applications utilizing their unique combination of properties. To date, however, problems of processability, mechanical properties and environmental stability remain to be solved. Clearly new polymers and/or new processes will be required to meet this challenge and to realize the bright economic potential that exists. This represents another opportunity for creative chemistry.

Two approaches currently seem open. The first is directed toward conjugated polymers. Here substituted systems may hold potential. Cyclopolymerization is a viable method of synthesis of substituted polyacetylenes. Likewise, chemical modification of polyacetylene itself may have potential. The preparation of polyacetylenes by elimination reactions from precursor polymers needs more attention. The second approach is toward non-conjugated polymers containing aromatic functionalities; little work has been done in this area. Much creative chemistry remains to be explored and it will most likely involve new monomers and polymers.

REFERENCES

1. H.W. Gibson, Polymer, in press (1983).
2. C.B. Duke and H.W. Gibson, Encyc. Chem. Tech., Wiley and Sons, New York, Vol. 18, pp. 755-794,1982.
3. G. Natta, G. Mazzanti, P. Corrandini, Atti Acad. Naz. Lincei, Rend. Cl. Sci. Fis., Mat. Nat. 25(8), 3,(1958).
4. D.J. Berets and D.S. Smith, Trans. Faraday Soc., 64, 823(1968).
5. T. Ito, H. Shirakawa, S. Ikeda, J. Polym. Sci., Polym. Chem. Ed., 12, 11(1974).
6. A.G. MacDiarmid, A.J. Heeger, Synth. Met. 1, 101(1979/1980), C.K. Chiang, A.J. Heeger, A.G. MacDiarmid, Ber. Bunsenges. Phys. Chem., 83, 407(1979) and references therein.
7. V. Enkelmann, W. Muller and G. Wegner, Synthetic Metals, 1, 185(1980); W.H. Watson, W.C. McMordie and L.G. Lands, J. Polym. Sci., 55, 137(1961); P.S. Woon and M.F. Farona, J. Polym. Chem. Ed., 12, 1749(1974).
8. G.B. Butler, G.C. Corfield and C. Aso, Prog. Polym. Sci., 4, 71(1975), G.B. Butler and R.J. Angelo, J. Amer. Chem. Soc., 79 3128(1958); G.B. Butler, A. Crawshaw and W.L. Miller, J. Amer. Chem. Soc., 80, 3615(1958).
9. J.K. Stille and D.A. Frey, J. Amer. Chem. Soc., 83, 1697(1961).
10. H.W. Gibson, F.C. Bailey, A.J. Epstein, H. Rommelmann and J.M. Pochan, J. Chem. Soc., Chem. Comm., 426(1980); H.W. Gibson, F.C. Bailey, A.J. Epstein, H. Rommelmann, S. Kaplan, J. Harbour X.-Q. Yang, D.B. Tanner and J.M. Pochan, J. Amer. Chem. Soc., in press (1983).
11. A.J. Epstein, H. Rommelmann, R. Fernquist, H.W. Gibson, M.A. Druy and T. Woerner, Polymer, 23, 1211(1982).
12. T. Akaishi, K. Miyasaka, K. Ishikawa, H. Shirakawa and S. Ikeda, J. Polym. Sci., Polym. Phys. Ed., 18, 745(1980); H. Haberkorn, H. Naarmann, K. Penzien, J. Schlag and P. Simak,

Synth. Metals, 5, 51(1982); P. Robin, J.P. Pouget, R. Comes, H.W. Gibson and A.J. Epstein, Phys. Rev. B, in press (1983).

13. J.M. Pochan, H.W. Gibson and J. Harbour, Polymer,23, 439(1982); unpublished data.

14. J.M. Pochan, D.F. Pochan and H.W. Gibson, Polymer, 22, 1367(1981); J.M. Pochan, H.W. Gibson and J. Harbour, Polymer, 23, 435(1982).

15. H.W. Gibson and J.M. Pochan, Macromolecules, 15, 242(1982).

16. R.I. Yakhimovich, E.A. Shilov and G.F. Dvorko, Dokl. Akad. Nauk SSSR, 166, 98 (1966) (Eng. Tr.).

17. L.A. Akopyan, G.V. Ambartsumyan, E.V. Ovakimyan and S.G. Matsoyan, Vysokomol. Soed., A19, 271(1977).

18. I thank Dr. John Spiewak of these laboratories for pointing this possibility out.

19. L.W. Shacklette, R.L. Elsenbaumer, R.R. Chance, H. Eckhardt, J.E. Frommer and R.H. Baughman, J. Chem. Phys., 75, 1919(1981) and references therein.

20. G.E. Wnek, J.C.W. Chien, F.E. Karasz and C.P Lillya, Polymer, 20, 1441 (1979); J.C.W. Chien, R.D. Gooding, F.E. Karasz, C.P. Lillya, G.E. Wnek and K. Yao, Org. Coat. Plast. Chem., 43, 886(1980); J.R. Reynolds, J.C.W. Chien, F.E. Karasz, R.D. Gourley, and C.P. Lillya, Conf. Phys. Chem. Cond. Polym., Les Arcs, France, Dec. 11–15, 1982, to appear in J. Physique.

21. T. Yamamoto, K. Sanechika, and A. Yamamoto, J. Polym. Sci. Polym. Lett. Ed.,18,9(1980); J.W. P. Lin and L. Dudek, J. Polym. Sci., Polym. Chem. Ed., 18, 2869(1980).

22. S.T. Wellinghoff, T. Kedrowski and H. Ishida, Conf. Phys. Chem. Cond. Polym. Les Arcs, France, Dec. 11–15, 1982, to appear in J. Physique.

23. G. Wolmershauser, R. Jotter and T. Wilhelm, Conf. Phys. Chem. Cond. Polym., Les Arcs, France, Dec. 11–15, 1982, to appear in J. Physique.

24. A. Diaz, Chemica Scripta, 17, 145(1981).

25. G. Kossmehl and G. Chatzitheodorou, Proc. IUPAC 28th Macromol. Symp., Amherst, Mass., July 12–16, 1982, p.419.

26. J.C.W. Chien, G.E. Wnek, F.E. Karasz and J.A. Hirsch, Macromolecules, 14, 479(1981).

27. W. Diets, P. Cukor, M. Rubner and H. Jopson, Ind. Eng. Chem. Prod. R and D, 20, 696(1981). M. Rubner and W. Deits, J. Polym. Sci., Polym. Chem. Ed., 20, 2043(1982).

28. T. Higashimura and T. Masuda, Proc. IUPAC 28th Macromol. Symp. Amherst, Mass., July 12–16, 1982, p.113; T. Higashimura, Y.X. Deng and T. Matsuda, Macromolecules, 15, 234(1982).

29. M.J. Kletter, A.G. MacDiarmid, A.J. Heeger, E. Faulques, S. Lefrant and P. Bernier, J. Polym. Sci., Polym. Lett. Ed., 20, 211(1982).

30. J.H. Edwards and W.J. Feast, Polymer, 21, 595(1980); Conf. Phys. Chem. Cond. Polym., Les Arcs, France, Dec. 11-14, 1982, to appear in J. Physique.

31. J.H. Edwards, W.J. Feast and D.C. Bott, private communication.

32. G. Leising, F. Stelzer, and H. Kahlert, Conf. Phys. Chem. Cond. Polym., Les Arcs, France, Dec. 11-14, 1982, to appear in J. Physique.

33. A.J. Heeger and A.G. MacDiarmid, Chemica Scripta, 17, 115(1981) and references therein.

34. G. Wegner, Angew. Chem. Int. Ed. Eng., 20, 361(1981); Proc. IUPAC 28th Macromol. Symp., Amherst, Mass., July 12-16, 1982, p.410; G. Wegner, V. Enkelmann, M. Monkenbusch, G. Wieners and G. Lieser, Conf. Phys. Chem. Cond. Polym., Les Arcs, France, Dec. 11-14, 1982, to appear in J. Physique.

35. C. Krohnke, V. Enkelmann and G. Wegner, Ang. Chem. Int. Ed. Eng., 19 912(1980).

36. V. Hadek, P. Zach, K. Ulbert, and J. Honzl, Collect. Czech. Chem. Comm., 34, 3139(1969); J. Honzl, K. Ulbert, V. Hadek, and M. Tlustakova, J. Polym. Sci. Part C, 16, 4465(1969); V. Hadek, J. Chem. Phys., 49, 5202(1968); K. Ulbert, J. Polym. Sci. Part C, 22,881(1969).

37. H. Block, M.A. Cowd and S.M. Walker, Polymer, 18 781(1977); H. Block, Adv. Polym. Sci. 33, 93(1979).

38. S. Lefrant, Conf. Phys. Chem. Conf. Polym., Dec. 11-14, 1982, Les Arcs, France, to appear in J. Physique; H. Kuzmany, ibid; G.P. Brivio and E. Mulazzi, ibid.

POLYMERIZATION OF BUTADIYNE:

POLYMER CHARACTERIZATION AND PROPERTIES

Arthur W. Snow

Naval Research Laboratory
Polymeric Materials Branch
Washington, D. C. 20375

INTRODUCTION

Butadiyne, H-C≡C-C≡C-H, as a polymerizable monomer, has received very little attention from polymer chemists although
its discovery dates back to Bayer[1] in 1885. This structurally
simple, highly reactive bifunctional molecule would be expected
to have been a monomer of considerable interest in the field of
polymer chemistry. Possibly, limited butadiyne stability may
account for the small amount of polymerization research. The
The compound is a liquified gas at room temperature (BP = 10°C),
discolors slowly in sealed vessels at 20°C and may explode if
heated. Storage and instability problems may be circumvented.
Prevention of explosion may be accomplished by addition of an
inert diluent such as butane[2]. The monomer may also be
stored in the form of a labile complex with N-methylpyrrolidone[3]. Its thermal condensation or polymerization
was briefly recorded as an observation by Bayer and described in
a little more detail by Müller[4] in 1925. Prevention of this
thermal polymerization has been the subject of several patents[5]
with methylene blue, pyridine and vinylpyridine claimed as inhibitors.

Polymerization has been attempted by conventional free
radical, cationic and anionic initiators[6]. Reactivity was observed, but a well defined polymer structure was not characterized. Butadiyne has been reported to form condensation copolymers by nucleophilic addition of diamines[7] or heterocyclic compounds[8]. From its structure as the simplest diacetylene, it
would be expected to figure prominently in solid state
diacetylene polymerizations (Eq. 1). In one instance Wegner does

399

$$ n \quad H-C{\equiv}C-C{\equiv}C-H \quad \longrightarrow \quad {\Big(}{\overset{H}{\underset{H}{C}}}-C{\equiv}C-C{\Big)}_n \qquad (1) $$

report the solid state 1,4-butadiyne polymerization, although
preparative details and properties other than electrical
conductivity were not disclosed[9]. Very recently, a single
crystal solid state polymerization by γ-irradiation at -78°C was
reported[10]. A highly colored brittle material was obtained but
no characterization or property data was presented. Solid state
polymerization to an undefined structure also occurs when
butadiyne is deposited on a cold AsF_5 film.[18]

In seeking new preparative routes to highly conjugated polymeric systems, we elected to investigate the polymerization
of butadiyne. In this presentation we summarize previously
unpublished work on a Ziegler-Natta polymerization of butadiyne
and some of the more current developments on the thermal vapor
deposition polymerization of this monomer.

EXPERIMENTAL

Butadiyne was prepared by the dehydrochlorination of
1,4-dichloro-2-butyne in aqueous potassium hydroxide-dioxane
solution[11]. The crude product was vacuum distilled to IR
purity[12]. This distillation is conducted from a -80°C trap to a
-195°C trap at 10^{-4} torr to remove an impurity identified as
2-chloro-1-butene-3-yne[13]. This less volatile impurity is
detectable in the infrared by stong bands at 850 and 1590 cm^{-1}
where butadiyne is transparent. Other bands assigned to
2-chloro-1-butene-3-yne are observed at 620, 920, 1015, 1230,
1260, 1703, 2118, 3142, 3182, 3197, 3328 cm^{-1}. By periodically
monitoring infrared spectra (100 torr, 10 cm gas cell) of the
distillate, we find 50 to 60% of the butadiyne has distilled
before impurity bands begin to appear. The infrared spectrum of
the purified butadiyne is identical to that of reference 12, and
PMR (10% $CDCl_3$ solution) did not detect proton impurities. Care
and precautions should be exercised in handing the butadiyne or
detonation may occur on mixing with oxygen and warming[14].

Ziegler-Natta catalysts were prepared in a nitrogen atmosphere by syringe or inert atmosphere transfer of the titanium
or zirconium alkoxide to toluene solvent followed by dropwise
syringe addition of the triethylaluminum. The metal alkoxide
concentration was 0.5 M and aluminum: titanium or zirconium ratio
was 4:1 unless stated otherwise. The C_2H_5, nC_3H_7, iC_3H_7, nC_4H_9,
$CH_2CH(C_2H_5)C_4H_9$ titanium tetraalkoxides, zirconium tetraisopropoxide and triethylaluminum were purchased from Ventron.
Titanium tetramethoxide was prepared by reaction of titanium
tetraisopropoxide with methanol[15]. The catalyst was magnetically
stirred for 1 hr. after mixing and transferred by syringe to the

reactor. The reactor was a three stage apparatus consisting of a central tubular vessel (300 ml) and two bulbs (600 ml and 50 ml) connected at the top by teflon valves. The large bulb was charged with 700 torr of monomer, and small bulb was charged with 20 ml dry degassed benzene. Catalyst was introduced to the central reactor and degassed 3 times under high vacuum. Polymerizations were conducted at $25^{\circ}C$ by coating the vessel walls with catalyst solution by shaking, followed by immediate opening of the monomer bulb valve. Reaction times were two to three hours. The reactor was then flooded with dry nitrogen, excess catalyst solution removed by syringe, and the reactor was evacuated. The product was washed by vacuum transfer of benzene into the central vessel and decantation back to the solvent bulb. The cycle was repeated until the supernate was colorless.

Butadiyne thermopolymer samples were prepared by placing films of substrate organic polymers (polyethylene (PE) 3.7 mil, polyvinylidene fluoride (PVF2) 1.7 mil, polytetrafluoroethylene (PTFE) 2.8 mil and poly(tetrafluoroethylene-co-hexafluoropropylene) (FEP) 10.5 mil) in an atmosphere of 700 torr butadiyne at $20^{\circ}C$ for 5 wk. The order of polymerization reactivity is PVF2 > PE > PTFE > FEP as determined by visual observation of color development and the corresponding weight percent increase (PVF2 81%, PE 61%, PTFE 23%, FEP 0.3%).

DSC and TGA data were obtained from 10 mg samples in a nitrogen atmosphere at a $10^{\circ}C$ per minute heating rate using a DuPont 990 Thermal Analyzer, 910 Differential Scanning Calorimeter and 951 Thermogravimetric Analyzer. Postpolymerization heat treatments were conducted for 1 hr. periods at selected temperatures between 120 and $470^{\circ}C$ in an evacuated, sealed pyrex tube. Transmission IR spectra were recorded on a Perkin-Elmer 267 grating spectrophotometer. ESR spectra were recorded on a Bruker ER200D spectrometer. Surface resistivity measurements were made using an interdigital electrode array consisting of 50 finger pairs 25 microns in width and separated by 25 microns[24]. The upper measurement limit of this apparatus is 10^{13} ohm/square.

ZIEGLER-NATTA POLYMERIZATION

Our initial objective was to prepare organic polymeric systems which may be modified to electrically conducting materials. The approach was to obtain the butadiyne polymer in a thin film form by taking advantage of the synthetic procedure, characterization techniques and structure-property insight that have been developed for polyacetylene[16]. The formation of a partially crosslinked polyvinylene system that may reduce or eliminate the requirement of charge carrier transfer by intermolecular chain contact was envisioned (Eq. 2).

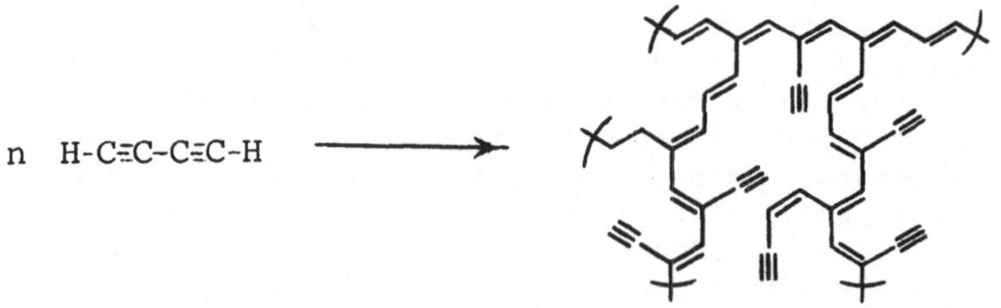

$$n \quad \text{H-C} \equiv \text{C-C} \equiv \text{C-H} \quad \longrightarrow \qquad \qquad \qquad \qquad (2)$$

This approach was considered promising from the following considerations. Like acetylene, butadiyne is a gas at room temperature and may be handled according to the procedure developed
by Shirakawa for polyacetylene films[16]. If the triple bonds of
butadiyne are considered to react independently, butadiyne, with
either end of the molecule available for polymerization, may be
the most susceptable substituted acetylenic monomer for polymerization. Sterically, the terminal acetylenic group is quite
small. The diolefin analogue, butadiene, when subjected to the
same Ziegler-Natta catalyst ($Ti(OC_4H_9)_4/Al(C_2H_5)_3$) as that of the
Shirakawa formulation, polymerizes to a stereoregular polymer
(syndiotactic 1,2-polymerization) of respectable molecular weight
($\bar{M}v = 5 \times 10^6$) in which crosslinking and gel formation occur at

Table I

Variations in Composition of the Ziegler-Natta Catalyst		
RELATIVE COMPOSITION	**ALKOXIDE**	**TRANSITION METAL**
$Al(C_2H_5)_3 : Ti(OC_4H_9)_4$	$4\ Al(C_2H_5)_3 : Ti(OR)_4$	
1 : 1	$-OR$	
2 : 1	$-OCH_3$	$Ti(O-\overset{\displaystyle |}{CH}-CH_3)_4$
4 : 1	$-OCH_2CH_3$	$\overset{\displaystyle |}{CH_3}$
8 : 1	$-OCH_2CH_2CH_3$	$Zr(O-\overset{\displaystyle |}{CH}-CH_3)_4$
10 : 1	$-O-\overset{\displaystyle |}{CH}-CH_3$ $\overset{\displaystyle |}{CH_3}$	$\overset{\displaystyle |}{CH_3}$
	$-OCH_2CH_2CH_2CH_3$	
	$-OCH_2-\overset{\displaystyle |}{CH}-CH_2CH_2CH_2CH_3$ $\overset{\displaystyle |}{CH_2}-CH_3$	

moderate degrees of conversion $(40\%)^{17}$. The polyacetylene film
forming technique with the Shirakawa formulated catalyst as well
as with other related catalysts (Table I) did not yield free
standing polybutadiyne films. Other formulations included
systematic variation in catalyst component quantities, size of
titanium alkoxide ligands and a change to a larger metal atom of
the same group. All catalysts had varying levels of
polymerization activity with acetylene. In cases where butadiyne
product was obtained, it was in the form of a very air sensitive
precipitate. Even after exhaustive washing and exclusion of air,
analysis indicated 30% by weight ash content which was presumably
due to bound or occluded catalyst. The carbon and hydrogen
contents were 56 and 6% respectively. Attempted
copolymerizations with deuterated acetylene were equally
discouraging. With the Shirakawa catalyst, a 1:1 butadiyne to
acetylene monomer ratio produced very fragile polymer film with
high oxygen sensitivity and high ash content, while a 1:7 ratio
yielded a film with marginally better mechanical strength and
lower oxygen sensitivity but still a high ash content (11% ash,
70.3% C, 6.8% H) after exhaustive washing and air exclusion.

VAPOR DEPOSITION POLYMERIZATION

During the course of investigating the Ziegler-Natta polymerization, it was observed that butadiyne undergoes a rapid
thermal polymerization in the liquid state at $0°C$ to precipitate
a highly colored polymer which has a brass metallic appearance at
high conversion. There is a small but significant body of literature on the melt and solution polymerization of disubstituted
(dimethyl and diphenyl) diacetylenes. Oligomers having a combination of polyene and polyacene structures have been proposed
as polymeric products (Eq. 3).[18]

What is unique about the butadiyne system, in addition to being
much more reactive than the substituted diacetylenes, is a
surface selective polymerization from the vapor phase and an
opportunity for characterization without interference from
pendent groups. The surface selectivity of this vapor deposition
polymerization on different substrates is in the order hydrocarbon (polyethylene) > fluorocarbon (teflon) > sodium chloride
aluminum oxide as measured by quantity of deposited polymer per

unit time on substrates of the same surface area[19]. Since the
vapor deposited polybutadiyne is highly insoluble, its deposition
on flexible organic polymer films is a particularly convenient
way of handling and characterizing the deposited polymer. Considering that the high degree of unsaturation of the C_4H_2 monomer
may also be a characteristic of the polymer and that such a
property is desirable for an electroactive polymer, a characterization study was undertaken.

POLYBUTADIYNE CHARACTERIZATION

The insolubility of polybutadiyne necessitated the use of
solid state characterization techniques and the tailoring of
sample preparation to a particular characterization experiment.
The characterization includes infrared, ultraviolet-visible,
electron spin resonance and x-ray diffraction spectroscopies,
thermogravimetric analysis and differential scanning calorimetry.

Infrared spectra were obtained from three different sample
preparations. The most convenient and best quality spectrum was
obtained in transmission using polyethylene and teflon supported
samples. The complementary infrared transmissions of these two
films allow a complete polybutadiyne spectrum to be represented
(Figure 1). Curiously, reflection spectra of these samples
displayed only bands assignable to the substrate polymer. The
second method involved a slow polymerization of a thin film onto
an NaCl surface over a thirty day period at $25^{\circ}C$. While the
color of the film made it easily visible to the eye, only a very
weak spectrum was obtainable by FTIR spectroscopy. This spectrum
displayed the same bands as those of Figure 1 except the 3315 cm^{-1}
band which, while still a strong absorption, was shifted to 3290
cm^{-1}. Attempts to obtain spectra of bulk polymer dispersed in
KBr or Nujol mull were unsuccessful because of small partical
scattering. A transmission spectrum of an unsupported film
flake, which had formed on the glass wall above a bulk sample
prepared by sealed tube polymerization, was very diffuse, but a
band centered more at 3290 cm^{-1} as opposed to 3315 cm^{-1} was
observed.

The spectrum in Figure 1 was used for analysis, and the band
assignments are presented in Table II. The bands at 3315, 640
and 2090 cm^{-1} are assigned to a terminal acetylenic functional
group. The two most intense bands at 3315 and 640 cm^{-1}
correspond to a stretching and bending deformation of the
acetylenic carbon hydrogen bond[20]. The 2090 cm^{-1} wavenumber
of the carbon triple bond stretching is indicative of a
monosubstituted acetylenic group that is conjugated with another
unsaturated structural unit. Within the normal $-C\equiv C-$ stretching
range (2250-2100 cm^{-1}), monosubstituted acetylenic bonds appear
at lower frequencies than disubstituted acetylenic bonds, and the
effect of conjugating a terminal triple bond with another triple
bond, double bond or benzene ring causes a further decrease in

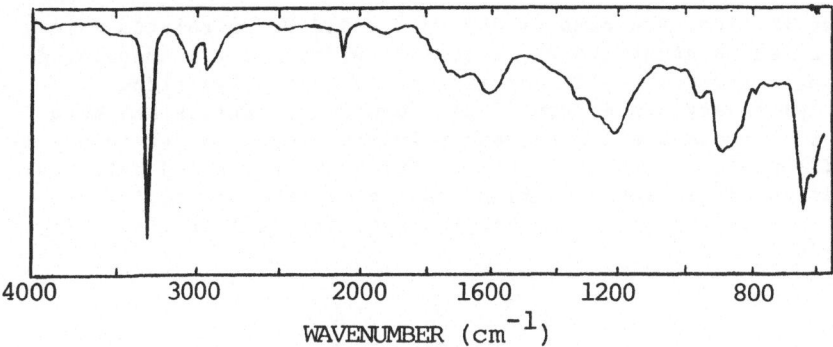

Figure 1.   Infrared Spectrum of Poly(butadiyne).

Table II

Infrared Band Assignments for Polybutadiyne

WAVENUMBER (cm^{-1})	ASSIGNMENT
3315	Terminal Acetylenic C-H stretching
3020	Aromatic or olefinic C-H stretching
2090	Terminal acetylenic C≡C stretching
1600	Aromatic or olefinic C=C stretching
1210	Aromatic or olefinic C-H in-plane bending
885	Aromatic or olefinic C-H out-of-plane bending
640	Terminal acetylenic C-H in-plane bending
605	Terminal acetylenic C-H out-of-plane

the frequency[21]. A benzene ring causes a further decrease in the
frequency . A pertinent example is mycomycin where an
appreciable orbital overlap over three conjugated bonds results
in a terminal -C≡C-stretching frequency of 2040 cm^{-1}[22]. It
should also be noted that the acetylenic C-H stretching band
exhibits very little frequency dependence on substituent structure but does shift with changes in physical state[21] which is
consistent with the shifting of the 3315 cm^{-1} band described in
the preceeding paragraph. The ≡C-H bending can exhibit a dependence on the symmetry of a substituent group. For example,
phenylacetylene displays two absorption maxima at 642 and 613 cm^{-1}
which correspond to in-plane and out-of-plane deformations
relative to the phenyl ring[21]. The two maxima at 640 and 605
suggest a similar structural situation in polybutadiyne. The
broad bands at 1600, 1210 and 885 cm^{-1} are assigned to the
polymer chain structure and interpreted to indicate a high degree
of unsaturation and poly-dispersity. In the context of this

polymerization, the band at 885 cm^{-1} could be correlated with a central acene structure or, alternately, with a trisubstituted benzene structure[21]. In describing the melt polymerized diphenyldiacetylene system[18c] the former correlation has been used to propose a block polyene-polyacene copolymer structure (I and II) while, in the present case with unsubstituted butadiyne and direct detection of a high density of terminal acetylenic groups, a substituted polyphenyl structure (III) is also being considered.

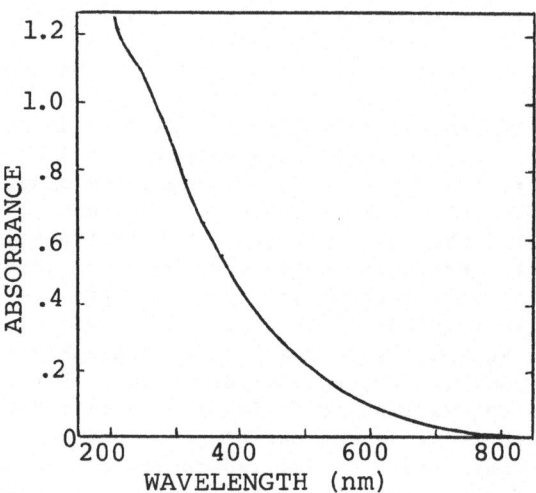

III

The possibility that the chromophores of structures I, II and III might be distinguishable by their visible-ultraviolet spectra was examined. As shown in Figure 2, the spectrum of a polybutadiyne thin film, obtained by vapor deposition polymerization onto a quartz cell, is a nearly featureless continuum from 800 to 190 nm. This suggests that neither structure predominates and is in general agreement with a polydispersity as indicated by the breath of the infrared spectrum's aromatic bands.

Figure 2.   Visible-Ultraviolet Spectrum of Poly(butadiyne).

Electron spin resonance studies have been reported elsewhere in detail[23] and will be briefly summerized here. Spectra were obtained from vapor deposited film and bulk polymerized samples. A remarkably intense singlet signal with a free electron g-value, 10 gauss line width and a spin density of 8×10^{19} spin/gram was observed. The spin density, line width and line shape were independent of temperature over a -150 to 25°C range and not changed by irradiation of the sample with ultraviolet or visible light. The signal is also unchanged when the polybutadiyne is exposed to oxygen or atmospheric moisture. When the polymer is prepared as a very thin film, the spectral shape changes from Lorentzian for a bulk sample to unsymmetrical for the thin film. This affect was attributed to a g-anisotropy. A shift in g-value was observed to be dependent on the nature of the substrate surface. The g-value shifts from 2.0027 for a bulk sample or one deposited on an impermeable surface such as Pyrex glass to 2.0030 for a teflon supported sample.

The polybutadiyne bulk polymer is non-crystalline, non-melting and of high thermal stability. The x-ray diffraction spectrum of a polybutadiyne bulk sample displayed no crystalline reflections, and DSC displayed no endothermic melting transitions. TGA in nitrogen atmosphere at a 10°C/min. heating rate showed only a 20% weight loss at 1100°C (Figure 3). If the bulk polymer is heated very rapidly in air or after exposure to air, a bright explosive flash occurs. When the experiment was conducted in an evacuated IR gas cell, methane and carbon dioxide were qualitatively detected as products. The density of the bulk polymer was measured to be 1.317 g/cm^3 by flotation in hexane-carbon tetrachloride.

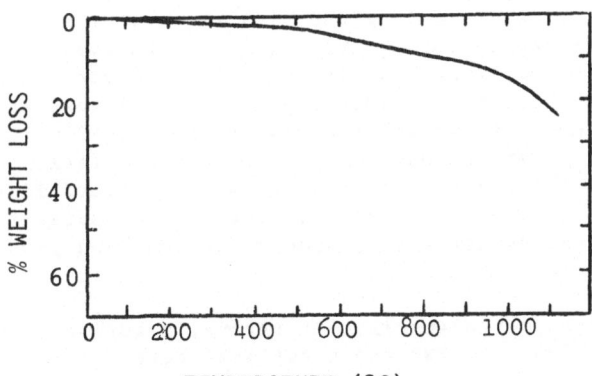

Figure 3.   TGA Thermogram of Poly(butadiyne).

Electrical Conductivity

Electrical measurements on polybutadiyne bulk samples and vapor deposited films gave resistance values greater than 20 MΩ for both bulk (1 mm probe spacing) and surface (interdigital electrode separation distance of 25 microns[24]) techniques. A doping experiment was conducted by depositing a film over chromium interdigital electrodes on a quartz surface followed by exposure to a 130 torr AsF_5 vapor pressure for three days at room temperature[25]. No measurable conductivities were observed (resistance > 20 MΩ). At the end of the three day period, a blistering of the film was observed. This was due to diffusion of the AsF_5 through the film, followed by a reaction with the glass to produce gaseous pockets of HF which detached the film from the glass wall. Other doping experiments involving exposure of teflon supported polybutadiyne samples to bromine, sulfuric acid, hydrogen peroxide, butyl lithium, sodium naphthalide and lithium ethylenediamine were also conducted. No significant conductivity was observed, and significant weight uptake was observed only with bromine where infrared spectra indicated a partial addition to the triple bonds.

Postpolymerization Reaction

We investigated the postpolymerization thermal treatment of vapor deposited polybutadiyne on the hypothesis that the pendant acetylenic groups of structure I would be converted to the ladder structure II. Such a structural speculation has accompanied a reported electrical conductivity increase of twelve orders of magnitude when the poly(diphenylbutadiyne) system was subjected to a 700°C heat treatment[18c]. The attractive features of our simpler polybutadiyne system are: (1) we could easily monitor reaction of the acetylenic group, (2) the higher reactivity of an unsubstituted acetylenic group suggested the reaction would occur at a lower temperature, (3) if the polybutadiyne could be converted to a moderately conducting material at reasonable temperatures, a combination with its ease of application to substrate organic polymers could have important practical consequences, and (4) the polyacene structure is theoretically predicted to be capable of very high electrical conductivities on the basis of ionization potential, band width and band gap calculations[26].

We studied the postpolymerization thermal reaction of polybutadiyne by DSC, TGA, IR, ESR and electrical conductivity measurements. The results are summarized as follows[27]. Polybutadiyne was vapor deposited onto thin films of polyethylene (PE), poly(vinylidene fluoride) (PVF2), poly(tetrafluoroethylene) (PTFE) and poly(tetrafluoroethylene-co-hexafluoropropylene) (FEP). DSC thermograms of these polymer samples displayed two interesting effects. The thermograms of the PTFE sample is presented in Figure 4. First was a very broad exotherm ranging from

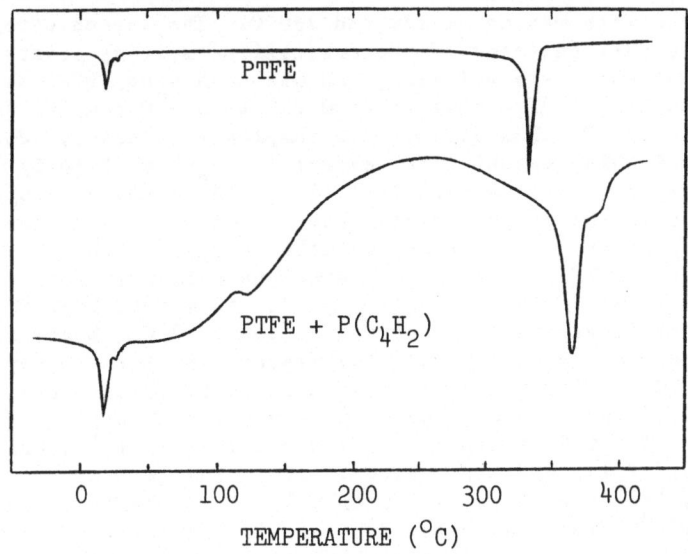

Figure 4.   DSC Thermograms of PTFE Supported Poly(butadiyne)
            and PTFE Control.

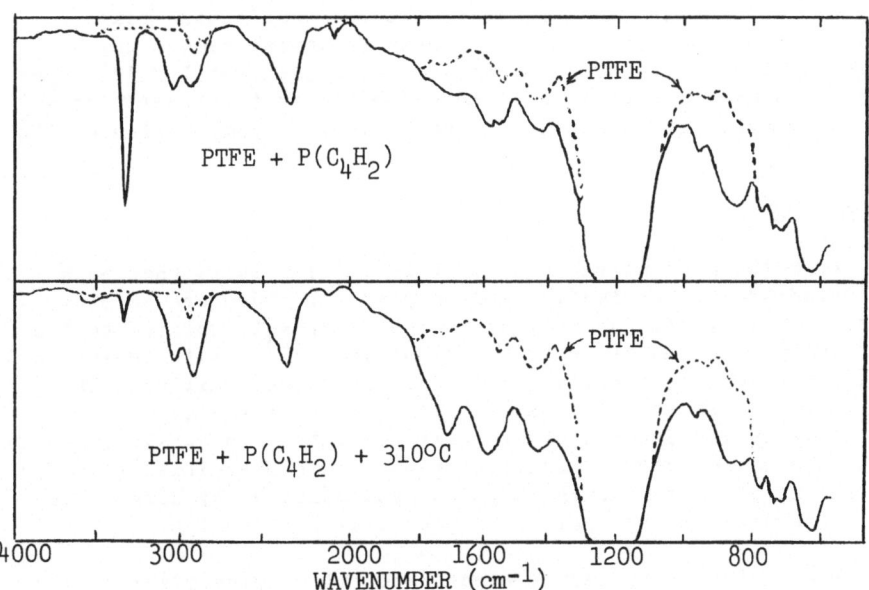

Figure 5.   IR Spectrum of PTFE Supported Poly(butadiyne) before
            and after 310°C Thermal Treatment.

100 to 400°C with maxima at 120 and 250°C. The second effect was
an unpredictable shifting in the crystalline melting points of
three of the substrate polymers; PE was unchanged, PVF2 shifted
from 163 to 113°C, PTFE shifted from 327 to 357°C and FEP shifted
from 267 to 273°C. TGA data on all samples were nearly identical
to the blank polymer substrates except for a very slightly lower
temperature of initial weight loss (10 to 20°C) and an excess
residue due to the polybutadiyne after decomposition of the
substrate polymer. Temperature induced changes in the IR
spectrum of a PTFE supported polybutadiyne film were monitored
over a 120 to 470°C temperature range for 60 minute heat treatment periods in vacuum at intervals of about 40°C. Spectral
changes occurred gradually with temperature and were mostly complete at 310°C (Figure 5). At that temperature, the acetylenic
bands at 3315 and 2090 cm^{-1} were one tenth their initial
intensity. The C–H absorptions at 3020 and 2935 cm^{-1} increased
by a factor of 1.2 and 2.3, respectively. The 1600 cm^{-1} aromatic
absorption increased by a factor of 2.7. Heat treatment above
410°C to 470°C caused a marked increase in the infrared back-
ground scattering, but no measurable weight loss in the sample.
We also noted no melting or flow above the PTFE melting point.
The characteristics of the ESR spectrum changed considerably.
The spin density increased from 8×10^{19} spin/gram to 2×10^{20}
spin/gram, and the linewidth narrowed from 12 to 1.5 gauss in
vacuum. The signal was no longer oxygen insensitive but displayed a strong, reversible pressure dependent oxygen broadening
from 1.5 gauss to 6 gauss on admission of air. The electrical
surface resistivity remained above the 10^{13} ohm/square measurement limit of the apparatus until heat treatment temperatures
above 400°C were reached. A surface resistivity of 7×10^{11} ohm/
square was measured after a 12 hour 470°C heat treatment. Heating was stopped at this temperature to avoid decomposition of the
PTFE.

DISCUSSION

From the preceeding results it is quite evident that both
the butadiyne polymerization and polymer structure are complex.
Currently, we think the polymerization is closely related to the
melt polymerization of substituted butadiynes[18]. With regard to
structure, we cannot write a well defined repeat unit with the
currently available data, but we can identify a specific functional group and examine close parallels with related systems
such as vinylacetylene, phenylacetylene and diphenyldiacetylene.
We will briefly examine these points and conclude by describing
some observations that relate to potential applications.

The ambient temperature vapor deposition polymerization of
butadiyne appears to involve an initial impregnation of the
substrate film by the butadiyne monomer followed by its polymerization. When a relatively thick (10 mil) teflon film on which
polybutadiyne has been deposited is cut and the cross-section

examined, a color gradient indicating a polybutadiyne concentration gradient is observed. The frequency dependence of the
acetylenic hydrogen stretching on the physical state (3315 cm^{-1}
in teflon vs. 3290 cm^{-1} on NaCl) and the ESR g-value (2.0030
in teflon vs. 2.0027 for a bulk sample) are consistant with
impregnation of the polybutadiyne. The DSC result of polybutadiyne induced variations in the crystalline melting point of
the substrate film can only be consistant with an impregnation of
the host polymer. Polybutadiyne deposition also causes changes
in the x-ray diffraction spectrum of the host polymer.

The structure of vapor deposited polybutadiyne, based on the
current amount of characterization, is an amorphous, paramagnetic, highly unsaturated polymer with a substantial quantity of
pendent, terminal acetylenic groups. The presence of the pendent
triple bond is seen quite clearly by the infrared absorptions at
3315, 2090 and 640 cm^{-1}. This indicates that, for many of the
butadiyne monomer residues, only one triple bond is consumed in
the polymer forming reaction. As to the polymer linking
structure, the broad infrared bands at 1600, 1210 and 885 cm^{-1}
are suggestive of a polyene (I) or polyaromatic structure (II or
III), although it is not possible to distinguish between them.
The 885 cm^{-1} band would correlate with the =C-H out-of-plane
deformation vibrations of a polyene, trisubstituted polyphenyl or
polyacene structure, the 1210 cm^{-1} band would correlate with the
=C-H in-plane vibrations and the 1600 cm^{-1} band with the C=C
stretching. The 3020 cm^{-1} =C-H stretching band is also consistent with the three structures. Such structures are proposed
for closely related systems such as poly(phenylacetylene), poly
(diphenyldiacetylene) and poly(vinylacetylene). The linear
phenyl substituted polyvinylene structure with 5 to 10 repeat
units has been proposed for thermally polymerized phenylacetylene[28]. This structure is reasonably well characterized by
hydrogenation and comparison of the infrared spectrum with that
of polystyrene, although some question exists as to the end group
identity and the source of an aliphatic C-H infrared band. The
thermal polymerization of diphenyldiacetylene is proposed to
yield an oligomeric structure composed of polyene and polyacene
blocks where the former structure may be converted to the latter
by postpolymerization heat treatment[18b,c]. The structure assignment is not based on direct detection of a polymer chain linking
chromophore or on chemical conversion to an identifiable
structure. It is based mostly on the proposition that a postpolymerization thermal treatment, which enhances the ultraviolet-
visible absorption without a corresponding molecular weight
change, must result from the interconversion between two such
structures. This reaction is also speculated to occur in the
structurally related vinylacetylene polymer (Eq. 4).[29] However,
this interconversion polymerization was not observed in either
the original poly(vinylacetylene) work[30] or a more current poly
(vinylacetylene) synthesis where the terminal acetylenic group
was protected by a trimethylsilyl group[31]. With regard to

(4)

the third polyphenyl structural possibility, it should be
noted that in both the phenylacetylene and diphenyldiacetylene
thermal polymerization systems the respective trimers,
1,3,5-triphenylbenzene[28a] and 1,2,4-triphenyl-3,5,6-tris
(phenylethynyl)benzene[18b] were isolated in small yield. It is
known that butadiyne is much more reactive, and a more facile
trimerization could easily be envisioned. Such a polymeric
structure was proposed for the $TiCl_2/AlCl_3$ catalyzed
cotrimerization polymerization of phenylacetylene and
diethynylbenzene[32]. However, the physical properties of this
system (solubility, infrared and ultraviolet spectra) are not a
good match with those of polybutadiyne.

    Our current hypothesis is that the butadiyne thermopolymer
is structurally related to melt polymerized substituted butadiyne
polymers. The absence of substituent groups allowed us to
make a direct observation of the gradual decrease in triple
bond concentration with increasing temperature of heat treatment.
This is consistent with the conversion of a polyene to a
polyacene structure, although it does not exclude the polyphenyl
structure. The presence of an aliphatic C-H band at 2935 cm^{-1}
and its intensity increase with heat treatment is paralleled by
the poly(diphenyldiacetylene) system[18b]. There is an oxygen uptake (4% O) on exposure to air and a strong ESR signal both of
which are also characteristic of the poly(diphenyldiacetylene)
system[18c]. These observations are not understood at this time.

    In seeking practical applications, we can cite observations
in three areas that may be of potential interest. First is the
prospect of coating polymeric materials either for protection or
for surface modification. It was observed in the process of
attempting to dope a polybutadiyne sample deposited on PTFE with
sodium naphthalide/THF that no weight change occurred after
several minutes immersion. However, a control sample, without
the polybutadiyne coating, was rapidly defluorinated loosing 30%
by weight. Along similar lines, it was thought that the polybutadiyne coating would stabilize polyacetylene to atmospheric
oxidative degradation which normally occurs over a period of
several days[33]. When a freshly prepared trans polyacetylene film
was exposed to an atmosphere of butadiyne, an uptake of 250 wt %
was observed over a two week period. In appearance the film
changed from metallic silver to brilliant blue. In texture the
film changed from flexible to rigid. Attempted subsequent doping
with iodine vapor resulted in no iodine uptake or measurable
electrical conductivity increase. Conversely, when the

polyacetylene film was first doped with iodine vapor to a high
electrical conductivity, then exposed to a butadiyne atmosphere,
no polybutadiyne deposition or conductivity change occurred. The
second area involves the modification of crystalline properties
of polymeric materials. In addition to crystalline melting point
changes, which can be quite large as was demonstrated with poly
(vinylidene fluoride), x-ray diffraction data indicates the d
spacing changes and crystallinity decreases. Finally, when
sufficient quantity of polybutadiyne is deposited in a substrate polymer and the polymer is heated well above its melting point, the sample does not flow or stick to supporting
surfaces, but maintains the integrity of its initial thin film
form. As an example, a polyethylene sample could be cycled
through a 250°C temperature without a change in sample dimensions
or weight. Current work is directed toward quantifying these
observations.

SUMMARY

The significant aspects of our polybutadiyne work may be
summarized as follows:

(1) Butadiyne does not have a susceptibility to Ziegler-
Natta polymerization comparable to acetylene.

(2) The vapor deposition polymerization of butadiyne on host
polymers involves a condensation and permeation of the substrate
prior to polymerization.

(3) The butadiyne thermopolymer structure is characterized
by IR, UV and ESR spectroscopy as having a polyconjugated chain
structure with pendant terminal acetylinic functional groups.

(4) The vapor deposited polybutadiyne has high thermal and
chemical stability and can reinforce permeated substrate
polymers.

REFERENCES

1. A. Bayer, Ber., 18, 2269 (1885).
2. F. Zobel and W. Hunsman, Ger. Patent 860,212 (1952),
    CA50:4510a.
3. N. Shachat, J. Org. Chem., 27, 2928 (1962).
4. F. G. Muller, Helv. Chim. Acta. 8, 826 (1925).
5. H. H. Nelson, U. S. Patents 2715101 (1955), 2861041 (1959);
    D. W. McDonald, U. S. Patent 2965565 (1960).
6. N. L. Desai, The Polymerization of Diacetylene, Ph.D.
    thesis, Pennsylvania State University, PA. (1965).
7. I. A. Chekulaeva, V. A. Ponomarenko, I. B. Bystrova and
    G. V. Talypina, Vysokomol. Soedin., Ser B9, 652 (1967)
    Vysokomol. Soedin., Ser. A12, 1180 (1970).

8. I. A. Chekulaeva, and V. A. Ponomarenko, Vysokomol. Soedin.,
    Ser. B16, 126 (1974).
9. G. Wegner, Makromol. Chem., 154, 35 (1972).
10. P. J. Russo and M. M. Labes, Chem. Comm., 53 (1982).
11. J. B. Armitage and E. R. Jones, J. Chem. Soc., 44 (1951).
12. A. V. Jones, Proc. Roy. Soc. (London), A211, 285 (1952).
13. K. K. Georgieff and Y. Richard, Can. J. Chem., 36, 1280 (1958).
14. R. Tedeschi and A. Brown, J. Org. Chem., 29, 2051 (1951).
15. F. Bischoff and H. Adkins, J. Amer. Chem. Soc., 46, 256 (1924).
16. T. Ito, H. Shirakawa and S. Ikeda , J. Polym. Sci., Polymer
    Chem., 12, 11 (1974).
17. D. H. Dawes, and C. A. Winkler, J. Poly. Sci., A2, 3029
    (1964)
18. M. G. Chauser, I. D. Kalikhman, M. I. Cherkashin and
    A. A. Berlin, Izv. Akad. Nauk. SSSR, Ser. Khim., 2421
    (1969) (b) A. A. Berlin, M. I. Cherkashin, M. G. Chauser
    and R. R. Shifrina, Vysokomol. Soed., A9, 2219 (1967)
    (c) M. G. Chauser, M. I. Cherkashin, M. Ya. Kushnerev,
    T. I. Protsuk and A. A. Berlin, Vysokomol. Soed.,
    A10, 916 (1968) (d) M. G. Chauser, I. D. Kalikhman,
    M. I. Cherkashin and A. A. Berlin, Vysokomol. Soed.,
    A12, 1022 (1970).
19. A. W. Snow, Nature, 292, 40 (1981).
20. R. A. Nyquist, and W. J. Potts, Spechrochim. Acta., 16, 419
    (1960).
21. GH. D. Mateescu, "Infrared Spectroscopy", Wiley–Interscience,
    New York (1972) Part II, Chapter 1.
22. W. D. Celmer, and I. A. Solomons, J. Amer. Chem. Soc., 75,
    1372 (1953).
23. A. W. Snow, Carbon 19, 467 (1981).
24. Apparatus designed by Dr. H. Wohltjen.
25. AsF$_5$ is one of the strongest doping agents for inducing high
    electrical conductivity in conjugated polymers. R. H.
    Baughman, J. L. Bredas, R. R. Chance, R. L. Elsenbaumer and
    L. W. Shaklette, Chem. Rev. 82, 209 (1982).
26. J. L. Bredas, R. R. Chance and R. H. Baughman, J. Chem.
    Phys., 76, 3673 (1982).
27. A. W. Snow and J. R. Griffith, IUPAC Macromolecular
    Symposium, Amherst, MA, 1982. Preprint, p. 432.
28. (a) Y. Okamoto, A. Gordon, F. Movsovicius, H. Hellman and
    W. Brenner, Chem. and Ind., 2004 (1961) (b) J. H. Lai,
    Macromolecules 10, 1253 (1977).
29. V. N. Salaurov, Yu. G. Kryazhev, T. I. Vakul'skaya, and
    M. G. Voronkov, Makromol. Chem., 175, 757 (1974).
30. C. C. Price and T. F. McKeon, J. Poly. Sci., 41, 445 (1959).
31. I. Kaneko, and N. Hagihara, J. Poly. Sci., Polymer Lett., 9,
    275 (1971).
32. W. Bracke, J. Poly. Sci., A-1, 10, 2097 (1972).
33. H. W. Gibson and J. M. Pochan, Macromolecules, 15, 249
    (1982) Osterholm, J. E., Levenson, C. L. and Yasuda, H. K.,
    J. Applied Poly. Sci., 27, 931 (1982).

SYNTHESIS OF COPOLYMERS OF m-DIISOPROPENYLBENZENE

AND m-DIMETHOXYBENZENE

Robert Alan Smith, Dennis B. Patterson
and Howard A. Colvin

The Goodyear Tire & Rubber Co
142 Goodyear Blvd
Akron, OH 44316

INTRODUCTION

The polymerization of diisopropenylbenzenes (DIPB's) was first reported over 25 years ago. As opposed to free radical initiated polymerizations, which produce cross-linked gels ionic initiated polymerizations produce soluble (uncrosslinked) polymer. Anioic techniques, for examples, can produce linear polymer in which only on unsaturation of each DIPB is consumed. The aromatic ring of each pendant group thus carries an unreacted isopropenyl group.[1]

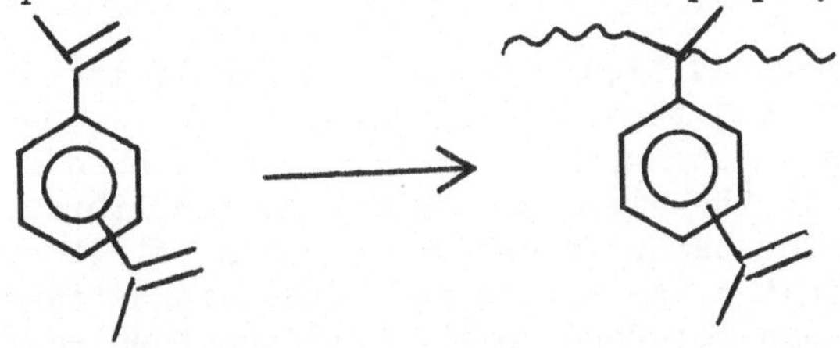

The cationic polymerization of DIPB's has been found to produce polymers containing predominantly a polyindane structure arising from a step-growth mechanism.[2] D'Onofrio, for example, showed that soluble polyindane is produced at polymerization temperatures above 70°C using Lewis acid initiation systems.[3]

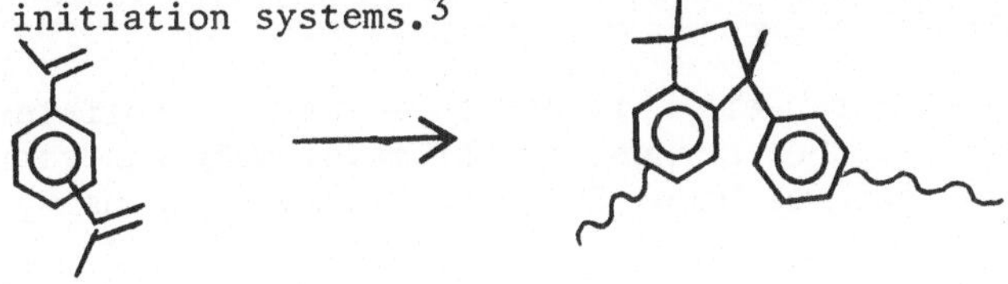

415

Direct cationic copolymerization of diisopropenylbenzenes with electron rich aromatics has only been reported using phenol[4] as co-monomer. The process is complicated by the simultaneous occurrence of propagation between successive m-DIPB units as well as alkylation of the electron rich aromatic by the m-DIPB carbenium ion. The occurrence of the propagation reaction leads to points of trifunctionality which produces branching and ultimately gelation. Consequently, alkylation must be maximized relative to propagation.

Alkylation is favored by increasing the ability of the aromatic ring to stabilize the positive charge resulting from reaction with the carbenium ion. Consequently, m-dimethoxybenzene, having two reactive positions located ortho/para to two methoxy groups should be an acceptable comonomer. This paper reports the first direct copolymerization of m-diisopropenylbenzene (m-DIPB) with m-dimethoxybenzene (m-DMB).

EXPERIMENTAL

All materials were used as received. Polymerizations were conducted in 30 ml vials which were capped with a Teflon lined self sealing gasket. Two different methods of monomer addition were used:

i) Monomer addition Method A - Stoichiometric amounts of m-DIPB and m-DMB were charged into a vial along with the solvent. The head space was flushed with nitrogen and the vial capped and equilibrated at the desired temperature in a constant temperature bath equipped with a motor driven rotor. A master solution of either Lewis or Bronsted acid was added and the vial replaced in the bath.

ii) Monomer addition Method B - One-half the stoichiometric amount of m-DIPB and the entire stoichiometric amount of m-DMB was placed in a vial along with the solvent. The head space was flushed with nitrogen and the vial was capped and equilibrated at the desired temperature in the bath. A master solution of either Lewis or Bronsted acid initiation system was added via syringe through the cap. The vial was returned to the bath and tumbled for 2 hours at the desired temperature at which time the remaining half of the m-DIPB was added via syringe.

The copolymers were purified by twice reprecipitating in methanol.

NMR spectra were obtained on a Varian Associates A360 60 MHz spectrometer. Gel permeation chromatography was carried out on a Water's Associates Model 302 High Pressure GPC using polystyrene

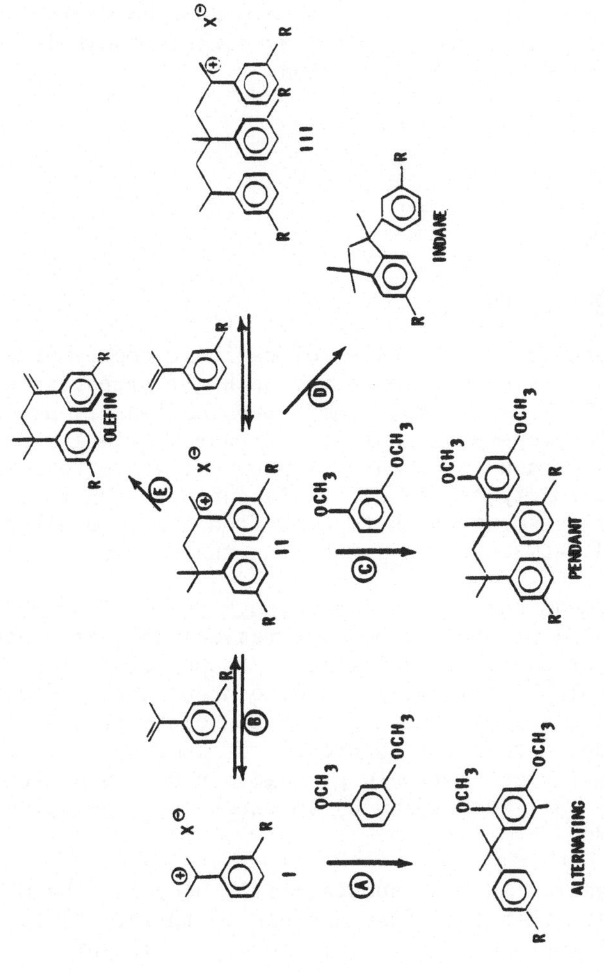

FIGURE 1.   Mechanistic origin of various enchainments in m-dimethoxy-
benzene/m-diisopropenylbenzene copolymerization.

equivalents. Weight average molecular weights were determined on
a Chromatix KMX-6 Low Angle Laser Light Scattering Spectrophotometer
(LALLS).

    Copolymer structure was determined using 60 MHz ^{1}H NMR spectros-
copy as will be outlined in a forthcoming publication.[5] Total m-DMB
content was obtained by comparing the total methoxy methyl resonance
area to the total aliphatic resonance area. The amounts of pendant
and alternating m-DMB residues were obtained by comparing the areas
of the magnetically unequivalent methoxy methyls located β (2.65 ppm)
or α (3.15 ppm) to a site of alkylation.

RESULTS AND DISCUSSION

Expected Copolymer Structure

    Figure 1 details the mechanism of cationic copolymerization of
m-DIPB and m-DMB showing the competing paths of propagation and
alkylation (R= ⟨, ↞). Simple cumyl-type cations I, produced by
protonation of isopropenyl groups, can either alkylate m-DMB (or
alkylated m-DMB) or add to another isopropenyl group to form dimer
cations II. The moiety formed by alkylation of m-DMB is a precursor
to an alternating segment which results from a second alkylation of
the incorporated m-DMB nucleus furthering chain growth.

    Type II cations can then react through any of four pathways.
Alkylation of m-DMB produces a pendant residue whereas proton loss
produces indane or olefin enchainments. Propagation to give trimeric
cations III (or higher homologs) can also occur but is much less
likely than with the simple cations I because the propagation rate
of α-methylstyrene derivatives decreases markedly with increasing
chain size.[6] Sufficient cationic propagation occurs at the temperature employed in this work (0-60°C) to cause rapid gelation in the
absence of m-DMB.

    Such propagation must be substantially overriden by m-DMB
alkylation (in addition to indane and olefin formation) to produce
the soluble copolymers which are described in this work.

    From the mechanism, four possible linakges can result:

        i) alternating copolymer (path A)
       ii) pendant copolymer (path C)
      iii) indane (path D)
       iv) olefin (path E)

Variation of Copolymer Structure With Experimental Parameters

Copolymerizations were conducted using various initiation systems in several solvents between 0 and 60°C.

Initiation System

The nature of the initiation system was found to be the most important variable affecting copolymer structure with respect to both m-DMB content and mode of m-DMB incorporation. Various Lewis and Bronsted acids were used and representative data are displayed in Table I.

Three different Lewis acids were examined. Boron trifluoride etherate used over a wide temperature range, provided soluble co-polymer with relatively high m-DMB content. Titanium tetrachloride used over the same temperature range produced soluble copolymers having lower amounts of m-DMB incorporated (c.f. samples 1 and 2). Stannic chloride produced gelled products under all attempted co-polymerization conditions. In general, Lewis acids produced co-polymers containing predominantly polyindane linkages with moderate amounts of m-DMB incorporated as monoalkylated, pendant residues. A typical copolymer NMR spectrum is shown in Figure 2a.

Three different Bronsted acids were examined as to their ability to initiate m-DMB/m-DIPB copolymerizations. Trifluoromethanesulfonic (triflic) acid and Super Filtrol (clay impregnated with sulfuric acid) were found to be effective initiators. Sulfuric acid alone did not initiate copolymerizations.

As can be seen in Table I the two Bronsted acids produced quite different copolymers. Triflic acid produced copolymers having the largest amount of m-DMB incorporated predominantly as alternating enchainments. Super Filtrol produced copolymers similar in structure to those produced using BF_3OEt_2. An NMR spectrum of a typical co-polymer prepared using CF_3SO_3H initiation is shown in Figure 2b. The much higher amounts of total m-DMB incorporation and m-DMB in-corporated by alternating enchainments are reflected in the larger total methoxy resonance (2.65 and 3.15 ppm) and predominance of the 2.65 ppm resonance over that at 3.15 ppm in spectrum b as compared to spectrum a, respectively.

Copolymerization Temperature

Copolymerizations were conducted between 0 and 60°C using BF_3OEt_2 and CF_3SO_3H initiation systems. The results are shown in Table II. Neither total m-DMB content nor the mode of m-DMB incor-poration (alternating or pendant) varied much over the temperature range used. Temperatures lower than 29°C could not be used with BF_3OEt_2 initiation in the solvents mentioned because of gelation.

TABLE I

Structure of m-Dimethoxybenzene/m-Diisopropenylbenzene Copolymers
Produced Using Various Initiation Systems by Monomer Addition Method A

Sample	Initiation System	Plzn Temp, °C	Plzn Solvent	m-DMB Content	
				Mole % Incorp.	Alternating/Pendant %
1	BF_3OEt_2[a]/"H_2O"	60	Chlorobenzene	28	11/89
2	$TiCl_4$[a]/"H_2O"	"	"	10	0/100
3	BF_3OEt_2[a]/"H_2O"	39	"	28	13/87
4	Super Filtrol[b]	60	"	27	27/73
5	CF_3SO_3H[c]	"	"	42	62/38

a.   0.27 M
b.   8.7 parts per hundred parts monomer (phm)
c.   .007 M

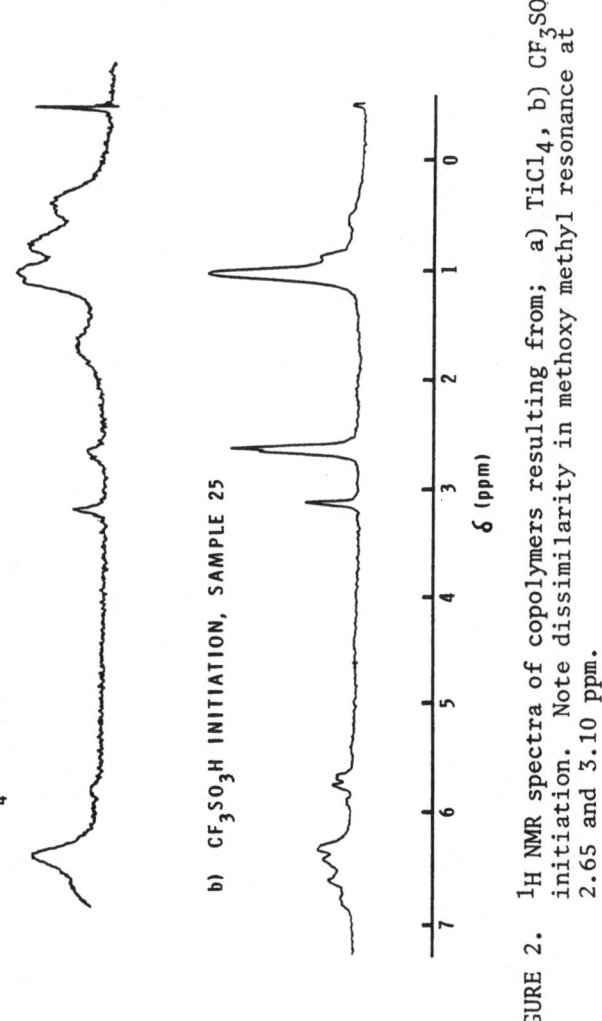

FIGURE 2. ^{1}H NMR spectra of copolymers resulting from; a) TiCl$_4$, b) CF$_3$SO$_3$H initiation. Note dissimilarity in methoxy methyl resonance at 2.65 and 3.10 ppm.

TABLE II

Variation of Copolymer Structure With Polymerization Temperature
In Chlorobenzene Solvent

Sample	Initiation System	Plzn Temp, °C	m-DMB Content Mole % Incorp.	Alternating/Pendant %
6	BF_3OEt_2[a]/"H_2O"	29	32	11/89
3	" "	39	28	13/87
8	" "	50	27	20/80
1	" "	60	28	11/89
9	CF_3SO_3H[b]	0	43	53/47
10	"	20	49	50/50
11	"	40	48	50/50
12	"	55	29	38/62

a. $[BF_3OEt_2] = .027$ Monomer addition method B
b. $[CF_3SO_3H] = .007$ "   "   "   A

Copolymerization Solvent

The structure of copolymers resulting from BF_3OEt_2 and triflic acid initiation in solvents of varying polarity is shown in Table III. Very little change in the structure of the copolymers occurred as the solvent polarity increased from heptane to toluene to chlorobenzene for each initiation system. The amount of m-DMB incorporated into the BF_3OEt_2 coinitiated systems decreases upon going from chlorobenzene to methylene chloride solvent at 60°C. Gelation problems prohibited use of methylene chloride as copolymerization solvent at 40°C with BF_3OEt_2.

Molecular Weight Characterization

An extensive GPC analysis of the copolymers was carried out. Comparison of weight average molecular weights obtained from the GPC chromatograms (polystyrene equivalents) with those from the absolute method, Low Angle Laser Light Scattering (LALLS), revealed the GPC values to be inaccurate.

Table IV displays representative $\overline{M}w$ values, determined by LALLS, for copolymers initiated with BF_3OEt_2 and CF_3SO_3H. As shown in the table, copolymers produced using BF_3OEt_2 generally have a higher weight average molecular weight than those produced using CF_3SO_3H.

Figure 3 compares chromatograms of copolymers produced using BF_3OEt_2 and CF_3SO_3H. The MWD of the copolymer produced using BF_3OEt_2 is considerably broader and displays multimodality. This likely reflects the higher degree of branching in these copolymers. The higher degree of branching in turn reflects a higher tendency for cations of Type II to be formed (Figure 1) and to proceed on to cations III (and higher homologs), which serve as branch points. Because of the breadth of the MWD of the copolymers produced using BF_3OEt_2 osmometric determination of $\overline{M}n$ was not attempted.

Correlation of Copolymer Structure with Copolymerization Mechanism

Figure 1 details the entire mechanism of the copolymerization. Protonation of an isopropenyl group by an acid of general structure HX(HX=CF_3SO_3H, BF_3OH_2, $TiCl_4OH_2$ etc) produces a simple cumyl-type cation I (R= $\langle$, $\longleftarrow$). Subsequent reaction occurs either by alkylation of m-DMB ultimately forming an alternating enchainment (path A) or by propagation to form carbenium ion II (path B). Subsequently this carbenium ion alkylates m-DMB to produce a pendant copolymer (path C) or loses a proton to form either indanyl or olefinic enchainments (paths D and E respectively).

As shown in Table I, copolymer structure is very dependent on the nature of the initiation system. Such dependency arises from the nature of X^- produced from each general acid HX:

TABLE III

Variation of Structure With Polymerization Solvent

Sample		Temp, °C	Solvent	Solvent Dielec. Const.	m-DMB Content	
					Mole % Incorp.	Alternating/ pendant
13	BF_3OEt_2[a,c]/"H_2O"	40	Heptane	1.9	30	9/91
14	"	"	Toluene	2.4	35	0/100
15	"	"	Chlorobenzene	5.7	33	0/100
1	"	60	"	"	28	11/89
16	"	"	Methylene Chloride	9.1	19	0/100
17	CF_3SO_3H[b,d]	20	Heptane	1.9	48	55/45
18	"	"	Toluene	2.4	50	53/47
19	"	"	Chlorobenzene	5.7	49	50/50

a. [m-DIPB] = [m-DMB] = 1.96, Monomer addition method B
b. "   "   "   ,   "   "   "   A
c. [BF_3OEt_2] = .027
d. [CF_3SO_3H] = .002

TABLE IV

Representative $\overline{M}w$ Values for M-DIPB/m-DMB Copolymers as Determined by LALLS

Sample[a]	Init System	Solvent	Temperature	$\overline{M}w \times 10^{-3}$
20[b]	$BF_3OEt_2/"H_2O"$	Chlorobenzene[d]	40	20.5
21[c]	"	"	"	43.6
22	"	"	"	36.7
23	CF_3SO_3H	Methylene Chloride	"	18.3
24	"	Heptane[e]	0	6.7
17	"	"	20	32.5
25	"	"	40	10.8
26	"	"	55	6.4

a.   Unless otherwise noted; $[BF_3OEt_2]=.027$, $[CF_3SO_3H]=.002$
b. $[BF_3OEt_2]=.014$
c. $[BF_3OEt_2]=.017$
d.   Monomer addition method B
e.   "   "   "   A

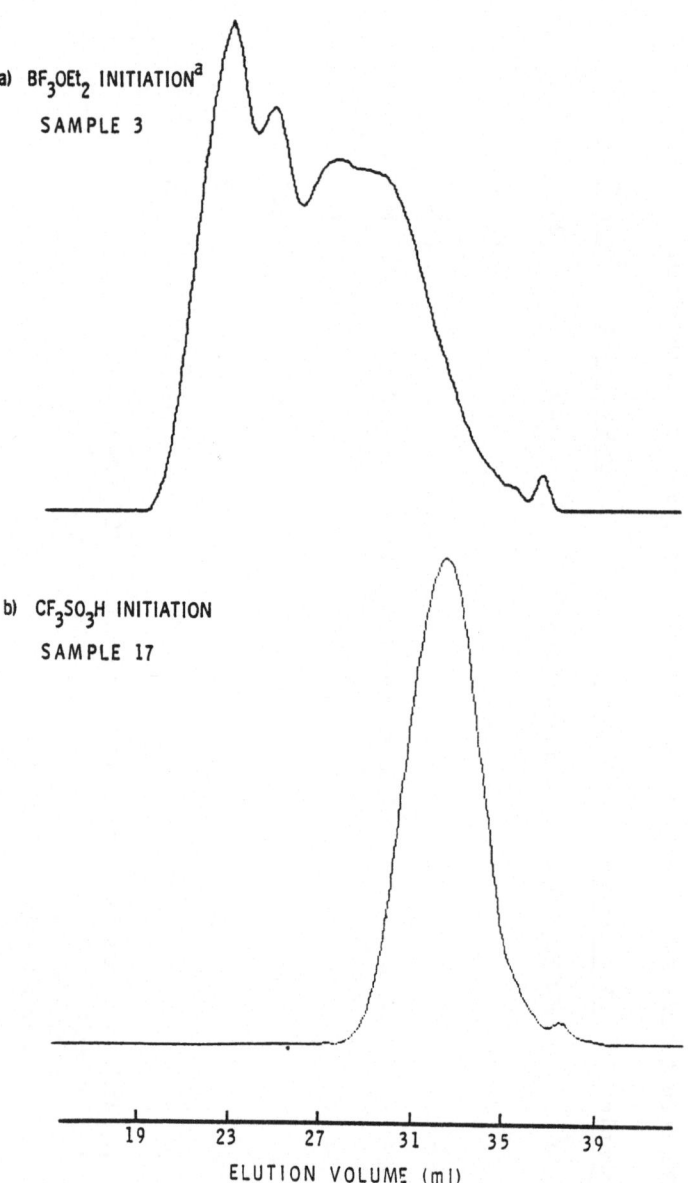

FIGURE 3.   GPC traces of copolymers prepared
using;   a) BF_3OEt_2, b) CF_3SO_3H
initiation.

The size and nucleophilicity of X^- affects the relative reactivity of the carbenium ion toward the various electron rich substrates in the system.[7] Thus X^- influences the propensity of carbenium ions I and II to alkylate m-DMB (paths A and C) or undergo other reactions (paths B, D and E).

As previously mentioned, copolymers produced using Lewis acid coinitiators, BF_3OEt_2 and $TiCl_4$, contain nearly exclusive pendant incorporation of m-DMB. Thus when $X^-=BF_3OH^-$ or $TiCl_4OH^-$ conversion of type I to type II carbenium ions completely predominates, or nearly so, over m-DMB alkylation, i.e. path B>>A.

The difference in composition between BF_3OEt_2 and $TiCl_4$ coinitiated copolymers arises from the different reaction preference of the respective dimeric ion pairs. Copolymers produced with BF_3OEt_2 usually contain $\sim$ 30 mole % m-DMB with the remainder consisting of predominantly indane enchainments indicating paths C and D to be of comparable importance. Titanium tetrachloride coinitiated copolymers contain a preponderance of indane enchainments showing path D to be predominant.

The two Bronsted acids used as initiators also produced copolymers of quite different structure from each other. Super Filtrol initiated copolymers contained moderate amounts of alternating copolymer indicating a larger (though still subordinate) contribution of path A as compared to path B with respect to Lewis acid coinitiated copolymerizations. Ion pair II displays comparable propensities to react through paths C and D in copolymerizations initiated with Super Filtrol as with BF_3OEt_2.

The structure of the counterion present in Super Filtrol initiated copolymerizations is unknown. However, the material consists of clay-bound sulfuric acid and presumably the counterion is sulfate bound to a silico-aluminate matrix.

Triflic acid copolymerizations result in significant amounts of alternating copolymer. The triflate counterion results in comparable rates of alkylation and propagation (paths A and B). Propagation yields ion pair II which predominantly alkylates m-DMB (path C) resulting in only minor amounts of indane enchainments (path D).

The reason for the exceptional ability of triflate counterion to mediate the copolymerization in favor of m-DMB enchainment is difficult to ascertain. A recent study of Hasegawa and Higashimura[8] on the cationic polymerization of divinylbenzene showed reduced propagation tendency relative to chain transfer when triflic acid initiation was used. This implies that increased m-DMB incorporation from triflic acid initiation is a result of reduced propagation rather than enhanced alkylation.

Tables II and III show that copolymer structure varies little with copolymerization solvent and temperature. Unfortunately, the range of solvent polarities used in the study was rather limited. The relatively polar methylene chloride could not be used below 50°C as gelation occurred, and the near explosive rate of the copolymerizations prevented it's use at higher temperatures.

These limitations notwithstanding, some variation in structure should be expected over the solvent and temperature conditions employed. Changing the temperature and solvent should affect the ionicity (degree of ionization) of the active center and hence alter it's reactivity

This insensitivity to solvent polarity and temperature may relate to early cationic copolymerization studies carried out by Overberger, Arnold and Taylor[9] and by Marvel and Dunphy.[10] These groups found reactivity ratios for the cationic copolymerization of styrene and α-methylstyrene with chlorostyrene to also be insensitive to solvent and temperature changes.

Subsequently, Overberger and coworkers[11,12] proposed that insensitivity of reactivity ratios of these experimental parameters resulted from preferential solvation of the propagating carbenium ion by the most polar component present in the system (either nitrobenzene from the mixed solvent system they employed or chlorostyrene monomer). The active center in m-DIPB/m-DMB copolymerizations could be preferentially solvated by free m-DMB thus screening any effect of changing bulk solvent. However, Overberger, Ekrig, and Tanner[11] noted no dependence of reactivity ratios on initiation system whereas m-DMB/m-DIPB copolymerizations are very dependent on the nature of the initiation system. Obviously, a more detailed study is required.

CONCLUSIONS

The direct cationic copolymerization of m-diisopropenylbenzene and m-dimethoxybenzene can be conducted using Lewis or Bronsted acid initiation in any of several different solvents over a range of temperatures. Two modes of m-DMB incorporation are possible.

    i)   Alternating enchainment resulting from attack of
        a cumyl type carbenium ion on a m-DMB nucleus.

   ii)   Pendant attachment resulting from attack of a
        dimeric carbenium ion on a m-DMB nucleus.

The relative amount of each mode of m-DMB incorporation is
dependent upon the initiation system. Copolymer structures ranging
from predominantly indane with a small amount of pendant m-DMB residues to a mixture of alternating m-DMB and pendant m-DMB enchainments
were obtained with different initiation systems. Copolymer structure
was insensitive to solvent or temperature changes.

The $\overline{M}w$ of copolymers produced using BF_3OEt_2 coinitiation was
generally found to be higher than when triflic acid initiation was
utilized. The molecular weight distribution, as determined by GPC,
was also much broader in the copolymers prepared using BF_3OEt_2 suggesting that copolymers prepared with this coinitiator are more
highly branched.

ACKNOWLDEGEMENT

The assistance of C. T. Enos , N. L. Dotson and D. J. Keith
for GPC and LALLS analysis; Dennis Romain for artwork and Rose Kordinak
for typing the manuscript is gratefully acknowledged.

REFERENCES

1. P. Lutz, G. Beinert, P. Rempp, Makromol. Chem. 183, 2727 (1982).
2. H. Brunner, A. L. L. Palluel and D. J. Walbridge, J. Polym. Sci.,
    28, 629 (1958).
3. A. A. D'Onofrio, J. Appl. Polym. Sci., 8, 521 (1964)
4. L. F. Sonnabend, U.S. Patent 3,004,953 (1961).
5. R. A. Smith, D. B. Patterson and C. T. Enos, J. Polym. Sci.,
    Polym. Chem. Ed. to be submitted.
6. S. Bywater in "The Chemistry of Cationic Polymerization,"
    P. H. Plesch Ed., Pergamon Press, New York 1963 , p. 311.
7. J. P. Kennedy, "Cationic Polymerization of Olefins: A Critical
    Inventory," Wiley-Interscience, New York, 1980, p. 24.
8. H. Hasegawa and T. Higashimura, Macromolecules, 13, 1350 (1980).
9. C. G. Overberger, L. H. Arnold and J. J. Taylor, J. Amer. Chem.
    Soc., 73. 5541 (1951).
10. C. S. Marvel and J. F. Dunphy, J. Org. Chem., 25, 2209 (1960).
11. C. G. Overberger, R. J. Ehrig and D Tanner, J. Amer. Chem. Soc.,
    76, 772 (1954).
12. C. G. Overberger and V. G. Kamath, J. Amer. Chem. Soc., 85, 446
    (1963).

PROGRESS IN RING-OPENING POLYMERIZATION OF CYCLIC ETHERS AND

CYCLIC SULFIDES

David A. Tirrell

Department of Chemistry
Carnegie-Mellon University
Pittsburgh, PA 15213

INTRODUCTION

Studies of ring-opening polymerization of cyclic ethers and cyclic sulfides are underway in many laboratories throughout the world. A search of eleven leading polymer science journals, made in the course of preparing the present review, revealed 109 papers on these subjects just in the period 1980-1982 (1). This chapter makes no attempt to be comprehensive in its treatment of this active area of research. Instead, discussion is limited to important advances of three kinds:

A. New Initiator Systems
B. New Methods of Synthesis of Block and Graft Copolymers and
C. New Reactive Polyethers and Polysulfides

Emphasis is placed on important developments which have occurred since 1980.

NEW INITIATOR SYSTEMS

The development of new polymerization initiators is one of the continuing objectives of polymer synthesis research. Two developments stand out among recent studies of this kind: the work of Crivello and coworkers on iodonium and sulfonium salt initiators, and that of Inoue's laboratory on metalloporphyrin systems. Crivello's research has produced several new classes of cationic initiators, some of which may be triggered by light, others by heat, and still others by reducing agents. Inoue's development of the metalloporphyrin initiators has allowed, for the first time, very general syntheses of living epoxide polymers and block copolyethers. Each of these developments is discussed below.

Iodonium and Sulfonium Salts

Crivello and Lam introduced the diaryliodonium salts (I) as photoinitiators for cationic polymerization in 1976 (2). The efficiency of photopolymerization was shown (not surprisingly) to be

$$\text{Ar} - \text{I}^+ - \text{Ar}' \quad \text{MX}_n^- \qquad\qquad\qquad \text{I}$$

highly dependent on the nature of the counterion, MX_n^-, with the complex metal halide ions BF_4^-, PF_6^-, AsF_6^- and SbF_6^- being the most thoroughly studied. (3). As is the rule in cationic polymerizations, initiators containing necleophilic anions such as the simple halides are unsuitable.

The development of photoinitiators for cationic polymerization is an important one, with both practical and fundamental applications. Practical uses are primarily in the area of light-induced curing of coatings, a process which is expected to be energy-conservative in comparison with conventional thermal curing (4). The advantages of using such photoinitiators in fundamental studies of cationic polymerization include the capability of generating the initiating species at controlled and variable rates, and the ability to vary the counterion while holding constant all other conditions of polymerization.

Careful studies of the photodecompositions of diaryliodonium salts by Crivello and Lam (2,3) have provided convincing evidence that the initiating species in these polymerizations are the Bronsted acids HMX_n, generated according to Eq. 1.

$$\text{Ar}_2\text{I}^+\text{MX}_n^- \xrightarrow{\ h\nu\ } [\text{Ar}_2\text{I}^+\text{MX}_n^-]^* \qquad\qquad \text{1a}$$

Major Pathway $\quad [\text{Ar}_2\text{I}^+\text{MX}_n^-]^* \longrightarrow \text{ArI}^+\cdot + \text{Ar}\cdot + \text{MX}_n^- \qquad \text{1b}$

$$\text{ArI}^+\cdot + \text{SH} + \text{MX}_n^- \longrightarrow \text{ArI}^+\text{H} + \text{S}\cdot + \text{MX}_n^- \qquad \text{1c}$$

$$\text{ArIH}^+ + \text{MX}_n^- \longrightarrow \text{ArI} + \text{H}^+\text{MX}_n^- \qquad\qquad \text{1d}$$

Minor Pathway $\quad [\text{Ar}_2\text{I}^+\ \text{MX}_n^-]^* + \text{SH} \longrightarrow \text{ArSH}^+ + \text{ArI} + \text{MX}_n^- \qquad \text{1e}$

$$\text{ArSH}^+ + \text{MX}_n^- \longrightarrow \text{ArS} + \text{H}^+\text{MX}_n^- \qquad\qquad \text{1f}$$

The diaryliodonium salts are highly stable in the absence of light, but decompose when irradiated at 313 or 365 nm with quantum yields of 0.2-0.3 (3). The rate of photodecomposition is independent of the structure of the counter anion, and is insensitive to temperature and to atmospheric oxygen (3). No thorough investigation of solvent effects has been reported, but decomposition rates appear to be identical in acetone and acetonitrile when quartz tubes are used and the irradiation source is a water-cooled Hanovia 450 W medium pressure mercury lamp (2). Decomposition in nitromethane is slower, probably as a result of the increased light absorption by this solvent.

Crivello and Lam have demonstrated the use of diaryliodonium salts as photoinitiators for polymerization of electron-rich olefins, cyclic ethers, cyclic sulfides, lactones and spiro orthoesters, but a vast majority of their published work concerns polymerization of substituted oxiranes, illustrating the potential of such systems in photocuring of epoxy resins. Such polymerizations can be quite fast; in the most favorable example, a 93% yield of polymer of $\bar{M}_n$ 10,700 was obtained from 3-vinylcyclohexene oxide after only 90 seconds of irradiation at room temperature with 4,4'-di-tert-butyldiphenyliodonium hexafluoroantimonate as initiator (3). The substituted salts are often preferred to the simple unsubstituted diphenyliodonium compounds for reasons of solubility (2). The use of diaryliodonium salts in combination with various dyes allows one to initiate cationic polymerization with visible light (5).

Crivello and Lam extended their pioneering studies of iodonium salt-initiated cationic polymerizations in a series of largely parallel investigations of photosensitive triarylsulfonium salts (6-10). These compounds decompose via a mechanism which is directly analogous to that shown in Eq. 1, with $C-S^+$ bond cleavage as the key first step. The most significant differences between the iodonium and sulfonium initiators appear to be the greater sensitivity of the behavior of the sulfonium salts to changes in aryl substitution (6), and the greater versatility of sensitization of the iodonium compounds (7). A single paper has appeared which describes photoinitiation by triarylselenonium salts (11).

Two additional classes of sulfonium salt initiators, also introduced by Crivello and Lam, are the dialkylphenacylsulfonium salts (II) (12,13) and the dialkyl-4-hydroxyphenylsulfonium salts (III) (13,14). The initiating species generated upon irradiation of such compounds is again the Bronsted acid HMX_n^-, but there is an important difference between these compounds and the diaryliodonium and triarylsulfonium salts discussed previously. Irradiation of compounds II and III results not in irreversible decomposition via carbon-heteroatom bond cleavage, but rather in reversible proton transfer from the excited sulfonium ion to produce the sulfonium ylid (IV or V):

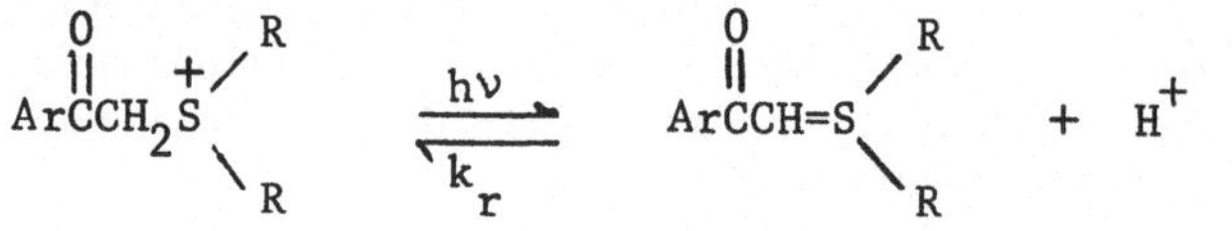

The reversibility of proton transfer limits to some extent the nature of the monomers which are subject to photopolymerization with these initiator systems, but epoxides, trioxane and vinyl ethers work very well (12,14). The sensitization of cationic polymerizations photoinitiated by compounds II and III has been reported by Crivello and Lee (13).

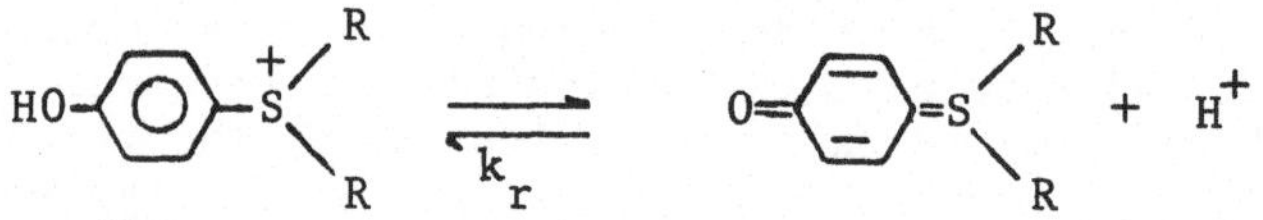

The most recent work of the General Electric group has addressed the development of diaryliodonium salts as thermal (15) or redox (16) initiators of cationic polymerization. These systems contain Cu^I (added as such or generated by reduction of added Cu^{II}), which serves to reduce the iodonium salt. Reduction of the iodonium salt produces the cationating agent (either H^+ or Ar^+) which initiates chain growth. The authors suggest eq. 2 as the mechanism of initiation (15).

$$ROH + 2Cu^{II}L_2 \xrightarrow[\text{or } \Delta]{\text{R.T.}} R=0 + 2Cu^IL + 2HL \qquad 2a$$

$$Ar_2I^+MX_n^- + Cu^IL \longrightarrow ArI + [ArCu^{III}LMX_n^-] \qquad 2b$$

$$[ArCu^{III}LMX_n^-] + \bigcirc \longrightarrow Ar-\overset{+}{\bigcirc}MX_n^- + Cu^IL \qquad 2c$$

or

$$[ArCu^{III}LMX_n^-] + ROH \longrightarrow ArOR + H^+MX_n^- + Cu^IL \qquad 2d$$

$$H^+MX_n^- + \bigcirc \longrightarrow H-\overset{+}{\bigcirc}MX_n^- \qquad 2e$$

The presence of phenoxy endgroups in polyethers prepared by thermal initiation of epoxides with diaryliodonium salts was verified by ultraviolet spectroscopy (15), but the authors do not comment on the fraction of such endgroups (i.e., on the relative importance of Eqns. 2c and 2e).

The key first step in Eq. 2 is the reduction of Cu^{II} to Cu^{I}. Crivello has suggested that this might be done in either of two ways: i) by adding to the system a strong hydroxylic reducing agent such as ascorbic acid (16), or ii) by allowing adventitious hydroxylic impurities in the monomer to serve as (much weaker) reducing agents (15). By selection of reducing agents of varying reducing power, one can achieve rapid room temperature redox cures (e.g., with ascorbate) or thermally-triggered cures (with no added reducing agent). Cyclohexene oxide, epichlorohydrin, tetrahydro-furan and trioxane all have been successfully polymerized using Cu-modified diaryliodonium salt initiator systems.

Metalloporphyrin Initiator Systems

Takeda and Inoue in 1978 reported a rapid polymerization of propylene oxide at room temperature in the presence of a catalyst prepared from equimolar amounts of tetraphenylporphyrin ($TPPH_2$) and diethylaluminum chloride (Et_2AlCl) (17). In 1981, the same laboratory expanded this initial report in a series of three papers (18-20) which described: i) the molecular weight distribution of polyethers prepared with the $TPPH_2/Et_2AlCl$ system (18), ii) the synthesis of block copolyethers by sequential monomer addition to living oxirane polymers (19), and iii) the structure of the grow-ing chain end in ethylene oxide and propylene oxide polymerizations initiated with $TPPH_2/Et_2AlCl$ (20). Most recently, this same catalyst system has been shown to afford cyclic carbonates from epoxides and CO_2 in the presence of 1-methylimidazole (21).

The significance of the development of this catalyst system is the control which it allows in the synthesis of polyethers of known molecular weights and narrow molecular weight distributions, and in the synthesis of block copolyethers. Although ethylene oxide is "well-behaved" in conventional anionic polymerizations, monomers such as propylene oxide (i.e., those with protons β to the oxirane ring) suffer severe molecular weight-limiting chain transfer via proton abstraction under anionic polymerization conditions. In constrast, Inoue and coworkers report a very good linear relation between the experimental number-average molecular weight ($\bar{M}_n$) and the initial ratio of monomer and catalyst concentrations (18,19), at least up to $[Monomer]_o/[Catalyst]_o = 800$. Typical polydispers-ity indices ($\bar{M}_w/\bar{M}_n$) are less than 1.1 for ethylene oxide or propylene oxide, and only slightly greater for 1,2-butene oxide (19). In block copolymerizations involving sequential monomer additions, very high (perhaps quantitative) blocking efficiencies are achieved

(19). Finally, the polymerizations are fast enough to be convenient;
for example, polyethylene oxide with a degree of polymerization of
400 is produced in 80% yield in 3 hours at room temperature (19).

NEW METHODS OF SYNTHESIS OF BLOCK AND GRAFT COPOLYMERS

Methods available for the synthesis of block and graft copolymers have been expanded by two important new ideas: i) that
the transformation of the mode of propagation (e.g., from anionic
to cationic) should allow the preparation of block copolymers from
monomers which do not share a common polymerization mechanism (22),
and ii) that graft copolymers might be assembled "branches first",
followed by a stitching together of the graft copolymer backbone
(23). The following discussion will describe recent applications
of these ideas to ring-opening polymerizations of cyclic ethers.

Block Copolymerization via Active Center Transformation

The synthesis of block copolymers of controlled structures is
most conventionally accomplished through the use of living anionic
polymerization. One can easily imagine, however, desirable block
copolymers derived from monomers which are inert to anionic polymerization conditions, or which do not share any common mode of
polymerization. In a recent series of papers (24-34), Richards
and coworkers have addressed this problem in a general way, and
have developed methods which convert one kind of active center
into another. Within the context of cyclic ether polymerizations,
Richards has focused on the preparation of block copolymers of
styrene and tetrahydrofuran (THF); several methods of accomplishing
this copolymerization are described in the following paragraphs.

The general problem here is of course the fact that styrene
is readily polymerized to a living polymer by anionic methods, but
not by cationic initiation, while for THF the situation is reversed.
Thus a plausible route to a styrene/THF block copolymer might consist of living anionic polymerization of styrene followed by conversion of the living carbanion into a cationating agent, or
alternatively, of a living cationic polymerization of THF followed
by transformation of the growing oxonium ion into a species capable
of initiating anionic polymerization of styrene. Richards and co-
workers have investigated both methods.

Polymerization of styrene with n-butyllithium using vacuum-
line techniques, followed by addition of excess Br_2 or xylylene
dibromide, produced reactive Br-terminated polystyrenes VI and VII
(24,25). Both contain benzylic bromine atoms, which are readily

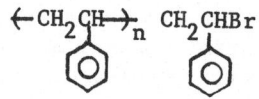   VI

$$\left(\text{CH}_2\text{CH}\right)_n \text{CH}_2\text{CHCH}_2 \text{—} \bigcirc \text{—} \text{CH}_2\text{Br} \qquad \text{VII}$$

precipitated as AgBr by treatment with $AgClO_4$ or $AgPF_6$. The carbocations produced in this way are effective initiators for cationic polymerization of THF (Eq. 3).

$$\text{PS—CHBr} \xrightarrow[\text{O}]{\text{AgClO}_4} \text{PS—CH-O}^{\oplus} \bigcirc \; ClO_4^- \quad \bigcirc^O \quad \text{PS—CH}\left[\text{O(CH}_2)_4\right]\text{—}$$
$$+ \text{AgBr} \qquad\qquad \underline{3}$$

Richards and coworkers have followed these conversions very carefully by gel permeation chromatography, and have reached the following conclusions: i) Preparation of VI and VII is best done by preliminary conversion of the living polystyryllithium to the corresponding Grignard reagent prior to addition of Br_2 or xylylene dibromide. Yields of VI and VII produced in this way are high (80-95%), though not quantitative. ii) The efficiency of the transformation is about 80% for either VI or VII under the most favorable conditions. iii) The only significant contaminant produced along with the block copolymer is homopolystyrene; no polytetrahydrofuran is found. iv) Under optimum conditions, block copolymers with $\bar{M}_w/\bar{M}_n = 1.04$ can be prepared. Although the overall yields of the block copolymers thus far obtained via these procedures are at best about 70%, it seems likely that higher yields will be realized after more extensive optimization of the transformation reactions.

The alternate route - the cation to anion transformation - has thus far been less successful. Living cationic polytetrahydrofuran, prepared with $AgPF_6$/RX initiation, has been terminated by the lithium salt of cinnamyl alcohol (26) (Eq. 4) or by n-butylamine (27) (Eq. 5). The first of these procedures is designed

$$\text{—}^+\!\!\bigcirc \; PF_6^- \xrightarrow[\bigcirc]{\text{LiOCH}_2\text{CH=CH}} \text{—OCH}_2\text{CH}_2\text{CH}_2\text{CH}_2\text{OCH}_2\text{CH=CH} \quad \underline{4}$$

$$\text{—}^+\!\!\bigcirc \; PF_6^- \xrightarrow{\text{RNH}_2} \text{—OCH}_2\text{CH}_2\text{CH}_2\text{CH}_2\text{NHR} \quad \underline{5}$$

to yield a carbanion by addition of butyllithium; the latter an amide anion by metallation with potassium metal film. In each case the termination reaction is reported to be quantitative, but subsequent initiation affords blocking efficiencies of only 20-30%.

Thus while the concept of the cation to anion transformation has been demonstrated, widespread use of the technique for the preparation of block copolymers will require the development of more efficient methods for accomplishing the transformation.

Graft Copolymer Syntheses via Macromolecular Monomers

Conventional syntheses of graft copolymers, beginning with the initial report of a graft copolymerization by Carlin and Shakespeare in 1947 (35), have consisted of assembly of the main polymer chain first, followed by grafting of the branches of the second monomer. In 1974, a patent by Milkovich (23) suggested a new method of graft copolymerization - the "macromer" technique - in which branches bearing one polymerizable endgroup are prepared first, and then the main chain is assembled by copolymerization of these macromolecular monomers with conventional monomers. The synthesis of macromers of well-defined molecular weight, molecular weight distribution and mono-functionality is a fascinating challenge for the organic polymer chemist, and the preparation and use of macromers has become an unusually active area of polymer synthesis research. Several recent reports of macromer syntheses which involve ring-opening polymerization of cyclic ethers are described in the following paragraphs.

A recent paper by Franta, Rempp and coworkers illustrates very nicely the two most general methods of macromer preparation: i) initiation of a transfer-free polymerization with a molecule containing a polymerizable group or ii) killing of a living polymerization with a terminating agent bearing a polymerizable group (36). Polymerization of ethylene oxide with the potassium salt of p-isopropenylbenzyl alcohol, followed by termination by acidic methanol or by benzyl chloride, produced ethylene oxide macromers VIII or IX, respectively. The macromers were characterized by gel

$$CH_2=C\overset{CH_3}{\underset{|}{}}-\bigcirc-CH_2O(CH_2CH_2O)_{n-1}CH_2CH_2OH \qquad\qquad VIII$$

$$CH_2=C\overset{CH_3}{\underset{|}{}}-\bigcirc-CH_2O(CH_2CH_2O)_{n-1}CH_2CH_2OCH_2-\bigcirc \qquad IX$$

permeation chromatography, NMR and UV spectroscopy, double bond titration, vapor pressure osmometry and light scattering. The average molecular weights obtained by these methods were in very good agreement, and confirmed the presence of one isopropenyl group per molecule. Number-average molecular weights were varied from 1000 to 5000. An alternate procedure developed by the same workers consisted of initiation of living ethylene oxide polymerixation by diphenylmethyl potassium or by the potassium salt of

2-methoxyethanol, followed by termination with methacryloyl chloride. Characterization as above again produced consistent average molecular weights. The yields of macromers obtained by both procedures were 70-80%, but this presumably reflects only product losses during isolation, since there is good agreement between the theoretical and experimental number-average molecular weights. The paper provides no experimental detail, and no results are reported for polymerization or copolymerization of the polyethylene oxide macromers.

Very similar techniques have been reported by Franta and Rempp (37,38) and by Asami and Takaki (39,40) for the synthesis of tetrahydrofuran macromers.

NEW REACTIVE POLYETHERS AND POLYSULFIDES

Our laboratory has recently reported the preparation of a new series of polymers which undergo nucleophilic displacement reactions with catalysis by backbone functional groups (41-45). Equation 6 shows this catalysis in a schematic way. The consequences of this

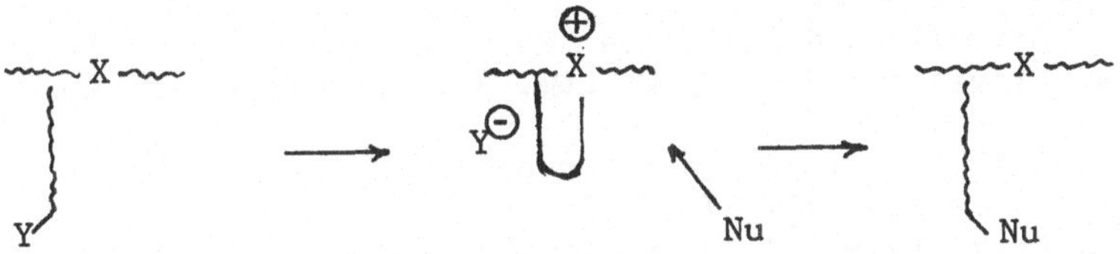

kind of catalysis include enhanced reaction rates and the formation of rearranged chain units.

A simple example of a polymer with this kind of molecular architecture is poly(chloromethylthiirane) (PCMT) (X). PCMT rearranges at room temperature in solution or in bulk, to produce an equilibrium copolymer (XI) of chloromethylthiirane (CMT) and 3-chlorothietane (3 CT) repeating units (Eq. 7).

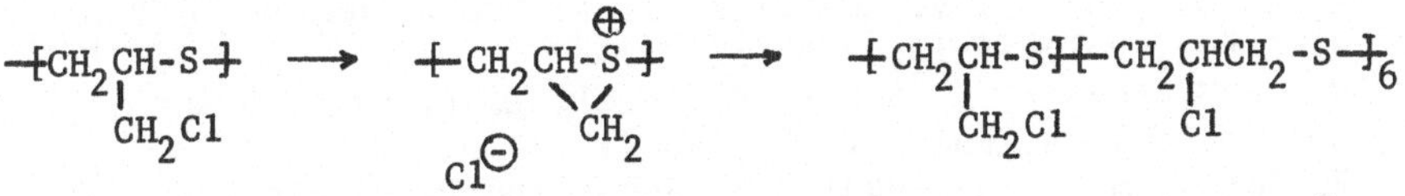

That this is in fact an equilibrium process is proven by the observation that the same equilibrium copolymer is obtained either from PCMT or from P3CT. We anticipate extension of this chemistry to polymers other than polysulfides, as well as applications in adhesives, in polymer modification and in enzyme immobilization.

440 D. A. TIRRELL

REFERENCES

1. This bibliography is available from the author.
2. J. V. Crivello and J. H. W. Lam, J. Polym. Sci., Symp. 56, 383 (1976).
3. J. V. Crivello and J. H. W. Lam, Macromolecules 10, 1307 (1977).
4. J. V. Crivello, CHEMTECH 10, 624 (1980).
5. J. V. Crivello and J. H. W. Lam, J. Polym. Sci., Polym. Chem. Ed. 16, 2441 (1978).
6. J. V. Crivello and J. H. W. Lam, J. Polym. Sci., Polym. Chem. Ed. 17, 977 (1979).
7. J. V. Crivello and J. H. W. Lam, J. Polym. Sci., Polym. Chem. Ed. 17, 1059 (1979).
8. J. V. Crivello and J. H. W. Lam, J. Polym. Sci., Polym. Chem. Ed. 17, 759 (1979).
9. J. V. Crivello and J. H. W. Lam, J. Polym. Sci., Polym. Chem. Ed. 18, 2677 (1980).
10. J. V. Crivello and J. H. W. Lam, J. Polym. Sci., Polym. Chem. Ed. 18, 2697 (1980).
11. J. V. Crivello and J. H. W. Lam, J. Polym. Sci., Polym. Chem. Ed. 17, 1047 (1979).
12. J. V. Crivello and J. H. W. Lam, J. Polym. Sci., Polym. Chem. Ed. 17, 2877 (1979).
13. J. V. Crivello and J. L. Lee, Macromolecules 14, 1141 (1981).
14. J. V. Crivello and J. H. W. Lam, J. Polym. Sci., Polym. Chem. Ed. 18, 1021 (1980).
15. J. V. Crivello, T. P. Lockhart and J. L. Lee, J. Polym. Sci., Polym. Chem. Ed. 21, 97 (1983).
16. J. V. Crivello and J. H. W. Lam, J. Polym. Sci., Polym. Chem. Ed. 19, 539 (1981).
17. N. Takeda and S. Inoue, Makromol. Chem. 179, 1377 (1978).
18. T. Aida, R. Mizuta, Y. Yoshida and S. Inoue, Makromol. Chem. 182, 1073 (1981).
19. T. Aida and S. Inoue, Macromolecules 14, 1162 (1981).
20. T. Aida and S. Inoue, Macromolecules 14, 1166 (1981).
21. T. Aida and S. Inoue, J. Am. Chem. Soc. 105, 1304 (1983).
22. D. H. Richards, Br. Polym. J. 12, 89 (1980).
23. R. Milkovich, M. T. Chiang, U. S. Patent 3,842,050 (1974).
24. F. J. Burgess, A. V. Cunliffe, J. R. MacCallum and D. H. Richards, Polymer 18, 719 (1977).
25. F. J. Burgess, A. V. Cunliffe, J. R. MacCallum and D. H. Richards, Polymer 18, 726 (1977).
26. M. J. M. Abadie, F. Schue, T. Souel, D. B. Hartley and D. H. Richards, Polymer 23, 445 (1982).
27. P. Cohen, M. J. M. Abadie, F. Schue and D. H. Richards, Polymer 23, 1105 (1982).
28. F. J. Burgess, A. V. Cunliffe, J. V. Dawkins and D. H. Richards, Polymer 18, 733 (1977).
29. T. Souel, F. Schue, M. Abadie and D. H. Richards, Polymer 18, 1292 (1977).

30. M. J. M. Abadie, F. Schue, T. Souel and D. H. Richards, Polymer 22, 1076 (1981).
31. C. H. Bamford, G. C. Eastmond, J. Woo and D. H. Richards, Polymer 23, 643 (1982).
32. F. J. Burgess, A. V. Cunliffe, D. H. Richards and D. C. Sherrington, J. Polym. Sci. Polym. Lett. Ed. 14, 471 (1976).
33. M. Abadie, F. J. Burgess, A. V. Cunliffe and D. H. Richards, J. Polym. Sci. Polym. Lett. Ed. 14, 477 (1976).
34. A. V. Cunliffe, G. F. Hayes and D. H. Richards, J. Polym. Sci. Polym. Lett. Ed. 14, 483 (1976).
35. R. B. Carlin and N. E. Shakespeare, J. Am. Chem. Soc. 68, 876 (1946).
36. P. Masson, G. Beinert, E. Franta and P. Rempp, Polym. Bull. 7, 17 (1982).
37. J. Sierra-Vargas, J. G. Zilliox, P. Rempp and E. Franta, Polym. Bull. 3, 83 (1980).
38. J. Sierra-Vargas, P. Masson, G. Beinert, P. Rempp and E. Franta, Polym. Bull 7, 277 (1982).
39. R. Asami, M. Takaki, K. Kita and E. Asakura, Polym. Bull. 2, 713 (1980).
40. M. Takaki, R. Asami and T. Kuwabara, Polym. Bull. 7, 521 (1982).
41. M. P. Zussman and D. A. Tirrell, Macromolecules 14, 1148 (1981).
42. M. P. Zussman and D. A. Tirrell, Polym. Bull. 7, 439 (1982).
43. M. P. Zussman and D. A. Tirrell, J. Polym. Sci. Polym. Chem. Ed., in press (1983).
44. D. A. Tirrell, M. P. Zussman, J. S. Shih and J. F. Brandt, in C. E. Carraher, Jr. and J. A. Moore, eds., Chemical Modification of Polymers, Plenum, in press.
45. J. S. Shih, J. F. Brandt, M. P. Zussman and D. A. Tirrell, J. Polym. Sci. Polym. Chem. Ed. 20, 2839 (1982).